AF379087

Optical Networks

Network Theory and Applications

Volume 6

Managing Editors:

Ding-Zhu Du, *University of Minnesota, U.S.A.*

and

Cauligi Raghavendra, *University of Southern California, U.S.A.*

The titles published in this series are listed at the end of this volume.

Optical Networks –
Recent Advances

Edited by

Lu Ruan

and

Ding-Zhu Du

Department of Computer Science and Engineering,
University of Minnesota,
Minneapolis, U.S.A.

KLUWER ACADEMIC PUBLISHERS
DORDRECHT / BOSTON / LONDON

A C.I.P. Catalogue record for this book is available from the Library of Congress.

ISBN 0-7923-7166-6

Published by Kluwer Academic Publishers,
P.O. Box 17, 3300 AA Dordrecht, The Netherlands.

Sold and distributed in North, Central and South America
by Kluwer Academic Publishers,
101 Philip Drive, Norwell, MA 02061, U.S.A.

In all other countries, sold and distributed
by Kluwer Academic Publishers,
P.O. Box 322, 3300 AH Dordrecht, The Netherlands.

Printed on acid-free paper

Printed in the Netherlands.

Contents

Foreword

With the rapid growth of bandwidth demand from network users and the advances in optical technologies, optical networks with multiterabits-per-second capacity has received significant interest from both researchers and practitioners. Optical networks deployment raises a number of challenging problems that require innovative solutions, including network architectures, scalable and fast network management, resource-efficient routing and wavelength assignment algorithms, QoS support and scheduling algorithms, and switch and router architectures. In this book, we put together some important developments in this exiting area during last several years. Some of the articles are research papers and some are surveys. All articles were reviewed by two reviewers.

The paper, "On Dynamic Wavelength Assignment in WDM Optical Networks," by Alanyali gives an overview of some issues in the analysis and synthesis of dynamic wavelength assignment policies for optical WDM networks and illustrates a new method of analysis. The paper by Ellinas and Bala, "Wavelength Assignment Algorithms for WDM Ring Architectures," presents two optimal wavelength assignment algorithms that assign the minimum number of wavelengths between nodes on WDM rings to achieve full mesh connectivity.

In the paper, "Optimal Placement of Wavelength Converters in WDM Networks for Parallel and Distributed Computing Systems," Jia *et al.* present the necessary and sufficient conditions that a subset of nodes guarantees load-wavelength assignability on general WDM networks. The optimal placement of converters on a class of WDM networks of special topologies are obtained. In the paper, "Allocation of Wavelength Converters in All-Optical Networks," Xiao and Leung discuss the problem of allocating wavelength converters in all-optical networks to maximize the performance. The authors describe three approaches to solve this allocation problem and the simulation-based optimization approach is shown to be widely applicable.

In the paper, "Lightpath Establishment in Wavelength-Routed WDM Optical Networks," Jue discusses the various issues related to the control and management of lightpaths in a wavelength-routed optical network, and presents some of the routing, wavelength assignment, and signaling protocols for establishing lightpaths in such a network. In the paper, "Connection Management for Wavelength-Routed Optical WDM Networks," Ramamurthy *et al.* study different connection management methods in wavelength routed optical WDM networks. The authors compare different implementation on connection management. Different RWA algorithms were discussed and compared with.

In the paper, "Dynamic Traffic Scheduling for QoS Support in WDM/TDM Networks with Arbitrary Tuning Latencies," Huang *et al.* propose a dynamic scheduling algorithm under single-hop WDM/TDM networks with QoS support and tuning latency consideration. Both CBR traffic and ABR traffic are supported.

The paper, "Multifiber WDM Networks," by Li and Somani provides an overview of the research work for multifiber WDM networks. Both static network design approaches and dynamic routing and wavelength assignment algorithms are reviewed and summarized.

Optical multistage interconnection network (MIN) is an important interconnecting scheme for communication and parallel computing systems. The paper by Pan *et al.*, " Recent Developments in Optical Multistage Networks," surveys the major challenges encountered and approaches adopted in the research of optical MINs.

The paper, "Multicast Routing in WDM Optical Networks," by Sreenath *et al.* discusses the multicast routing in various WDM networks: broadcast-and-select networks, linear lightwave networks, and wavelength routed networks.

The paper by Wang and Dixit, "Architecture and Analysis of Terabit Packet Switches Using Optoelectronic Technologies," lists the recent activities on optical packet switches and presents several architectures of a terabit-level packet switch using optoelectronic technologies.

We wish to thank all authors and all reviewers for their valuable efforts that made this book a reality. We hope this collection of articles would serve as a useful reference in study of optical networks.

Lu Ruan
Ding-Zhu Du

Optical Networks - Recent Advances
L. Ruan and D.-Z. Du (Eds.) pp. 1 - 17
©2001 Kluwer Academic Publishers

On Dynamic Wavelength Assignment in WDM Optical Networks

Murat Alanyali
Department of Electrical and Electronics Engineering
Bilkent University, Ankara, 06533 Turkey
E-mail: `alanyali@ee.bilkent.edu.tr`

Contents

1   Introduction

Optical fiber has been used as the physical medium for high rate data transmission since late 1960s. Early applications of optical fiber communications modulated data onto a single optical carrier frequency that is commonly referred to as a wavelength. The carried data rate is therefore limited by the speed of electronics that generate the signal, thereby grossly underutilizing the tens of THz of useful bandwidth available on the fiber[1]. The *Wavelength Division Multiplexing (WDM)* technology now allows multiplexing

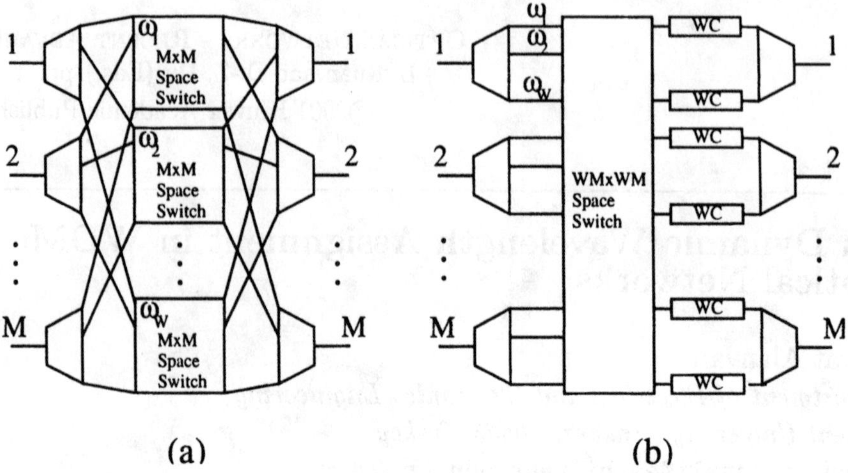

Figure 1: Structure of (a) a wavelength selective (WS) optical switch, and (b) a wavelength converter (WC) optical switch.

several optical carriers on the same fiber, opening up the available potential. Current deployments of WDM are based on point-to-point links. This entails demodulating the optical signal at each switching node and carrying out switching electronically. Remarkable increase in the carried data rates due to the WDM technology make these switches inherent bottlenecks in the network. One promising approach to solve this problem is optical switching which entails switching entire wavelengths without any electronic processing.

Optical switches have varying capabilities with respect to wavelength conversion. At one extreme, a *wavelength selective (WS)* switch offers no conversion capability: the signal on a particular wavelength cannot be modulated onto a different wavelength as it passes through the switch. Hence in a network employing exclusively WS switches, an end-to-end connection is established by using the same wavelength on each link of its path. This condition is referred to as the *wavelength continuity* constraint. Figure 1.a gives the structure of a WS switch with M input and M output fibers. Throughout the paper it is assumed that the same set of wavelengths is available on all fibers. At the other extreme, a *wavelength converter (WC)* switch can remodulate the data on any inbound wavelength onto any outbound wavelength. WC switches remove the wavelength continuity constraint and therefore result in better utilization of the fiber capacity. For example in the network of Figure 2 with two wavelengths w_1, w_2 on each fiber, if fiber

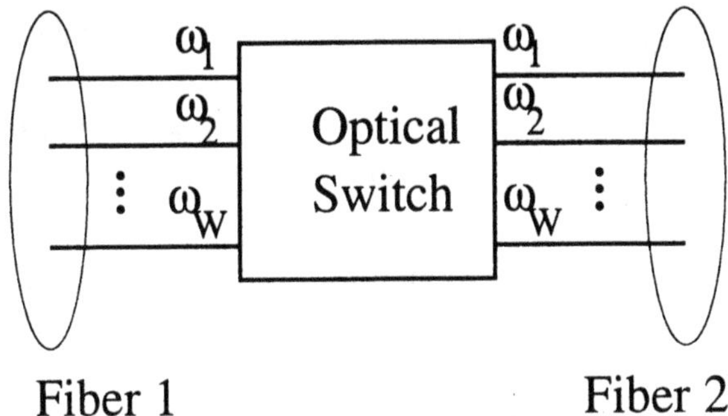

Figure 2: A simple topology with one optical switch and two fibers.

1 carries one connection which fully utilizes w_1 and fiber 2 carries another connection which fully utilizes w_2, a third connection that uses both fibers can be established only if the optical switch is a WC. The architecture of a WC switch is given in Figure 1.b. The wavelength converters in the figure ideally have the capability of providing conversion between any pair of wavelengths. However currently obtained conversion performance strongly depends on the specific wavelength pair, and technology is yet to mature to allow full conversion capability[2]. This practical consideration gives rise to the *limited wavelength converter* switch, which stands somewhere between WS and WC models, and provides conversion only between specified wavelength pairs. The architecture of the switch is the same as that of a WC switch except for the reduced capability of the converters.

A main theme of interest in recent research on WDM networks has been to identify effective routing and wavelength assignment strategies for end-to-end connections. Networks that comprise exclusively WC switches have immediate analogues in classical circuit switched networks, hence many issues about these networks are relatively well understood. Networks with limited wavelength converter or WS switches form a distinct class due to the wavelength continuity constraint, and appear to require more sophisticated analysis tools.

This paper focuses on dynamic wavelength assignment in WDM optical networks, and gives an overview of some recent results. The aim of the paper is to give a perspective on the subject and to illustrate a new method of analysis. Thus the articles cited here do not constitute an exhaustive list of references on wavelength assignment, however throughout the paper the reader is frequently referred to relevant survey papers for this purpose. It is assumed here that each end-to-end connection occupies a full wavelength, and that connections are routed over predetermined paths. The efficiency of an assignment strategy is considered in two categories depending on the performance metric and a corresponding modeling approach. The first approach assumes a probabilistic structure for the connection set-up/tear-down process: set-up requests arrive according to Poisson processes, and connections are torn-down after exponentially distributed holding times. The model therefore inherently involves blocking of requests due to capacity constraints, and the aim in wavelength assignment is to minimize the blocking probability. While the Poisson request arrival model seems possibly suitable for the optical networks of the future, contemporary optical networks serve aggregated traffic with inherently different statistics. In addition, blocking is highly undesirable in these networks due to the very high data rates carried on a single wavelength. In view of these the second approach adopts a deterministic analysis and seeks the minimum number of wavelengths required assure no blocking for a given class of dynamic scenarios.

The outline of the paper is as follows. Section 2 concerns probability of blocking. In Section 2.1 networks with WC switches are considered, and some results from classical circuit switching theory are cited. Wavelength assignment is not an issue here, however the cited results have implications in networks with WS switches. These networks are the subject of Section 2.2. A brief overview of simple assignment heuristics proposed in the literature is given, and some recent work on an asymptotically exact model is discussed. Section 3 summarizes some recent results on nonblocking WDM optical networks. We conclude by identifying some open issues in Section 4.

2   Loss Networks

We consider a network with J edges, each edge representing a fiber with W wavelengths. The network serves a number of connection types, each of which is identified by the set of edges in its route. Connection requests of different types arrive according to statistically independent Poisson processes, with rate ν_r for connection type r. If the network has sufficient free capacity to accommodate a new connection request, the request is granted and the connection holds one wavelength on each fiber on its route for the duration of its holding time. It is assumed that wavelength reassignment is not allowed: each connection keeps the same wavelength assignment as it remains in the network. Holding times are independent and have exponentially distribution with mean 1. Requests that cannot be granted are blocked and lost.

2.1 Networks without wavelength continuity constraint

If all network nodes are WC switches and the wavelength continuity constraint is not in effect, then a connection request is accepted if each link on its route has a free wavelength. Blocking in such networks has been studied extensively in the context of circuit switched telephone networks, and this section cites some results thereof. Let R denote the collection of connection types, and the routes be represented by the incidence matrix $A = [A_{j,r}]_{J \times R}$, where $A_{j,r} = 1$ if link j belongs to route of type r, and $A_{j,r} = 0$ otherwise. Hence $r = \{j \in J : A_{j,r} = 0\}$. Denote by $n_t(r)$ the number of type r connections in the network at time t, and define the vector $n_t = (n_t(r) : r \in R)$. Let $\underline{W}$ denote a $|J|$ dimensional vector whose entries are W. The process $(n_t : t \geq 0)$ is Markovian with state space $S = \{n \in Z_+^R : An \leq \underline{W}\}$. The equilibrium distribution of this process, θ, is given by[3]

$$\theta(n) = G^{-1} \prod_{r \in R} \frac{\nu_r^{n(r)}}{n(r)!} \qquad n \in S.$$

IIere G is a normalizing constant which makes θ a probability vector.

The equilibrium distribution θ leads immediately to the blocking probability for type r connections as

$$B_r = \sum_{n: An > \underline{W} - Ae_r} \theta(n), \tag{1}$$

where e_r denotes a binary vector whose entries are zero except the rth entry. However the task of evaluating this expression is $\#P$ complete[4]; thus one usually has to seek accurate methods to approximate blocking probabilities.

An appealing independence assumption gives rise to a well-known reduced load approximation: Consider the hypothetical situation in which links blocked requests in a mutually independent manner. Let L_j denote the probability that link j has no free wavelengths. The rate of connection requests of each type at a link i would then be thinned by a fraction $1 - L_j$ at every other link j on the route of that type. In particular, the process of connection requests seen by link i would be a Poisson process with rate

$$\rho_i = \sum_{r:i \in r} \nu_r \prod_{j \in r - \{i\}} (1 - L_j). \tag{2}$$

In turn, equation (1), applied to a single link, would imply that

$$L_i = \left(\sum_{n=0}^{W} \frac{\rho_i^n}{n!} \right)^{-1} \frac{\rho_i^W}{W!}.$$

(3)

The reduced load approximation is obtained by using

$$E_r = 1 - \prod_{i \in r}(1 - L_i)$$

to approximate the blocking probability B_r. Equations (2)–(3) admit a unique solution and the approximation is well defined.

The reduced load approximation outlined above is surprisingly effective for networks employing fixed routing. The approximation gives asymptotically exact results in a limiting regime in which the number of wavelengths W and the request arrival rates ν_r increase in proportion[3]. Variants of the method have been shown to perform well in the finite case also[5]. The reader is referred to [3] for an excellent discussion of reduced load approximations and further insight on blocking in classical loss networks.

2.2   Networks with wavelength continuity constraint

Quantifying the increase in blocking due to the wavelength continuity constraint has been a common theme in much of recent research. Several facts make this a challenging task. Determining the exact capacity of networks with WS switches appears difficult: whether a fixed set of connections can be established simultaneously in the network is a NP-complete problem [6]. For the dynamic case, a closed form expression for the steady state distribution can seldom be obtained for nontrivial topologies or reasonable wavelength assignment policies.

Addressing the fundamental limitations due to wavelength continuity entails having deeper insight on wavelength assignment. In a network of WS switches a wavelength may be shared by two connections only if the connections do not use the same fiber. Thus the wavelengths are reused spatially. A number heuristic wavelength assignment strategies proposed in the literature have been inspired by the informal observation that maximizing wavelength reuse lead to high utilization of network resources. Two such strategies are *First-Fit* [6], which selects the available wavelength with the smallest index, and *Most-Used* [7], which selects an available wavelength that is most utilized within the entire network. Both strategies implicitly promote packing connections into a small number of wavelengths so as to reserve free capacity for potential new requests.

The *Max-Sum* [8] strategy, which has a slightly different flavor than the previous two, selects a wavelength so as to reduce an explicit measure of blocking potential in the resulting state of the network. Simulations indicate that Max-Sum performs better than First-Fit and Most-Used strategies [8]. This is not totally unexpected since Max-Sum mimics a myopic optimization with regard to the future. In principle progressively better strategies can be obtained by extending the optimization horizon further. This entails formulating stochastic dynamic programs, however the dimension of the state space invalidates this approach for typical problems. One such formulation is adopted in [9] where policy iteration[10] is employed to improve the blocking performance of any given strategy. Here the cost-to-go functions of the original strategy are evaluated via simulations.

Majority of the analytic approaches employed in performance evaluation of networks of WS switches involve the assumption of link independence. This argument assumes that links have statistically independent states, and wavelengths on the same fiber have statistically independent occupancies. Clearly this is not the case under the general model of the present section. Typical applications of the link independence assumption do not entail a fixed point argument (see the list of references in [11]), hence these are fundamentally not analogous to the reduced load approximations of the previous section.

In principle a generalized form of the reduced load approximations may be incorporated in the analysis of the networks considered in this section by treating each fiber as W parallel links with unit capacity. The flexibility of choosing among possible wavelengths in the original network then corresponds to alternate routing in the classical circuit switched counterpart. The asymptotic exactness of the reduced load approximation does not extend to networks with alternate routing[13], instead one can employ the generalized reduced load approximation that is formulated in [14]. This technique does not offer hard guarantees on accuracy, and the involved computational complexity tends to be high for large networks[15]. See [15], [16] and [17] for applications of this technique.

While the link independence assumption permits an explicit analysis and captures some salient features of routing and wavelength assignment, it fails to answer somewhat deeper questions due to the inherent oversimplification. [18] addresses this shortcoming by considering a different approximate model that accounts for links correlations. Analysis of the model reveals that the blocking performance depends on the ratio of the average route length to the average number of links shared by two intersecting routes. This result confirms earlier observations that the wavelength continuity constraint has a smaller impact in ring topologies compared to mesh topologies. For detailed discussions of heuristic algorithms and approximate models, the reader is referred to the survey articles [11, 19].

Significant insight can be gained on the network behavior by using approximate dynamic models that are exact in some asymptotic sense. The rest of this section concerns one such model due to [20] that provides information about the dynamic as well as the steady state behavior of the network. Consider a network whose topology is given by Figure 2. Let W, ν_1, ν_2, ν_3 be positive numbers and let γ be a scaling factor so that each fiber has $\lfloor \gamma W \rfloor$ wavelengths and type r connections arrive at rate $\gamma \nu_r$. Each type 1 (respectively type 2) connection require one wavelength from fiber 1 (respectively fiber 2). Each type 3 connection requires one wavelength from each fiber, furthermore these wavelengths should be identical. The Most-Used wavelength assignment strategy is adopted: a type 1 (respectively type 2) connection is assigned an available wavelength that is already occupied on fiber 2 (respectively fiber 1), if such a wavelength exists at the time of arrival. If all available wavelengths are idle in both fibers, then one such wavelength is assigned arbitrarily.

For each time t and type $r = 1, 2, 3$ define

$$
\begin{aligned}
X_t(r) &= \text{number of type } r \text{ connections in the network at time } t, \\
X_t(4) &= \text{number of wavelengths available on both fibers at time } t,
\end{aligned}
$$

and let $X_t = (X_t(1), X_t(2), X_t(3), X_t(4))$. See Figure 3 for an illustration of these variables. Under Most-Used wavelength assignment the process $(X_t : t \geq 0)$ is Markovian. Its normalized version $(X_t/\gamma : t \geq 0)$, where division by γ is understood to be componentwise, converges in probability as γ grows towards infinity. The limit, denoted by $(x_t : t \geq 0)$, complies with the following differential equalities:

$$
\begin{aligned}
\dot{x}_t(1) &= \nu_1 \pi_{x_t}(m(1) + m(3) > 0) - x_t(1) & (4) \\
\dot{x}_t(2) &= \nu_2 \pi_{x_t}(m(2) + m(3) > 0) - x_t(2) & (5) \\
\dot{x}_t(3) &= \nu_3 \pi_{x_t}(m(3) > 0) - x_t(3) & (6) \\
\dot{x}_t(4) &= (2W - x_t(1) - x_t(2) - x_t(3) - 2x_t(4)) - \nu_3 \pi_{x_t}(m(3) > 0) \\
&\quad - \nu_1 \pi_{x_t}(m(1) = 0, m(3) > 0) - \nu_2 \pi_{x_t}(m(2) = 0, m(3) > 0). & (7)
\end{aligned}
$$

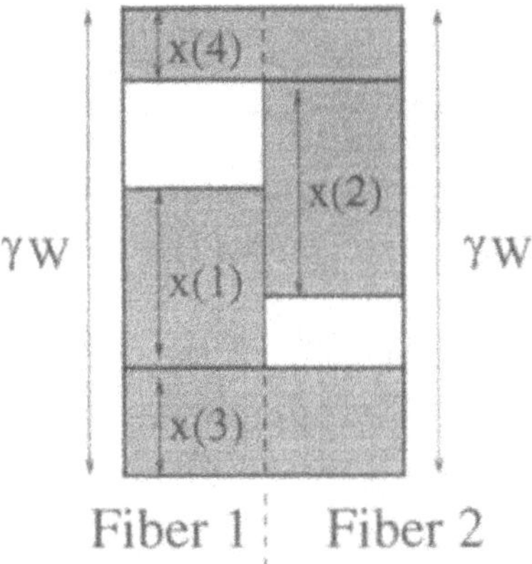

Figure 3: A typical state of the network of Figure 2.

Here π_{x_t} denotes an equilibrium distribution of a Markov process $(m(1), m(2)$ $m(3))$ that mimics the free wavelength process in the network. Informally, $m(1)$ (respectively $m(2)$) represents the number of wavelengths that are available *only* on fiber 1 (respectively fiber 2), and $m(3)$ represents the number ber wavelengths that are available on both fibers. Note that the actual free wavelength process is not Markovian: its instantaneous transition rates at time t depend on the value X_t. These rates are thus proportional to γ, and in the limit of large γ the free wavelength process settles to equilibrium before the value of the normalized process X_t/γ changes. The resulting equilibrium distribution is captured by the process $(m(1), m(2), m(3))$ whose transition rates are given by Table 1. Here and in the rest of the paper I denotes the indicator function, which is 1 if the condition in the argument holds, and 0 otherwise. Since the total number of wavelengths is unbounded in the limit, the coordinates of the process $(m(1), m(2), m(3))$ may take on the value ∞. Thus this process is inherently reducible and possesses multiple equilibrium distributions. Solving for the trajectories (4)–(7) entails identifying the right distribution at every t, which is a nontrivial procedure.

Equations (4)–(7) can be interpreted as follows. Equation (4) indicates that type 1 connections are admitted to the network at a rate proportional to the probability that there is a free wavelength on fiber 1. The total rate of termination of type 1 connections is proportional to the total number of such connections, since each connection terminates independently of the others. Equations (5) and (6) are interpreted similarly for type 2 and type 3 connections respectively. The number of wavelengths that are free on both

Jump	Rate
(1,0,0)	$x_t(1) + x_t(2) + x_t(3) + x_t(4) - W$
(0,1,0)	$x_t(1) + x_t(2) + x_t(3) + x_t(4) - W$
(0,0,1)	$2W - (x_t(1) + x_t(2) + x_t(3) + 2x_t(4))$
(-1,0,0)	$\nu_1 I\{m(1) > 0\}$
(0,1,-1)	$\nu_1 I\{m(1) = 0, m(3) > 0\}$
(0,-1,0)	$\nu_2 I\{m(2) > 0\}$
(1,0,-1)	$\nu_2 I\{m(2) = 0, m(3) > 0\}$
(0,0,-1)	$\nu_3 I\{m(3) > 0\}$

Table 1: Transition rates of the process $(m(1), m(2), m(3))$.

fibers increase either when a type 3 connection terminates, or when a type 1 (respectively type 2) connection terminates while its assigned wavelength is free on fiber 2 (respectively fiber 1). This is captured by the first term in equation (7). This number decreases upon a type 3 connection arrival or upon a type 1 or type 2 arrival while all available wavelengths are those that are free on both fibers, as indicated by the remaining terms in equation (7).

We next examine the solutions of the limiting model (4)–(7). Figures 4.a–4.d depict the trajectories $(x_t : t \geq 0)$ starting from an empty system, for four different sets of parameters. We concentrate on the steady state system behavior under the following classification:

1) Underloaded network $(\nu_1 + \nu_3 \leq W, \ \nu_2 + \nu_3 \leq W)$: In the underloaded case the capacity constraint due to finite W is not binding in the considered limit. Thus connections of each type $r = 1, 2, 3$ are virtually undisturbed by the existence other types, and their normalized number converge exponentially to ν_r in steady state. Furthermore the normalized number of wavelengths that are available on both fibers converge to $\min_{r \in \{1,2\}} (W - \nu_r - \nu_3)$, the largest possible value. Hence in the particular topology considered here, Most-Used assignment suffices to pack the utilized wavelengths as closely as possible.

2) Partially overloaded network $(\nu_1 + \nu_3 > W, \ \nu_2 + \nu_3 \leq W)$: In the underloaded fiber, in this case fiber 2, there is virtually no contention for capacity and therefore there is no blocking. In turn the normalized number of type 2 connections evolve as in the underloaded case considered above, converging to ν_2 in equilibrium. Type 3 connections however are subject to blocking since they share the overloaded fiber 1 with type 1 connections.

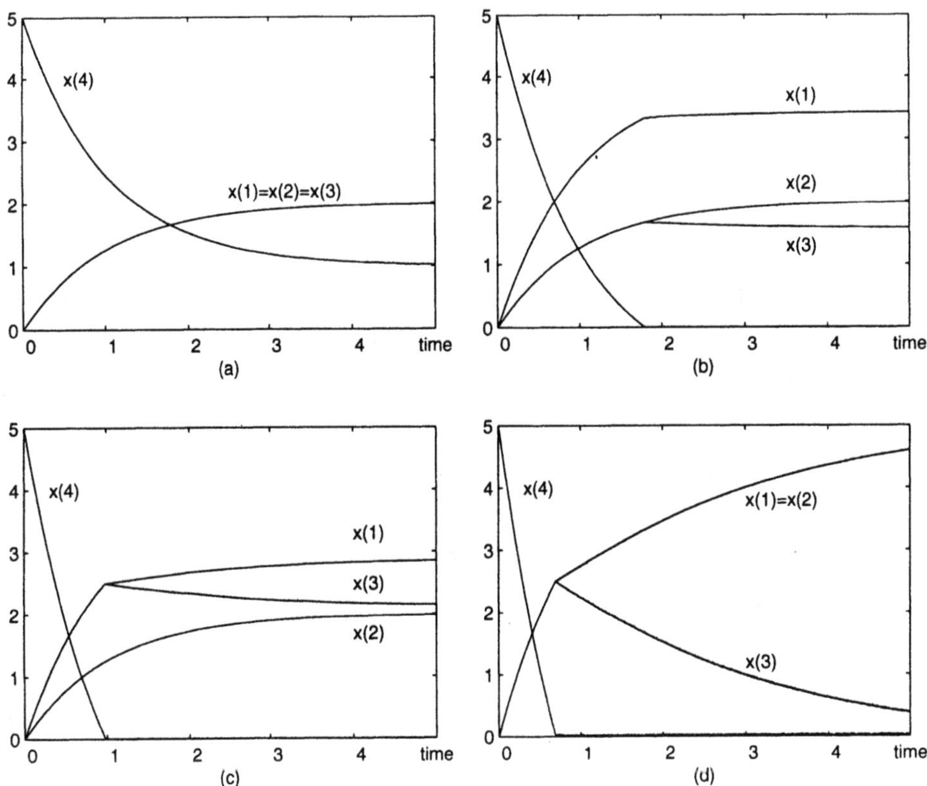

Figure 4: Limiting trajectories for four sets of system parameters that lead to qualitatively different behavior: (a) $\nu_1 = 2$, $\nu_2 = 2$, $\nu_3 = 2$, $W = 5$, (b) $\nu_1 = 4$, $\nu_2 = 2$, $\nu_3 = 2$, $W = 5$, (c) $\nu_1 = 4$, $\nu_2 = 2$, $\nu_3 = 4$, $W = 5$, (d) $\nu_1 = 5$, $\nu_2 = 5$, $\nu_3 = 5$, $W = 5$. All considered cases start with empty system.

On the one hand, type 3 connections have a disadvantage against type 1 connections, since they may not use the wavelengths occupied by type 2 connections on fiber 2. On the other hand, these wavelengths are selected with priority by type 1 connections, due to the Most-Used assignment rule.

The trajectory $(x_t : t \geq 0)$ converges to the limit point

$$x_\infty = (\nu_1 \pi(m(1) + m(3) > 0) \,, \ \nu_2 \,, \ \nu_3 \pi(m(3) > 0) \,, \ 0)$$

where π is the equilibrium distribution of the ergodic Markov process with values in two-dimensional nonnegative integers, and transition rates as given by Table 2. A fairly accurate approximation to x_∞ is obtained via an in-

Jump	Rate
(1,0)	ν_2
(-1,0)	$\nu_1 I\{m(1) > 0\}$
(0,1)	$W - \nu_2$
(0,-1)	$\nu_3 I\{m(3) > 0\} + \nu_1 I\{m(1) = 0, m(3) > 0\}$

Table 2: Transition rates of the Markov process $(m(1), m(3))$ that determines the limit point of the partially overloaded system under Most-Used assignment.

Jump	Rate
(1,0)	ν_2
(-1,0)	$\nu_1 I\{\hat{m}(1) > 0\}$
(0,1)	$W - \nu_2$
(0,-1)	$(\nu_3 + \nu_1 \pi(m(1) = 0))I\{\hat{m}(3) > 0\}$

Table 3: Transition rates of the Markov process $(\hat{m}(1), \hat{m}(3))$ that is used in obtaining the approximate limit point (8). Note that $\pi(m(1) = 0) = 1 - (\nu_2/\nu_1)$ by Table 2.

dependence assumption that is analogous to those used in the analysis of loss networks with state dependent routing[15]. The approximation entails replacing π with the equilibrium distribution of the Markov process that has transition rates given by Table 3, and results in the approximate limit point

$$(\nu_2 + (W - \nu_2)\frac{\nu_1 - \nu_2}{\nu_3 + \nu_1 - \nu_2} \, , \, \nu_2 \, , \, (W - \nu_2)\frac{\nu_3}{\nu_3 + \nu_1 - \nu_2} \, , \, 0). \qquad (8)$$

This point reflects an operating regime in which a fraction ν_2/ν_1 of all type 1 arrivals are assigned wavelengths that are occupied by type 2 connections, and the remaining fraction $(W - \nu_2)/W$ of all wavelengths on fiber 1 is shared between the type 3 and excess type 1 connections, in proportion with the respective arrival rates ν_3 and $\nu_1 - \nu_2$. Though intuitive, the approximation (8) is not exact in that it slightly underestimates the fair share of type 3 connections.

   3) Totally overloaded network $(\nu_1 + \nu_3 > W, \, \nu_2 + \nu_3 > W)$: Let us start with the case when $\nu_2 < \nu_1 \leq W$. There is competition for free wavelengths

on both fibers, however type 3 connections face a harsher contention on fiber
1. In turn enough of type 3 arrivals are knocked out by type 1 connections
so that type 2 connections virtually enjoy the perception of an underloaded
fiber. Therefore in steady state the normalized number of type 2 connections
in the network converge to ν_2. The argument presented for the partially
overloaded case above now applies, and it implies that the steady state
normalized numbers are again given by the vector x_∞ above. This discussion
holds as ν_2 increases towards ν_1, in particular if $\nu_1 = \nu_2 \leq W$ the limit point
is $x_\infty = (\nu_1, \nu_1, W - \nu_1, 0)$. Note that as ν_1, ν_2 increase towards W, the
number of type 3 connections carried by the network vanishes. In the case
when $\min\{\nu_1, \nu_2\} \geq W$ the network operates around the normalized state
$(W, W, 0, 0)$, and type 3 connections are virtually locked out of the network.
This is an extreme manifestation of the fairness issue that has been raised
in several works[11].

The asymptotic exactness of the model (4)–(7) is essentially a functional
law of large numbers, hence the error in the approximation is $o(\gamma)$. For
the unnormalized system, the model should be regarded as a qualitative
description of the expected system dynamics. In the particular context
considered here, this description yields insight on the transient and static
behavior of the network, which is not captured via link independence or
reduced load approximations.

From a practical point of view, perhaps the most significant of the above
results is that the Most-Used strategy has near optimal performance in utilizing the available wavelengths in the considered topology. In particular,
in the considered limit of large capacities and load, the Most-Used strategy has the best possible underload condition in which there is virtually no
blocking in the network. It is interesting to see if this property extends to
more general network topologies.

3 Nonblocking Networks

In this section we give a brief summary of results on requirements for no
blocking. Earlier work on these networks concerned determining the minimum number of wavelengths to satisfy static demand patterns. A popular
static pattern is a permutation, in which each node is the origin and the
destination of exactly one connection. Bounds on the necessary number of
wavelengths required to establish a permutation routing for general topologies are provided in [21, 22]. Similar results are given in [23] for bounded

degree topologies, and a randomized algorithm is given to route the connections with close to minimum number of wavelengths with high probability.

The number of wavelengths required to satisfy dynamic traffic patterns has been recently studied for ring topologies[24]. Namely a ring network with N WS nodes is considered. The traffic on the ring comprises end-to-end connections whose routes are predetermined in advance. It is shown that if the routes and the connection set-up/tear-down times are arbitrary subject to the condition that at any time no fiber carries more than L active connections, then a reservation based assignment strategy guarantees no blocking with roughly $L + L\log_2 N$ wavelengths. With regard to a lower bound, there exists an explicit set-up/tear-down sequence for which any wavelength assignment algorithm would need at least $0.5L\log_2 N$ wavelengths for no blocking. The First-Fit strategy is shown to guarantee noblocking with $2.52L\log_2 N + 5L$ wavelengths. For the case of limited wavelength conversion, [24] provides an explicit construction of a conversion scheme in which each wavelength can be converted to one of two specified wavelengths. With this construction the number of required wavelengths reduce to $L\log_2 \log_2 L + 4L$.

Results of [24] indicate that satisfying dynamic connection patterns on the ring via WS switches proves to be very demanding on network resources: Clearly L wavelengths suffice to guarantee no blocking with WC switches. Static patterns do not require more than $2L - 1$ wavelengths even with WS switches[24]. However dynamic patterns under the wavelength continuity constraint result in significant underutilization of wavelengths as summarized above. On the positive side, even a small degree of limited wavelength conversion helps reduce the wavelength requirement considerably, in agreement with results reported elsewhere[11, 19].

4   Conclusion

This paper concerned some current issues in the analysis and synthesis of dynamic wavelength assignment policies for optical WDM networks. We conclude by providing a brief list of open issues. Accurate analysis methodologies are needed to shed more light on the network behavior under the wavelength continuity constraint, and possibly to lead to identifying effective assignment strategies. Limited wavelength conversion appears to be a promising solution to increase wavelength utilization and deserves further investigation. Effective wavelength conversion and assignment methods, as well as fundamental limits on the gains obtained by limited conversion are unknown for general topologies. Finally, in current WDM systems link capacities are increased by using additional fibers. The crosspoint complexity per wavelength of a WS switch increases with the number of incident fibers, indicating possible improvement in wavelength utilization. Numerical studies[19] support this observation, and warrant further investigation of networks with multiple fibers per link.

References

[1] P. E. Green, *Fiber optic networks*, (Prentice-Hall, 1993).

[2] J. M. H. Elmir, and H. T. Muftah, All-optical wavelength conversion: technologies and applications in DWDM networks, *IEEE Communication Magazine* Vol. 38 No. 3 (2000) pp. 86–92.

[3] F. P. Kelly, Loss networks, *Annals of Applied Probability*, Vol. 1 (1991) pp. 319–378.

[4] G. Louth, M. Mitzenmacher, and F. P. Kelly, Computational complexity of loss networks, *Theoretical Computer Science*, Vol. 125 (1994) pp. 45–59.

[5] S. Chung and K. W. Ross, Reduced load approximations for multirate loss networks, *IEEE Transactions on Communications*, Vol. 41, No. 8 (1993) pp. 1222–1231.

[6] I. Chlamtac, A. Ganz, and G. Karmi, Lightpath communications: an approach to high bandwidth optical WANs, *IEEE Transactions on Communications*, Vol. 40 No. 7 (1992) pp.1171–1182.

[7] A. Mokhtar and M. Azizoglu, Adaptive wavelength routing in all-optical networks, *IEEE/ACM Transactions on Networking*, Vol. 6. No. 2 (1998) pp. 197–206.

[8] S. Subramaniam and R. A. Barry, Wavelength assignment in fixed routing WDM networks, *Proceedings of IEEE ICC* (1997) pp. 406–410.

[9] E. Hyytia and J. Virtamo, Dynamic routing and wavelength assignment using first policy iteration, *Fifth Symposium on Computers and Communications ISCC'00* (2000).

[10] S. M. Ross, *Introduction to stochastic dynamic programming*, (Academic Press, 1983).

[11] E. Karasan and E. Ayanoglu, Performance of WDM transport networks, *IEEE Journal of Selected Areas in Communications*, Vol.16 No. 7 (1998) pp. 1081–1096.

[12] E. Karasan and E. Ayanoglu, Effects of wavelength routing and selection algorithms on wavelength conversion gain in WDM optical networks, *IEEE/ACM Transactions on Networking*, Vol. 6 No. 2 (1998) pp. 186–196.

[13] P. J. Hunt and T. G. Kurtz, Large loss networks, *Stochastic Processes and their Applications*, Vol. 53 (1994) pp. 363–378.

[14] F. P. Kelly, Routing and Capacity Allocation in Networks with Trunk Reservation, *Mathematics of Operations Research*, Vol. 15 (1990) pp. 771–792.

[15] S. P. Chung, A. Kashper, and K. W. Ross, Computing Approximate Blocking Probabilities for Large Loss Networks with State-Dependent Routing, *IEE/ACM Transactions on Networking*, Vol. 1 (1993) pp. 105–115.

[16] D. Mitra, R. J. Gibbens, and B. D. Huang, State-Dependent Routing on Symmetric Loss Networks with Trunk Reservations – I, *IEEE Transactions on Communications*, Vol. 41 (1993) pp. 400–411.

[17] D. Mitra, and R. J. Gibbens, State-Dependent Routing on Symmetric Loss Networks with Trunk Reservations, II: Asymptotics, Optimal Design, *Ann. Oper. Res.*, Vol. 35 (1992) pp. 3–30.

[18] R. A. Barry and P. A. Humblet, Models of blocking probability in all-optical networks with and without wavelength changers, *IEEE Journal of Selected Areas in Communications*, Vol. 14 No. 5 (1996) pp. 858–867.

[19] J. M. Yates, M. P. Rumsewitz, and J. P. R. Lacey, Wavelength converters in dynamically reconfigurable WDM networks, *IEEE Communications Surveys*, http://www.comsoc.org/pubs/surveys, (1999).

[20] M. Alanyali, *Asymptotically exact models for some wavelength assignment strategies in a two-hop optical network*, (Technical Report, Electrical and Electronics Eng. Dept., Bilkent University), (2000).

[21] A. Aggarwal, A. Bar-Noy, D. Coppersmith, R. Ramaswami, B. Schieber, and M. Sudan, Efficient routing and scheduling algorithms for optical networks, *Proceedings of ACM-SIAM SODA*, (1994) pp. 412–423.

[22] R. A. Barry and P. A. Humblet, On the number of wavelengths and switches in all-optical networks, *IEEE Transactions on Communications*, Vol. 42 No. 2/3/4 (1994) pp. 583–591.

[23] P. Raghavan and E. Upfal, Efficient routing in all-optical networks, *Proceedings of 26th ACM Symp. Theory of Computing*, (1994) pp. 134–143.

[24] O. Gertsel, G. Sasaki, S. Kutten, and R. Ramaswami, Worst-case analysis of dynamic wavelength allocation in optical networks, *IEEE/ACM Transactions on Networking*, Vol. 7 No. 6 (1999) pp.833–845.

OPTICAL NETWORKS - RECENT ADVANCES
L. Ruan and D.-Z. Du (Eds.) pp. 19 - 45
©2001 Kluwer Academic Publishers

Wavelength Assignment Algorithms for WDM Ring Architectures

Georgios Ellinas
Tellium Inc.
2 Crescent Place, Oceanport, NJ 07757
E-mail: gellinas@tellium.com

Krishna Bala
Tellium Inc.
2 Crescent Place, Oceanport, NJ 07757
E-mail: kbala@tellium.com

Contents

1   Introduction

Self-healing wavelength-division-multiplexing (WDM) rings are the leading candidate architectures for high capacity local exchange carrier (LEC) networks [1, 2, 3, 4, 5, 6], because of the survivability capabilities they provide and the fact that their capacity can be shared by all the nodes connected to a ring. Self-healing ring (SHR) architectures provide restoration capability for a single link and network switch failure through their ring topology and simple but fast automatic protection switching schemes. These architectures are best suited to the implementation of high capacity Local Area Networks (LAN's), interoffice, or university campus networks [7].

SONET SHR architectures using high-speed electronics have been some of the preferred carrier architectures. SONET ring networks consist of SONET multiplexing equipment interconnected together to form a closed loop. These network architectures, protect against service interruptions by having two facility paths between SONET ADM's. If a failure occurs, there exists enough redundancy to re-route the affected signals to their destinations without any service interruption [8].

Alternatively, WDM SHR networks can also be used. Even though in SONET SHR's the network capacity is limited to the fiber transmission rate or the Add/Drop Multiplexer (ADM) processing capability, WDM SHR's provide sufficient capacity for growth by multiplexing a number of wavelengths for transmission through the same medium. WDM ring architectures can provide substantially larger bandwidth (10 or more network nodes at OC-3 - OC-192 rate/node [9]), by accessing the entire bandwidth of the fiber, thus permitting very high aggregate bit rates. The structure of the network nodes is also simplified since connections that pass through these nodes stay in the optical domain and are passively routed with no need to be converted to the electrical domain. These networks enable future growth of services, while offering enhanced protection. Several WDM SHR architectures have been proposed and a number of them have been experimentally demonstrated in a laboratory environment [6, 7, 10, 11, 12].

This report focuses on 2-fiber WDM ring architectures where no protection capability is available and 4-fiber WDM Shared Protecion Ring (SPRING) [13, 14] architectures. In both architectures 2 fibers (one in the clockwise direction and one in the counter-clockwise direction) are used to carry the working traffic around the ring. A 4-fiber WDM SPRING architecture is similar in concept to the 4-fiber SONET bi-directional line switched (BLSR) ring architecture. A 4-fiber WDM SPRING architecture is a bi-directional ring with four fibers connecting each adjacent pair of nodes. For the 4-fiber SPRING architecture two fibers (one in each direction) are used as working fibers while the other two (one in each direction) are used for protection. The working fibers carry all the traffic during the failure free period while the protection fibers carry traffic only when a failure occurs. Since there are two working fibers available and the connections are always routed via shortest paths, every working fiber carries only half of the traffic on the ring.

To recover from link or node failures, the 4-fiber SPRING architecture uses Automatic Protection Switching (APS) loop-back systems [13, 14, 15] utilizing the optical protection switches that are located at each network node. When a fiber link is cut (severing all working and protection lines in a segment of the ring that physically interconnects two network nodes), the two network nodes at the endpoints of the failed link engage their protection switches to interconnect the working to the protection fibers. The traffic then propagates on the protection fibers, in a direction away from the failure until it reaches the other side of the failed link. At that point, it switches back to the original working fiber and as a result the failure is restored. A similar procedure takes place when a network node fails with the protection switches at the edges of the failed node engaging as a response to the failure. Only traffic that passes through a failed node can be restored, whereas traffic that originates or terminates at that node is lost.

In this work, each node communicates with each other node on the ring, on a single wavelength, through shortest path routes. The main assumptions are that enough wavelengths are available for full mesh connectivity on the ring, and that each connection uses only one wavelength. The routing and wavelength assignment problem for the single self-healing ring is approached in this chapter from the static point of view. Given the aforementioned assumptions, the matrix algorithm described later in this work calculates the wavelengths needed for each connection in advance and if full mesh connectivity is required, it assigns wavelengths to all connections as specified by the algorithm. Similarly, if partial connectivity is required, when a connection request arrives, the algorithm finds the shortest path route and assigns the wavelength specified for this connection. Since enough wavelengths are available for full mesh connectivity and each connection does not require more than one wavelength, this matrix-driven method will always work for both (static) full mesh connectivity or dynamic (on the fly) partial connectivity.

Section 2 discusses the number of wavelengths required for full mesh connectivity in such rings, differentiating between the cases of odd and even number of nodes on the ring. Section 3 presents a *modular* wavelength assignment algorithm for the odd and even cases which assigns wavelengths to all connections on the ring so that no Color Clash[1] (CC) constraint violation occurs [16, 17, 18, 19, 20, 21, 22]. Section 4 presents a simpler more elegant approach of finding the wavelength assignment using a *matrix* approach. The scalability problem using the matrix approach is analyzed in Section 5 and conclusions follow in Section 6.

2 Wavelengths Required for Full Mesh Connectivity

The first part of the analysis calculates the number of wavelengths required to achieve full mesh connectivity in a ring with N nodes (each node i is connected to the other $(N-1)$ nodes on the ring through shortest path routes). The network architecture is a 2-fiber ring, so the direction chosen for each connection (clockwise or counter-clockwise) is the determining factor for the shortest path route. Since the network is symmetric, all the analysis performed for one direction can be applied to the other direction as well. The analysis differentiates between the cases of odd and even number of nodes attached to the ring :

Odd number of nodes on the ring ($N = N_{odd}$) :

To find the minimum number of wavelengths required for full mesh connectivity, while observing the CC constraint, a calculation of the maximum number of connections that will share a single fiber in a single direction is required [23]. This, in turn will be equal to the minimum number of wavelengths required to "color" all connections without violating the CC constraint. Since the shortest route is always chosen, the longest path for interconnecting any two nodes will be equal to $(\frac{(N_{odd}+1)}{2} - 1)$ segments of the ring (segment is defined as a link between two adjacent nodes on the ring). Therefore, at most $\frac{(N_{odd}+1)}{2}$ nodes will share a common fiber. The minimum number of wavelengths, W_{odd}, required for full connectivity is thus given by [14] :

$$W_{odd} = \left(\begin{array}{c} \frac{(N_{odd}+1)}{2} \\ 2 \end{array} \right) = \frac{\left(\frac{(N_{odd}+1)}{2}\right)!}{\left(2!\left(\frac{(N_{odd}-3)}{2}\right)!\right)} = \frac{(N_{odd}^2 - 1)}{8} \tag{1}$$

---

[1]The Color Clash constraint states that optical signals simultaneously sharing a single fiber are required to have different wavelengths.

It has to be shown that the minimum number of wavelengths required, W_{odd}, is sufficient to create all connections while not violating the CC constraint. Note that in the presence of wavelength conversion, this is also the sufficient number of wavelengths in the network.

Even number of nodes on the ring ($N = N_{even}$) :

Again, the maximum number of connections that will share a single fiber in a single direction has to be calculated. When choosing the shortest route, the longest path for interconnecting any two nodes in this case will be equal to $(\frac{N_{even}}{2})$ segments of the ring. Therefore, at most $(\frac{(N_{even}+2)}{2})$ nodes will share a common fiber. A sufficient number of wavelengths, W_{even}, required for full mesh connectivity is thus given by :

$$W_{even} = \left(\begin{array}{c} \frac{(N_{even}+2)}{2} \\ 2 \end{array} \right) = \frac{(\frac{(N_{even}+2)}{2})!}{(2!(\frac{(N_{even}-2)}{2})!)} = \frac{(N_{even}+1)^2 - 1}{8} \qquad (2)$$

W_{even} represents a sufficient number of wavelengths necessary to achieve full mesh connectivity if wavelength conversion is allowed. From equations 1 and 2, it can be seen that the N_{even} case requires as many wavelengths as the "next odd" ($N_{even} + 1$) case. As it will be shown later though, for the even case a smaller number of wavelengths is sufficient even without wavelength conversion.

While equation 2 assumes that all shortest path connections are allocated in a single direction, a new expression can be found by taking into consideration the fact that when there are even number of nodes on the ring, the shortest path for the longest connections can be in either direction (clockwise or counter-clockwise) [23]. The analysis differentiates between the case where $(\frac{N_{even}}{2})$ is an even number, i.e., the number of nodes on the ring is divisible by 2 and 4, (N^e_{even}) and the case where $(\frac{N_{even}}{2})$ is an odd number, i.e., the number of nodes on the ring is divisible by 2 but not divisible by 4, (N^o_{even}). For these cases, full mesh connectivity can be achieved with a minimum number of wavelengths given by equations 3 and 4 respectively [23, 24, 25]:

$$W^e_{even} = \frac{(N^e_{even})^2}{8} \qquad (3)$$

$$W^o_{even} = \frac{(N^o_{even})^2 + 4}{8} \qquad (4)$$

Again, it has to be shown that the number of wavelengths given by these equations is sufficient for full mesh connectivity in the network while not violating the CC constraint. The proofs of equations 1, 3 and 4 which are derived from the modular wavelength assignment scheme can be found in [26].

A calculation of the number of wavelengths required for full mesh connectivity in a 2-fiber WDM Unidirectional Path Protection Ring (UPPR) architecture with one working and one protection fiber is presented for comparison purposes. It can be easily seen that for the 2-fiber WDM UPPR architecture, the number of wavelengths required, W_{UPPR}, is

$$W_{UPPR} = \binom{N}{2} = \frac{N(N-1)}{2} \tag{5}$$

for both the N_{odd} and N_{even} cases. As expected, the 4-fiber SPRING architecture will always require a smaller number of wavelengths than the number required by the 2-fiber UPPR architecture in order to achieve full mesh connectivity without violating the CC constraint.

3   Modular Wavelength Assignment Algorithm

This section presents a modular wavelength assignment scheme that assigns wavelengths to all connections on the ring so that no CC constraint violation occurs. It also shows how the problem scales when two (or four) new nodes are added to the ring at a time, and how the scalability approach can be used to prove that the number of wavelengths required to color all connections is equal to the minimum number of wavelengths found in equations 1, 3 and 4. Again, the assignment scheme differentiates between the cases of odd and even number of nodes on the ring :

Case 1 : Odd number of nodes on the ring ($N = N_{odd}$) :

The modular algorithm depends on the wavelength assignment of a ring with ($N_{odd} - 2$) nodes to obtain the wavelength assignment of a ring with N_{odd} nodes. However, to obtain the assignment for the ($N_{odd} - 2$) case it requires knowledge of the assignment for the ($N_{odd} - 4$) case and so on. Thus, the algorithm starts from the simple case of $N_{odd} = 3$ and continues to find the wavelength assignment for $N_{odd} = 5, 7, 9$, etc. It maintains the previous wavelength assignment and uses "new" wavelengths to accommodate the connections created due to the addition of the two new nodes on the ring. Therefore, solving the wavelength assignment problem also solves the scalability problem (when two new nodes are added to the network at a time), while maintaining the same wavelength assignment for the connections between already existing nodes on the ring. The modular wavelength assignment algorithm is as follows [27, 28] :

(1). Start with a 3-node ring ($M = 3$).

(2). Assign wavelength λ_1 to the 1-hop connections on the ring that interconnect these three nodes together.

(3). Add two new nodes (denoted as (M+1) and (M+2)), one before node 1 and one after node $\frac{(M+1)}{2}$. The ring is now separated into two areas. Area 1 includes nodes 1,2, ..., $\frac{(M+1)}{2}$ and Area 2 includes nodes $\frac{(M+3)}{2}$, ..., M.

(4). $\frac{(M+1)}{2}$ extra wavelengths are used for the new connections on the ring (connections on the clockwise direction only). The new wavelengths are labeled λ_i, λ_{i+1}, ..., $\lambda_{(i+\frac{(M+1)}{2})}$, where $i = (j + 1)$ and j is the maximum number of wavelengths used in the M-node ring. The wavelength assignment is then as follows :

(4a). Area 1 : Assign wavelength λ_i from node $(M + 1)$ to node 1, wavelength $\lambda_{(i+1)}$ from node $(M + 1)$ to node 2, ..., wavelength $\lambda_{(i+\frac{(M+1)}{2})}$ from node $(M + 1)$ to node $\frac{(M+1)}{2}$. Assign wavelength λ_i from node 1 to node $(M + 2)$, wavelength $\lambda_{(i+1)}$ from node 2 to node $(M + 2)$, ..., wavelength $\lambda_{(i+\frac{(M+1)}{2})}$ from node $\frac{(M+1)}{2}$ to node $(M + 2)$.

(4b). Area 2 : Assign wavelength λ_i from node $\frac{(M+3)}{2}$ to node $(M + 1)$, wavelength $\lambda_{(i+1)}$ from node $\frac{(M+3)}{2} + 1$ to node $(M + 1)$, ..., wavelength $\lambda_{(i+\frac{(M-1)}{2})}$ from node M to node $(M + 1)$. Similarly, assign wavelength λ_i from node $(M + 2)$ to node $\frac{(M+3)}{2}$, wavelength $\lambda_{(i+1)}$ from node $(M + 2)$ to node $\frac{(M+3)}{2} + 1$, ..., wavelength $\lambda_{(i+\frac{(M-1)}{2})}$ from node $(M + 2)$ to node M.

(4c). Wavelength $\lambda_{(i+\frac{(M+1)}{2})}$ is assigned from node $(M + 2)$ to node $(M + 1)$ in order to interconnect the two new nodes.

(5). Rename all nodes on the ring (sequentially) as $1, 2, ..., M$, (where M now includes the two new nodes ($M = M + 2$)).

(6). Return to Step (3). Repeat until desired number of nodes has been reached.

Figure 1 shows an implementation of the algorithm when the network scales from $N_{odd} = 3$ to $N_{odd} = 5$, and finally to $N_{odd} = 7$ nodes.

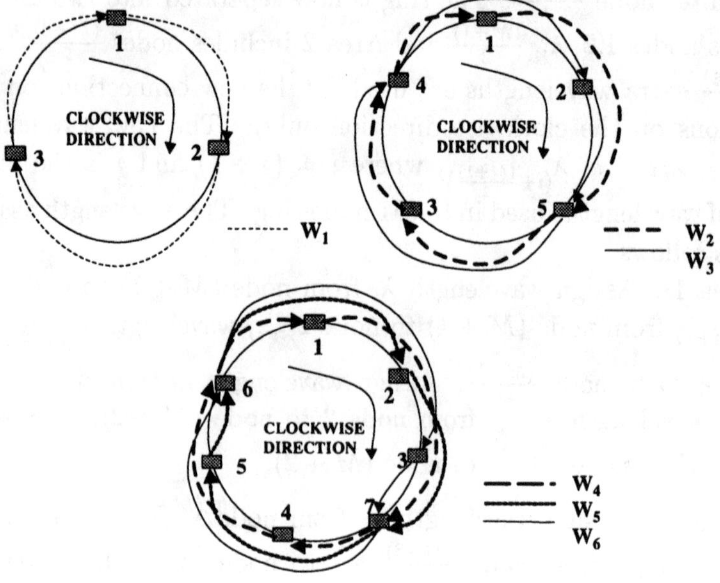

Figure 1: Wavelength assignment for the new connections created when the ring scales from $N = 3$ to $N = 5$ and then to $N = 7$ nodes

Clearly, the modular wavelength assignment scheme presented above provides the framework for the proof of equation 1 [26].

Case 2 : Even number of nodes on the ring ($N = N_{even}$) :

For the case of even number of nodes on the ring, a wavelength assignment algorithm similar to the one in Case 1 is used. It now depends on the assignment of a ring with ($N_{even} - 4$) nodes to obtain the wavelength assignment for a ring with N_{even} nodes. Thus, the algorithm starts from the simple case of $M = i$ and continues to find the wavelength assignment for rings with $M = (i + 4), (i + 8), (i + 12)$, etc. Again, the algorithm maintains the previous wavelength assignment and uses "new" wavelengths to accommodate the connections created due to the addition of four new nodes on the ring.

Two different variations of the same algorithm are used for N^e_{even} and N^o_{even} nodes on the ring. In the N^e_{even} case, four nodes are added to the ring at each time, whereas in the N^o_{even} case, the number of nodes on the ring is again incremented by four but this procedure is done in two phases. In the first phase, the number of nodes on the ring is incremented by two and the wavelength assignment for the new connections is made following a similar wavelength assignment scheme as in the N^e_{even} case. Two more nodes are then added to the ring and the same wavelength assignment process is repeated.

Figure 2 shows the addition of four nodes at a time and the new wavelength assignments when the ring expands from $N^e_{even} = 4$ to $N^e_{even} = 8$ nodes. The detailed wavelength assignment algorithms are not explicitly presented here but can be found in [26].

4   Matrix Wavelength Assignment Algorithm

A simpler and more graceful wavelength assignment algorithm that assigns wavelengths to all connections in order to achieve full mesh connectivity on the ring while avoiding violating the CC constraint is presented in this section. A matrix approach was found to work the best for this problem [23]. The columns of the matrix represent the nodes on the ring and the rows represent the wavelengths required for each connection from each node to every other node on the ring. The values placed in the matrix represent the connections, on one direction only, from each node to every other node (representing half of the total number of connections). By appropriately filling-in the matrix, while following some simple rules, an optimal wave-

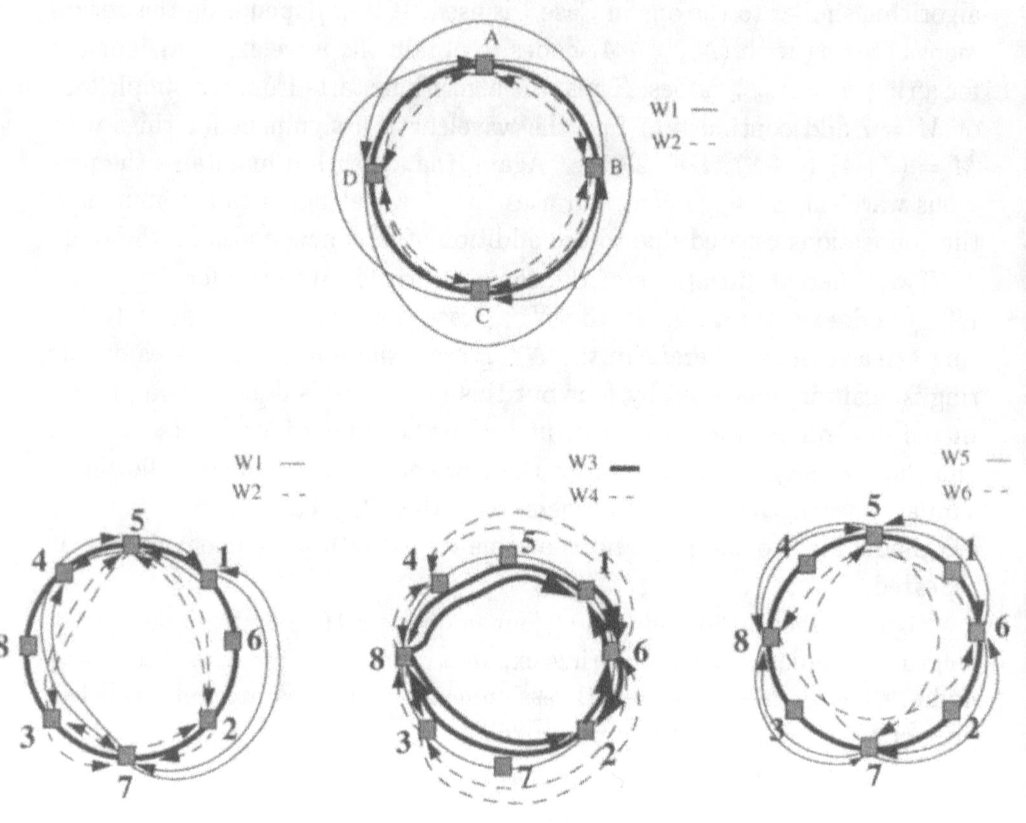

Figure 2: Wavelength assignment for the new connections created when the ring scales from $N = 4$ to $N = 8$ nodes

length assignment can be obtained. An explanation of how the algorithm works for the case of odd and even number of nodes attached to the ring is presented below:

Case 1 : Odd number of nodes on the ring ($N = N_{odd}$) :

Matrix Definitions for $N = N_{odd}$:

- The matrix is of size $\{W_{odd} \times N_{odd}\}$ where W_{odd} is the number of wavelengths required (calculated using equation 1) and N_{odd} is the number of nodes on the ring.

- The values placed in the matrix range from 1 to L_{max}, where L_{max} represents the longest connection between any two nodes on the ring through the shortest path route. This number equals : $L_{max} = \frac{(N_{odd}-1)}{2}$.

- A value K in position (i, j) states that node j is connected to node $(j + K)mod(N_{odd})$ using wavelength i.

- The symbol X in position (i, j) states that wavelength i is not used for any of the connections starting from node j.

The values placed in the matrix represent the connections, on one direction only, from each node to every other node (representing half of the total number of connections). In order to fill-in the matrix, specific rules have to be followed. These rules are :

Matrix Rules :

- In a row : After inserting value K in the matrix, $(K - 1)$ matrix locations following value K are left empty (including wrap-around), i.e., if value K is placed at (i, j) (wavelength λ_i connects node j to node $(K + j)mod(N_{odd})$), then locations $(i, ((j + 1)modN_{odd}))$, ..., $(i, (j + (K - 1))modN_{odd})$ will be empty (X's are placed at these locations). These locations cannot be used for a different connection, since there will exist two connections that have intersecting paths and use the same wavelength. However, this is not allowed since it would violate the CC constraint.

- In a column : No two entries K_1 and K_2 can exist where $K_1 = K_2$. Since each value represents the number of hops in a connection, by having the same value in a single column, a node repeats the same connection using different wavelengths. This contradicts the original assumption that there is a single connection between any two nodes with a dedicated wavelength assigned to that connection.

- In a column : The whole range of values $1, ..., L_{max}$ is used. Since these entries correspond to the "connection hops" , (i.e., if value 2 appears in column 1, it signifies a connection between node 1 and node 3), this guarantees that each node is connected with all other nodes (on one direction only), and thus full mesh connectivity is supported. Since this is a bi-directional ring, a node communicates with the remaining $\frac{(N_{odd}-1)}{2}$ nodes on the fiber going in the opposite direction.

- In a row : Summation of all the values in a row equals N_{odd}. This ensures that each wavelength is fully utilized since it is used on all segments of the ring exactly once.

These four rules ensure that the CC condition is not violated, there exist no multiple connections between nodes, all connections are assigned a distinct wavelength and each wavelength is fully utilized. Clearly, if the matrix defined above can be filled following these rules, an optimal wavelength assignment supporting full mesh connectivity (since the size of the matrix implies that the minimum number of wavelengths required, W_{odd} is used) can be obtained.

A simple way to fill-in the matrix is by cyclically shifting the matrix entries by one for each column, i.e., placing entries $(1, 2, ..., L_{max})$ in the first column, entries $(L_{max}, 1, 2, ..., (L_{max} - 1))$ in the second column and so on. Figure 3(a) displays a matrix for odd size ring networks ($N_{odd} = 7$). In the case of $N_{odd} = 7$ nodes, $W_{odd} = 6$ wavelengths are required and the longest connection is $L_{max} = 3$. The entries for the columns of the matrix follow the cyclic shift described above. After the matrix is calculated, the wavelength assignment is simple to obtain (Figure 3(b)). A second example is shown in Figure 4 for the case of $N_{odd} = 11$. The corresponding wavelength assignment graph is omitted because of the large number of wavelengths ($W_{odd} = 15$) required for full mesh connectivity. It is easy to recognize though that given the corresponding matrix, the wavelength assignment becomes a trivial exercise.

As previously mentioned, the matrix specifies the wavelength assignment on one direction. The wavelength assignment on the other direction will be identical. This does not violate the CC constraint since the connections on

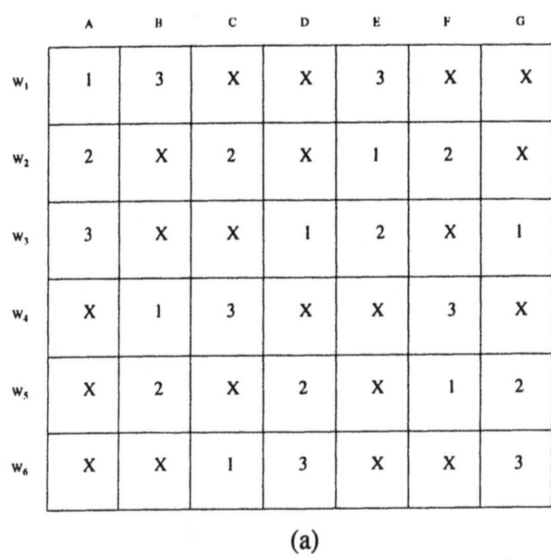

	A	B	C	D	E	F	G
W_1	1	3	X	X	3	X	X
W_2	2	X	2	X	1	2	X
W_3	3	X	X	1	2	X	1
W_4	X	1	3	X	X	3	X
W_5	X	2	X	2	X	1	2
W_6	X	X	1	3	X	X	3

(a)

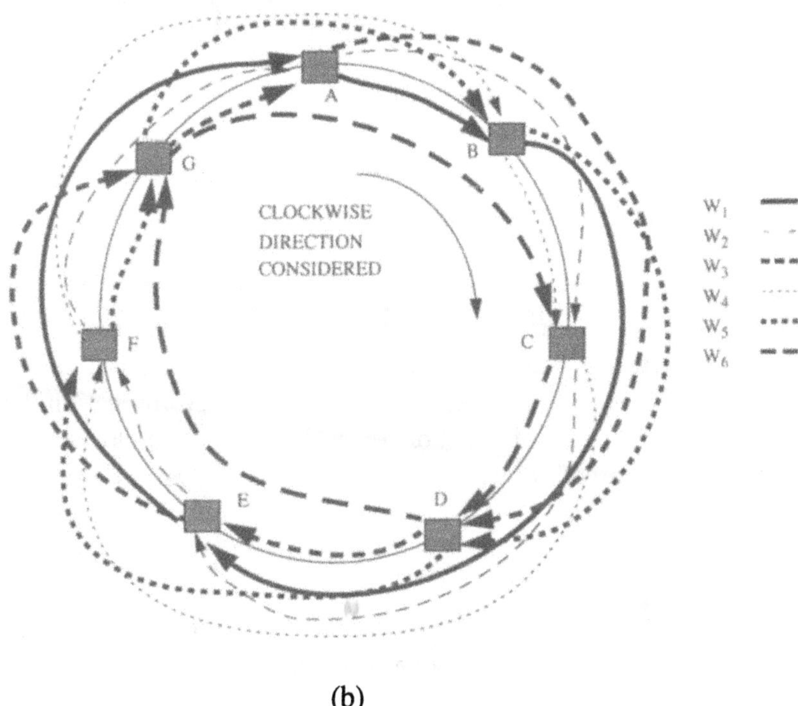

(b)

Figure 3: (a) Wavelength assignment matrix for $N = 7$ nodes on the ring (b) Wavelength assignment for the ring with $N = 7$ nodes as specified by the wavelength assignment matrix

	A	B	C	D	E	F	G	H	I	J	K
w_1	1	5	X	X	X	X	5	X	X	X	X
w_2	2	X	4	X	X	X	1	4	X	X	X
w_3	3	X	X	3	X	X	2	X	3	X	X
w_4	4	X	X	X	2	X	3	X	X	2	X
w_5	5	X	X	X	X	1	4	X	X	X	1
w_6	X	1	5	X	X	X	X	5	X	X	X
w_7	X	2	X	4	X	X	X	1	4	X	X
w_8	X	3	X	X	3	X	X	2	X	3	X
w_9	X	4	X	X	X	2	X	3	X	X	2
w_{10}	X	X	1	5	X	X	X	X	5	X	X
w_{11}	X	X	2	X	4	X	X	X	1	4	X
w_{12}	X	X	3	X	X	3	X	X	2	X	3
w_{13}	X	X	X	1	5	X	X	X	X	5	X
w_{14}	X	X	X	2	X	4	X	X	X	1	4
w_{15}	X	X	X	X	1	5	X	X	X	X	5

Figure 4: Wavelength assignment matrix for $N = 11$ nodes on the ring

the other direction are carried on a different fiber.

Case 2 : Even number of nodes on the ring ($N = N_{even}$) :

If N is an even number, when all connections are allocated in one direction only, the minimum number of wavelengths required for full mesh connectivity is shown in equation 2. As it was demonstrated in Section 2 though, by choosing the direction of the shortest path route for the longest connections, the minimum number of wavelengths required is given by equations 3 and 4. The analysis differentiates between the N_{even}^e and N_{even}^o cases :

Case 2a : $N = N_{even}^e$:

A wavelength assignment scheme similar to the one used in the $N = N_{odd}$ case is used here as well. The longest connections are allocated first, using a new allocation scheme presented below, and $\frac{N_{even}^e}{4}$ wavelengths are assigned to these connections. The rest of the connections are allocated using a modified matrix approach.

(a) For all the longest connections, a new wavelength assignment scheme is defined that assigns wavelength λ_i for connections (j, k) and (k, j), on one direction as well as connections $(j+1), (k+1)$ and $(k+1), (j+1)$ on the other direction, wavelength λ_{i+1} for connections $(j+2), (k+2)$ and $(k+2), (j+2)$, on one direction and connections $(j+3), (k+3)$ and $(k+3), (j+3)$ on the other direction and so on. In general, the wavelength assignment algorithm alternates directions for alternating nodes on the ring and assigns to both pairs of connections (a connection pair is a pair $(i,j), (j,i)$) the same wavelength.

Because of the wavelength re-use, only as many wavelengths as half the number of connections ($\frac{N^e_{even}}{4}$) are needed to color all the longest connections without violating the CC constraint. The total number of wavelengths required to color all connections will then be equal to $W^e_{longest}$ (number of wavelengths to color the longest connections using the scheme described above) in addition to the number of wavelengths required to color the remaining connections (W^e_{remain}). To calculate W^e_{remain}, the maximum number of connections that will share a single fiber needs to be determined. The second longest path for interconnecting any two nodes will be equal to ($\frac{N^e_{even}}{2} - 1$) segments of the ring, therefore at most $\frac{N^e_{even}}{2}$ nodes will share a common fiber. The minimum number of wavelengths required for all but the longest connections, W^e_{remain}, is thus given by :

$$W^e_{remain} = \left(\begin{array}{c} \frac{N^e_{even}}{2} \\ 2 \end{array} \right) = \frac{(\frac{N^e_{even}}{2})!}{(2!(\frac{(N^e_{even}-4)}{2})!)} = \frac{(N^e_{even} - 1)^2 - 1}{8}$$

and $W^e_{even} = W^e_{longest} + W^e_{remain} = \frac{N^e_{even}}{4} + \frac{((N^e_{even}-1)^2-1)}{8} = \frac{(N^e_{even})^2}{8}$ (eq. 3)

In the presence of wavelength conversion this is the sufficient number of wavelengths in the network. The minimum number of wavelengths required, W^e_{even}, is also sufficient to create all connections while not violating the CC constraint, even in the absence of wavelength conversion [26].

(b) The rest of the connections are assigned wavelengths using the following modified matrix approach :

Matrix Definitions for $N = N^e_{even}$:

1. The matrix is of size $\{W^o_{prev} \times N^e_{even}\}$ where W^o_{prev} is the number of wavelengths for the previous odd case, calculated using equation 1, and N^e_{even} is the number of nodes on the ring.

2. The values placed in the matrix range from 1 to ($L_{max} - 1$), where L_{max} represents the longest connection between any two nodes on the ring. The

number $(L_{max}-1)$ represents the longest connection between any two nodes in the "previous odd" $(N^o_{prev} = N^e_{even} - 1)$ case.

3. A value K in position (i,j) states that node j is connected to node $(j + K)mod(N^e_{even})$ using wavelength λ_i.

4. The symbol X in position (i,j) states that wavelength λ_i is not used for any of the connections starting from node j.

The values placed in the matrix represent all but the longest connections, on one direction only, from each node to every other node. The same matrix rules as in Case 1 apply here as well. Clearly, filling-in matrix $\{W^o_{prev} \times N^e_{even}\}$ following these rules, together with the new wavelength assignment scheme for the longest connections, corresponds to an optimal wavelength assignment supporting full mesh connectivity. The new matrix is now filled as follows :

Step 1 : The previous odd matrix is filled using the cyclic shift method.

Step 2 : A new column is added to the matrix at any arbitrary location.

Step 3 : For each row, starting from the new column, the algorithm traverses the row towards the left (including wrap-around), until it encounters the first number entry (other than an X). The new column is now filled using two simple rules :

Rule 1 : If the number entry encountered, K, is less than the longest connection entry for the previous odd case $(L_{max}^{prev(odd)})$, place an (X) at the new column and increment K by 1.

Rule 2 : If the number entry encountered, K, equals $L_{max}^{prev(odd)}$, denote by the number Q the number of X's after the new column. The algorithm then places the number $M = Q + 1$ at the new column and K now becomes $K = K - Q$.

Figure 5(a) shows such a matrix for the $N^e_{even} = 8$ case. In the matrix, all connections except the longest ones are allocated using 6 wavelengths (as many wavelengths as for the previous odd $(N^o_{prev} = 7)$ case). The allocation of half of the longest connections is shown in Figure 5(b) (clockwise direction). The identical assignment is valid for all the other connections in the counter-clockwise direction.

Case 2b : $N = N^o_{even}$:

The wavelength assignment scheme allocates first the longest connections, and $(\frac{N^o_{even}-2}{4} + 1)$ wavelengths are assigned to these connections. The rest

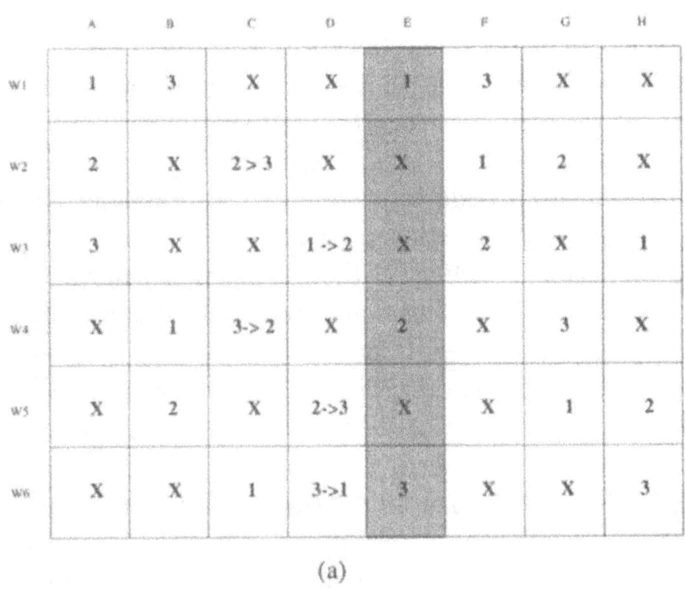

(a)

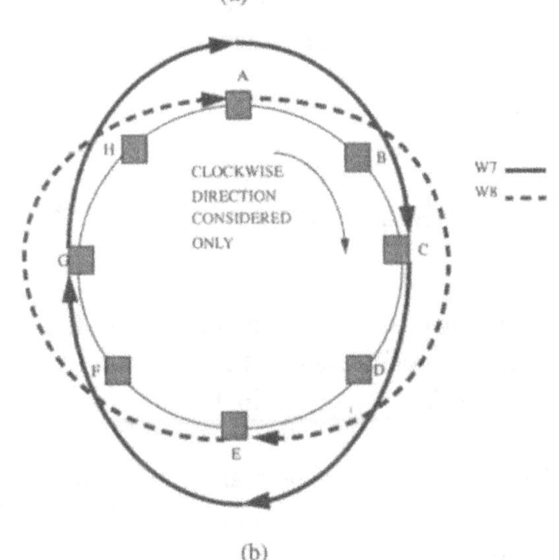

(b)

Figure 5: (a) Wavelength assignment matrix for $N = 8$ nodes on the ring (b) Wavelength assignment for the longest connections for a ring with $N = 8$ nodes.

of the connections are allocated using the same matrix approach as in the $N = N_{even}^e$ case.

The number of longest connections will again be equal to $\frac{N_{even}^o}{2}$, an odd number now. The new wavelength allocation algorithm for the longest connections assigns the same wavelength to the longest connections in a single direction for "pairs of nodes" now, i.e., if it assigns wavelength λ_i for a longest connection from node k to node j, it also assigns the same wavelength for the longest connection from node j to node k. Since $\frac{N_{even}^o}{2}$ is an odd number, after assigning all the "connection pairs", it needs to assign one more connection. This connection can be any of the remaining connections. There are $\frac{(N_{even}^o-2)}{2}$ pairs. Half of them are assigned to the clockwise direction and the other half to the counter-clockwise direction. This results in a requirement of $\frac{(N_{even}^o-2)}{4}$ wavelengths. For the last ("unpaired") connection an additional wavelength is needed. The total number of wavelengths required to color all the connections on the ring will then be equal to $W_{longest}^o$ (number of wavelengths to color the longest connections) in addition to the wavelengths required to color the remaining connections. The latter equals to the number of wavelengths needed for the "previous odd" case ($N_{prev}^o = N_{even}^o - 1$). That number of wavelengths is given by equation 1 as $W_{remain}^o = \frac{((N_{prev}^o)^2-1)}{8}$. Thus,

$$W_{even}^o = W_{longest}^o + W_{remain}^o = \left(\frac{N_{even}^o-2}{4} + 1\right) + \frac{((N_{even}^o-1)^2-1)}{8} = \frac{(N_{even}^o)^2+4}{8}$$
(eq. 4)

In the presence of wavelength conversion this is the sufficient number of wavelengths in the network. Again, the minimum number of wavelengths required, W_{even}^o, is sufficient, even in the absence of wavelength conversion [26].

Figure 6(a) shows the wavelength assignment matrix for $N_{even}^o = 6$. In addition, the longest connections assignment is shown in Figure 6(b) for the clockwise direction. As before, the identical assignment is valid for all the other connections in the counter-clockwise direction.

5   Scalability Using the Matrix Approach

If the ring size increases, for the new ring, the wavelength allocation algorithms presented in Section 4 can again find the new wavelength assignments for full mesh connectivity. This section presents the case of adding a *single*

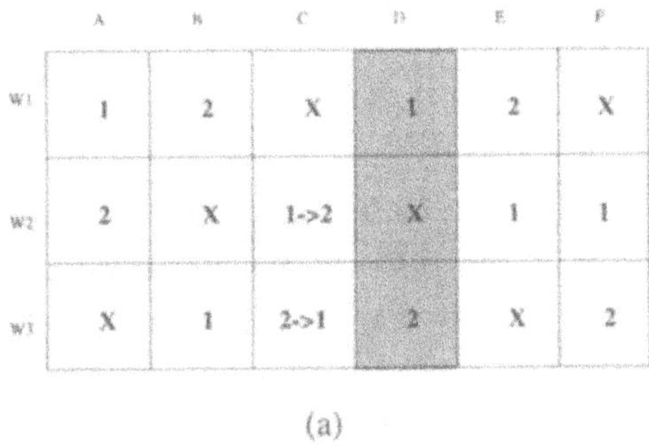

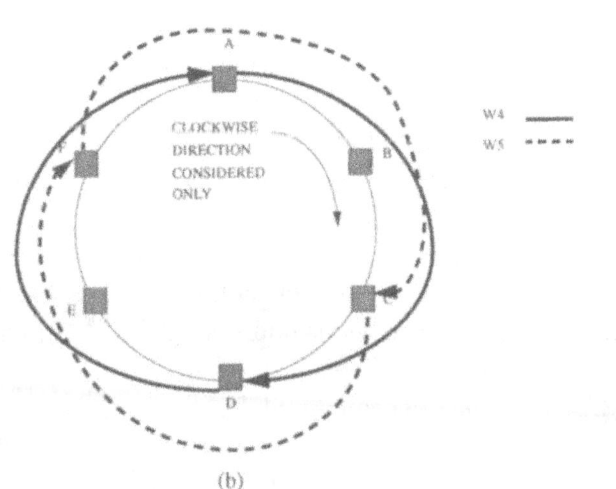

Figure 6: (a) Wavelength assignment matrix for $N = 6$ nodes on the ring (b) Wavelength assignment for the longest connections for a ring with $N = 6$ nodes.

node at a time at *any location* on the ring. It demonstrates that when the size of the ring increases by one, the matrix approach can be used to scale the network while re-assigning wavelengths to only a small number of the already existing connections [29]. The analysis differentiates between scaling rings with odd and even number of nodes. For both cases, the matrix scales from $\{W \times N\}$ to $\{W' \times (N+1)\}$. A short explanation on how the new matrix is filled follows :

Case 1 : N' = N + 1 (N = odd number) :

This new method scales a matrix with an odd number of nodes to a matrix with an even number of nodes while adding the new node at *any location* on the ring.

The matrix is expanded from $\{W \times N\}$ ($W = \frac{N^2-1}{8}$) to $\{W' \times N'\}$. The number of columns is increased by 1 and the number of rows is increased to correspond to the new number of wavelengths required, W' ($W' = \frac{(N')^2}{8}$ or $W' = \frac{(N')^2+4}{8}$). For each row, from the new column, the algorithm traverses the row towards the left (including wrap-around) until it encounters the first entry (other than an X). The first W rows of the new column are then filled following rules 1 and 2 below. The rest of the rows are filled with the longest connection entries using a simple algorithm that places entry L_{max} at location (i,j) if entry L_{max} is not used in column j, and location (i,j) is empty (no X entry) (Figure 8).

Rule 1 : If the number entry encountered, K, is less than the longest connection entry for the previous odd case ($L_{max}^{prev(odd)}$), place an (X) at the new column and increment K by 1.

Rule 2 : If the entry encountered K, equals $L_{max}^{prev(odd)}$, the algorithm places the number $M = (Q+1)$ at the new column and entry K becomes $K' = (K - Q)$. The number Q denotes the number of X's that follow the new column for that specific row.

Figure 7 show the required full mesh connectivity and the wavelength assignment for a 5-node ring. Figure 8 shows the new matrix when the number of nodes on the ring increases from $N = 5$ to $N' = 6$. Column F is the new column and rows $W_4 - W_5$ are the new rows added to the matrix. The resulting wavelength assignment is optimal while at the same time maintaining the same wavelength assignment for all but three of the existing connections on the ring.

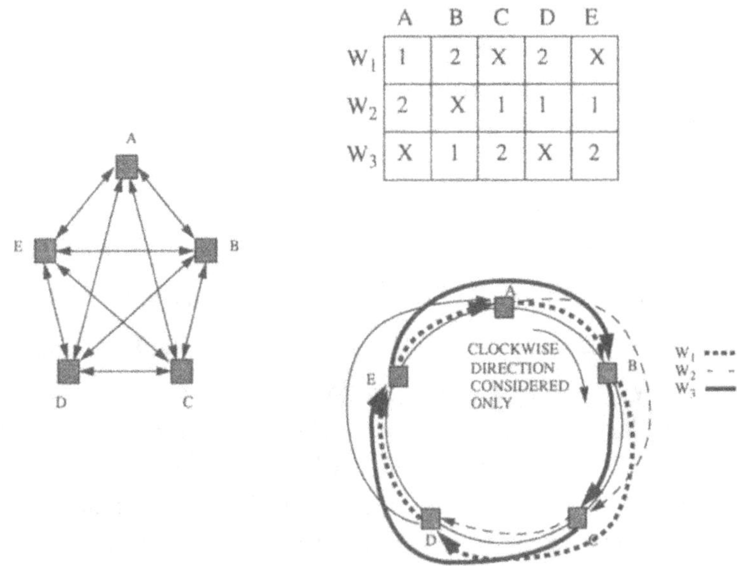

Figure 7: Wavelength assignment matrix for the 5-node ring and the wavelength assignment for full mesh connectivity as specified by the matrix

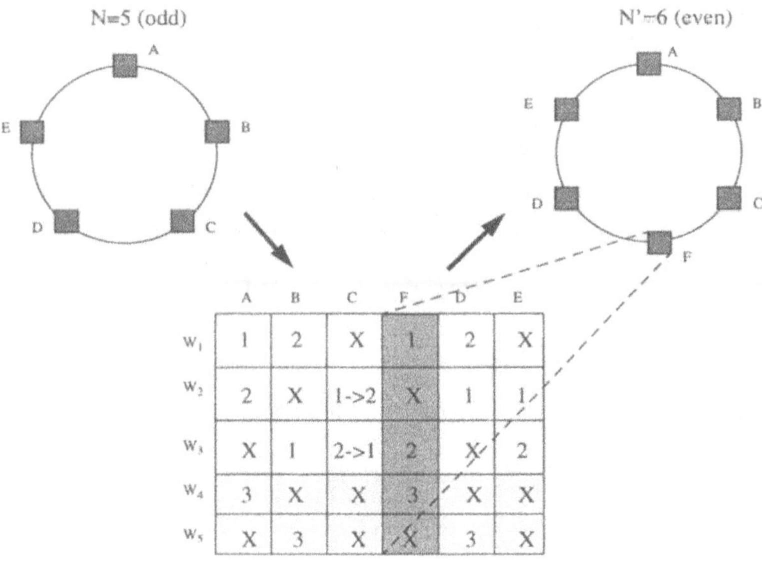

Figure 8: Wavelength assignment matrix for a ring that scales from $N = 5$ to $N' = 6$ nodes

Case 2 : N' = N + 1 (N = even number) :

The analysis also considers the case of scaling a matrix with an even number of nodes to a matrix with an odd number of nodes. As before, the new column (representing the new node) can be added at *any location* in the matrix.

The matrix is now expanded from $\{W \times N\}$ to $\{W' \times N'\}$ ($W' = \frac{(N')^2-1}{8}$). For the new column, for rows $1, ..., (\frac{(N-1)^2-1}{8})$, the algorithm follows Rule 1 as defined above. For rows $(\frac{(N-1)^2-1}{8}) + 1, ..., W$ (where W equals to the number of wavelengths for $N = N_{even}$ ($W = \frac{(N)^2}{8}$ if $N = N^e_{even}$ or $W = \frac{(N)^2+4}{8}$ if $N = N^o_{even}$)) the algorithm follows Rule 2. For rows $(W+1), ..., W'$ the algorithm places in the new column the connection entries that are not used in the rows above for that column in a sequential order. The rest of the row entries are filled using a simple algorithm that places entry K at location (i, j) if entry K is not used in column j, and location (i, j) is empty (no X entry) (Figure 9).

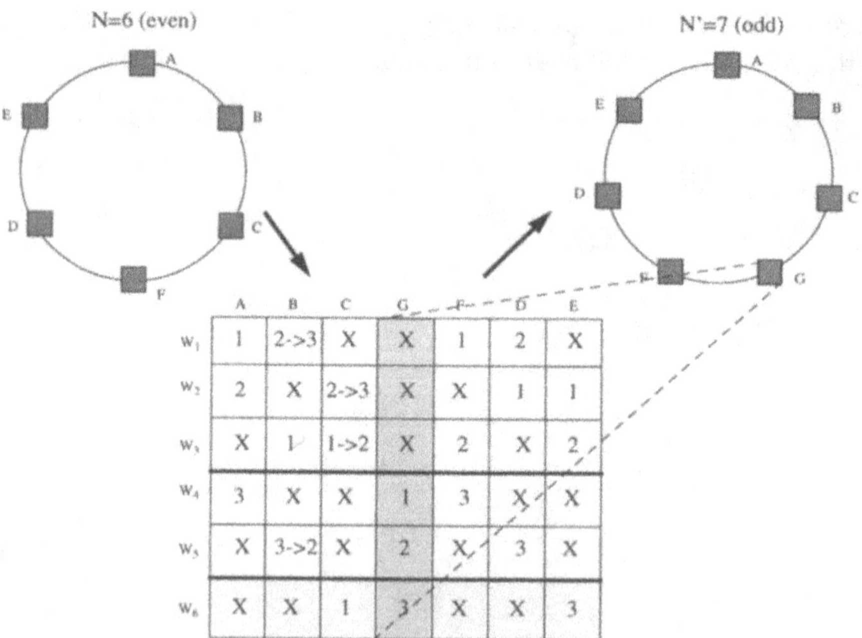

	A	B	C	G	F	D	E
W_1	1	2->3	X	X	1	2	X
W_2	2	X	2->3	X	X	1	1
W_3	X	1	1->2	X	2	X	2
W_4	3	X	X	1	3	X	X
W_5	X	3->2	X	2	X	3	X
W_6	X	X	1	3	X	X	3

Figure 9: Wavelength assignment matrix for a ring that scales from $N = 6$ to $N' = 7$ nodes

Figure 9 shows an example of the new matrix when the number of nodes on the ring increases from $N = 6$ to $N' = 7$. Column G is the new column and row W_6 is the new row added to the matrix.

6 Conclusions

This chapter presents 2-fiber multiwavelength rings with full mesh connectivity among their nodes. The number of wavelengths that is sufficient to achieve full mesh connectivity in these networks through shortest path routing is calculated for both the $N = N_{odd}$ and $N = N_{even}$ cases. Two optimal wavelength assignment algorithms are proposed that assign the minimum number of wavelengths between nodes on the ring to achieve the required connectivity while observing the CC constraint. These algorithms are based on a modular and a matrix approach respectively. The algorithm which is based on a matrix approach is found to be simpler and much easier to implement. After the entries of the matrix are calculated and the matrix is completed, the assignment of the wavelengths for each connection on the ring becomes a trivial exercise. If enough wavelengths are available for full mesh connectivity on the ring, and each node on the ring communicates with each other node on the ring with a single wavelength, the matrix algorithm will always work for both (static) full mesh connectivity or dynamic (on the fly) partial connectivity.

Scalability is also examined in this chapter. The modular approach demonstrates that it is possible to add two nodes at specified locations on the ring, without disturbing the existing wavelength assignments. Furthermore, it is shown that the matrix approach allows for a single node to be added to the ring network at a time, with only minimal disturbance to the existing wavelength assignments. The new node can be placed at any position on the ring. If the network scales by more than one node, the process is simply repeated until full mesh connectivity is achieved. Thus, assuming that enough wavelengths are available, the ring can scale to any size.

Finally, if not enough wavelengths are available for full mesh connectivity or a node connection requires more than one wavelength, the wavelength assignment will then have to be carried out dynamically. In that case, some of the node connections may be blocked. The goal then becomes to find an algorithm which can minimize the number of blocked connections on a ring with limited resources.

References

[1] A.S. Acampora, The scalable lightwave network, *IEEE Communications Mag.* Vol. 32 No. 12 (1994) pp. 36-42.

[2] S.B. Alexander, R.S. Bondurant, D.Byrne, V.W.S. Chan, S.G. Finn, R. Gallager, B.S. Glance, H.A. Haus, P. Humblet, R. Jain, I.P. Kaminow, M.J. Karol, R.S. Kennedy, A. Kirby, H.Q. Le, A.A.M. Saleh, B.A. Schofield, H.H. Shapiro, N.K. Shankaranarayanan, R.E. Thomas, R.C. Williamson and R.W. Wilson, A precompetitive consortium on wideband all optical networks, *IEEE/OSA Journal of Lightwave Technology* Vol. 11 No. 5/6 (1993) pp. 714-735.

[3] C.A. Brackett, Dense wavelength division multiplexing networks: Principles and applications, *IEEE Journal on Selected Areas in Communications* Vol. 8 No. 6 (1990) pp. 948-964.

[4] C.A. Brackett, A.S. Acampora, J. Sweitzer, G. Tangonan, M.T. Smith, W. Lennon, K.C. Wang and R. H. Hobbs, A scalable multiwavelength multihop optical network: A proposal for research on all-optical networks, *IEEE/OSA Journal of Lightwave Technology* Vol. 11 No. 5/6 (1993) pp. 736-753.

[5] G.R. Hill, P.J. Chidgey, F. Kaufhold, T. Lynch, O. Sahlen, M. Gustavsson, M. Janson, B. Lagerstrom, G. Grasso, F. Meli, S. Johansson, J. Ingers, L. Fernandez, S. Rotolo, A. Antonielli, S. Tebaldini, E. Vezzoni, R. Caddedu, N. Capanio, F. Testa, A. Scavennec, M.J. O'Mahony, J. Zhou, A. Yu, W. Sohler, U. Rust and H. Herrmann, Multi-wavelength transport network: A transport network layer based on optical network elements, *IEEE/OSA Journal of Lightwave Technology* Vol. 11 No. 5/6 (1993) pp. 667-679.

[6] R.E. Wagner, R.C. Alferness, A.A.M. Saleh and M. S. Goodman, MONET: Multiwavelength optical networking, *IEEE/OSA Journal of Lightwave Technology* Vol. 14 No. 6 (1996) pp. 1349-1355.

[7] M. Sharma, H. Ibe and T. Ozeki, WDM ring network using a centralized multiwavelength light source and add-drop multiplexing fibers, *IEEE/OSA Journal of Lightwave Technology* Vol. 15 No. 6 (1997) pp. 917-929.

[8] J.K. Conlisk, Topology and survivability of future transport networks, *In Proc. IEEE Globecom.*, pp. 826-834, Dallas, TX, November 1989.

[9] S. Wagner and T.E. Chapuran, Multiwavelength ring networks for switch consolidation and interconnection, *In Proc. IEEE Int'l Conf. Commun.*, pp. 1173-1179, Chicago, IL, June 1992.

[10] F. Arecco, F. Casella, E. Iannone, A. Mariconda, S. Merli, F. Pozzi and F. Veghini, A transparent, all-optical, metropolitan network experiment in a field environment: The "PROMETHEO" self-healing ring, *IEEE/OSA Journal of Lightwave Technology* Vol. 15 No. 12 (1997) pp. 2206-2213.

[11] B. Glance, C.R. Doerr, I.P. Kaminow and R. Montagne, Optically restorable WDM ring network using simple add/drop circuitry, *IEEE/OSA Journal of Lightwave Technology* Vol. 14 No. 11 (1996) pp. 2453-2456.

[12] P.A. Perrier, S. Ruggeri, A. Noury, P. Gavignet, S. Gauchard, V. Havard, L. Berthelon, H. Fevrier and J. Dupraz, 4-channel, 10-Gbit/s capacity, self-healing WDM ring network with wavelength add/drop multiplexers, *In Proc. IEEE/OSA Optical Fiber Commun. Conf.*, San Jose, CA, February 1996.

[13] K. Bala, Multiwavelength optical network architecture, *In 7th Workshop on Very High Speed Networks* Maryland, July 1996.

[14] A.F. Elrefaie, Multiwavelength survivable ring network architectures, *In Proc. IEEE Int'l Conf. Commun.*, pp. 1245-1251, Geneva, Switzerland, May 1993.

[15] T-H. Wu and R.C. Lau, A class of self-healing ring architectures for SONET network applications, *In Proc. IEEE Globecom.*, pp. 444-451, San Diego, CA, December 1990.

[16] K. Bala, T.E. Stern and K. Bala, Algorithms for routing in a linear lightwave network, *In Proc. IEEE Infocom.*, pp. 1-9, Bal Harbor, FL, April 1991.

[17] S. Baroni and P. Bayvel, Wavelength requirements in arbitrary connected wavelength-routed optical networks, *IEEE/OSA Journal of Lightwave Technology* Vol. 15 No. 2 (1997) pp. 242-252.

[18] R.A. Barry and P.A. Humblet, On the number of wavelengths and switches in all-optical networks, *IEEE Transactions on Communications* Vol. 42 No. 2/3/4 (1994) pp. 583-591.

[19] I. Chlamtac, A. Ganz and G. Karmi, Purely optical networks for Terabit communication, *In Proc. IEEE Infocom.*, pp. 887-896, Ottawa, Canada, April 1989.

[20] K.C. Lee and V.O.K. Li, A wavelength convertible optical network, *IEEE/OSA Journal of Lightwave Technology* Vol. 11 No. 5/6 (1993) pp. 962-970.

[21] B. Mukherjee, WDM-based local lightwave networks - Part I: Single-hop systems, *IEEE Network Mag.*, Vol. 6 No. 3 (1992) pp. 12-27.

[22] R. Ramaswami and K.N. Sivarajan, Routing and wavelength assignment in all-optical networks, *IEEE/ACM Transactions on Networking* Vol. 3 No. 5 (1995) pp. 489-500.

[23] G. Ellinas, K. Bala and G.K. Chang, A novel algorithm for wavelength assignment in 4-fiber WDM self-healing rings, *In Proc. IEEE Int'l Conf. Commun.*, Atlanta, GA, June 1998.

[24] D. Hunter, Optical mesh routing in four-fibre WDM rings, *Electron. Letters* Vol. 34 (1998).

[25] G. Wilfong, Minimizing wavelengths in an all-optical ring network, *In 7th International Symposium on Algorithms and Computation* pp. 346-355, 1996.

[26] G. Ellinas, Fault Restoration in optical networks: General methodology and implementation, Ph.D. Thesis, Columbia University, New York, NY, 1998.

[27] G. Ellinas, K. Bala and G.K. Chang, Algorithms for wavelength assignment in 2 and 4-fiber self-healing rings, Technical Report TM-25821, Bell Communications Research, Red Bank, NJ, September 1997.

[28] T.E. Stern and K. Bala, *Multiwavelength Optical Networks*, (Addison-Wesley, 1999).

[29] G. Ellinas, K. Bala and G.K. Chang, Scalability of a novel wavelength assignment algorithm for WDM shared protection rings, *In Proc.*

IEEE/OSA Optical Fiber Commun. Conf., pp. 363-364, San Jose, CA, February 1998.

OPTICAL NETWORKS - RECENT ADVANCES
L. Ruan and D.-Z. Du (Eds.) pp. 47 - 74
©2001 Kluwer Academic Publishers

Dynamic Traffic Scheduling for QoS Support in WDM/TDM Networks with Arbitrary Tuning Latencies

Nen-Fu Huang
Department of Computer Science
National Tsing Hua University, Hsinchu 300, Taiwan, R.O.C.
E-mail: nfhuang@cs.nthu.edu.tw

Te-Lung Liu
Department of Computer Science
National Tsing Hua University, Hsinchu 300, Taiwan, R.O.C.
E-mail: tlliu@cs.nthu.edu.tw

Ching-Fang Hsu
Department of Computer Science
National Tsing Hua University, Hsinchu 300, Taiwan, R.O.C.
E-mail: cfhsu@cs.nthu.edu.tw

Contents

1   Introduction

With the rapid growth of development of multimedia applications, the trend of accommodating a large variety of traffic with various quality-of-service (QoS) requirement has emerged. Optical networks based on wavelength division multiplexing (WDM) technology are capable of providing several terabits per second bandwidth; such attractive characteristics also makes WDM be considered as the most promising approach to achieve this goal. In order to efficiently exploit the extraordinary high bandwidth offered by optical fiber, WDM divides the bandwidth into a few wavelengths and there is no interference between transmissions on different wavelengths. Generally, there are two architectures, single-hop and multi-hop [17][18]. In a single-hop system, communications between a source node and a destination node complete in one hop. Conversely, in a multi-hop system, it is not necessary to satisfy the constraint of direct communication path; passing through one or more intermediate nodes is allowed.

Among several proposed physical topologies for WDM network, the star topology with a star coupler operating is the most widely discussed one under a LAN environment. In this topology, the star coupler merges incoming signals on all wavelengths and then broadcasts the combined information to all stations connected to it, as shown in Fig. 1. Each end station is equipped with one or more transmitter(s) and receiver(s). Roughly speaking, according to the tuning capability, a transmitter (receiver) can be classified into one of two categories, tunable transmitter (receiver) or fixed transmitter (receiver). A tunable transmitter (receiver) is capable of tuning across different wavelengths while a fixed transmitter (receiver) always works on a

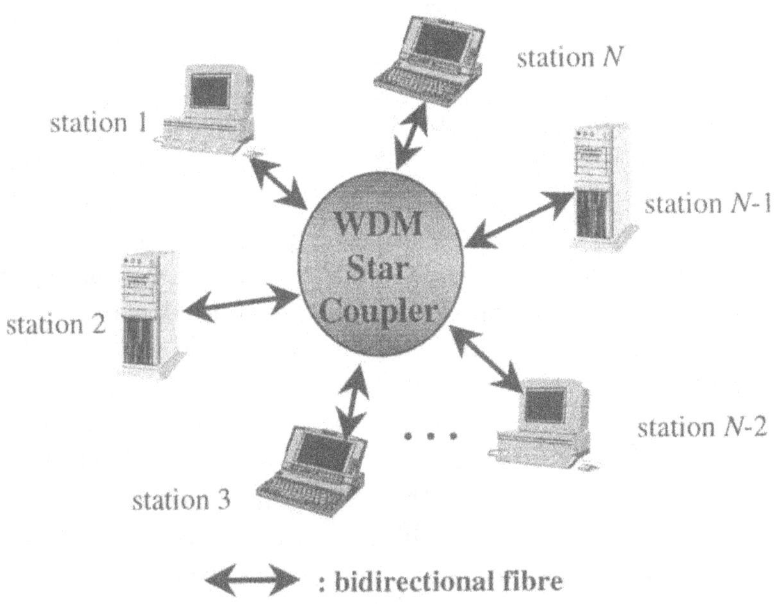

Fig. 1: WDM star-coupled network

fixed wavelength.

The tunability of transceivers mainly depends on tuning range and tuning delay. The former specifies the number of wavelengths that a transceiver can tune to and the latter is the required latency time for a transceiver to tune from one wavelength to another. With current technology, ideal transceivers that may tune across all wavelengths in a negligible time are still under design [1][6][21].

Taking the tunability of transceivers into account, there are four combinations of hardware components: fixed transmitters and fixed receivers (FT-FR), tunable transmitters and fixed receivers (TT-FR), fixed transmitters and tunable receivers (FT-TR), and tunable transmitters and tunable receivers (TT-TR) [17]. Lowest cost is the main merit of the FT-FR technique while a TT-TR system is more flexible than others. For provisioning higher flexibility, the following discussions are based on a WDM star-coupled network with TT-TR (tunable transmitters and tunable receivers) technique.

During the past ten years, many access protocols have been proposed for single-hop WDM star networks. Owing to invoking the concept of random access, early protocols easily suffer from the problem of low performance [10][16][17][18]. In order to provide more efficient coordination between source and destination, time division multiplexing (TDM) technology, which is broadly used in satellite communication, is investigated [8][9]. As a matter of fact, most proposed protocols and scheduling algorithms are based on the WDM/TDM systems afterwards [1][2][4][8][9][12][14][19]-[23].

In general, on the researches of traffic scheduling problem on WDM networks, there are two traffic types: static and dynamic. For the static traffic, traffic load between each source/destination pair is pre-determined. Because the network size and traffic demands are given and fixed, a matrix representation is a good way to keep the information. Such a matrix is usually called *traffic demand matrix*. The entry on the i-th row and j-th column is the amount of required traffic from station i to station j. On the contrary, dynamic traffic varies by time. We can not predict what the next request is or how much traffic is emitted from some station. Unlike static traffic, dynamic traffic is much closer to the real-world behavior; however, it is much more complex to deal with problems on the assumption that demands arrive dynamically. Hence, most proposed scheduling algorithms are based on a given traffic demand matrix [1][2][4][8][9][12][19]-[21][23] instead of dynamically generated traffic [22]. Generally speaking, the main goal of the former category is minimizing the schedule length or the tuning duration [8] (switching duration [12][23]) after decomposing the traffic demand matrix into several slot matrices [8] (switching matrices [12][23]).

In traditional scheduling algorithms, there is no consideration of QoS requirements [1][4][8][9][19]-[21]. With the popularity of multimedia applications, more and more researches set provisioning different QoS levels as one of their main goals [12][14][15][22][23]. In [12] and [23], both isochronous and asynchronous traffic are considered under a WDM star network with multiple transceivers while only isochronous traffic is scheduled in [22]. A scheduling mechanism for the assignment of guaranteed bandwidth (GBW) is proposed in [14] and two classes of traffic, real-time and nonreal-time, are processed in [15].

Because of the limitation of hardware components, the tuning latency is still considerable. Nevertheless, for the sake of simplicity, it was not taken into consideration in most literatures although such simplification is not practical [3][12][14][19][22][23]. In [22], the tuning latency is assumed to be incorporated into a time slot. Following this assumption, the bandwidth utilization will drop rapidly with the increasing of the latency time. In recent years, more and more discussions broke up this assumption and revealed the impacts that the tuning latency may bring [1][2][4][7][9][20][21].

This paper proposes a dynamic traffic scheduling algorithm in single-hop WDM/TDM networks with arbitrary tuning latencies to support guaranteed

QoS. To furnish different levels of QoS, two classes of traffic are considered: constant bit rate (CBR) and available bit rate (ABR). An effective bandwidth normalization scheme for ABR traffic is also derived for more accurate allocation. Two slot allocation policies are also suggested to allocate the time slots to connections and evaluated by simulations.

The rest of this paper is organized as follows. System assumptions and formal problem definition are presented in Section 2. In Section 3, the proposed dynamic traffic scheduling algorithm is introduced. The simulation model and results are shown and discussed in Section 4 and Section 5 concludes this paper.

2 Assumptions and Problem Definition

2.1 Assumptions

2.1.1 Network Model

WDM star topology using a broadcast-and-select star coupler is considered. Assume there are $W + 1$ wavelengths on each fiber link, λ_i, $0 \leq i \leq W$, where λ_0 is dedicated to the control channel and other W wavelengths are data channels. Transmission in the network operates in a time-slotted fashion. Synchronization problem is beyond the scope of this paper, hence we assume that all stations are synchronized at time slot boundaries and a time slot t is located in time interval $[t\text{-}1, t]$. Data are formatted in fixed-size cells, where one cell can be sent on one time slot. As to the control channel λ_0, every time slot is further divided into N minislots, where N is the number of stations. Each minislot is pre-assigned to one station and each station exactly possesses only one minislot, i.e. the mapping relation between N stations and N minislots is one-to-one. For example, the i-th minislot belongs to station i. In this way, no negotiation for distributing minislots to stations is needed.

Assume each station is equipped with a pair of fixed transceivers (control channel) and a pair of tunable transceivers (data channel). The scheduling problem is focused on the data channel and therefore for the rest of this paper, the term "transmitter" and "receiver" means those of the data channel. We also assume that all transceivers are tunable across all data channels and the tuning latencies are not negligible. Note that tuning latency is a quantity of time. For simplicity, the *normalized* tuning delay δ, expressed in units of cell duration instead of units of time, is used [1][21]. The value of δ is determined by three factors: the transmission rate, the actual tuning time, and the cell size. This paper will focus on the impacts brought by tuning latency and therefore assume all transceivers are tunable over all wavelengths. Here we assume all transceivers require the same normalized latency time, δ time slots, for tuning from one wavelength λ_i to another wavelength λ_j, $i \neq j$.

Thus, we say that a schedule is *feasible* if the following constraints are all satisfied:

(1) For each station i, at most one transmission is allowed within the same time slot.

(2) For each station i, at most one reception is allowed within the same time slot.

(3) For all transmitters, the time period between any two consecutive transmissions on different wavelengths can not be shorter than δ time slots.

(4) For all receivers, the time period between any two consecutive receptions on different wavelengths can not be shorter than δ time slots.

2.1.2 QoS Parameters

With QoS provisioning, both CBR and ABR traffic types are considered. For simplicity, traffic streams are referred to virtual circuits (VC's). As periodically recurring is the most significant characteristic of CBR VC's, the period of a traffic schedule containing CBR traffic is called as *cycle length*, denoted by L which will be explained later.

The behavior of a CBR VC is usually characterized by the peak cell rate (PCR). In addition to PCR, the minimum cell rate (MCR) is also used to characterize the behavior of an ABR VC. In this paper, we use a 2-tuple $< c, d >$ to describe cell rate, where c is the maximum number of slots that can arrive in any d slots. For CBR transmission, d is also the relative deadline, i.e. a cell of a CBR VC must be sent before slot $t + d$ if it arrives in slot t [22]. Since there is no delay or jitter constraint for ABR traffic, deadlines are meaningless for an ABR VC and the 2-tuple $< c, d >$ just means that $\frac{c}{d} \times L$ slots within L-slot period should be assigned to it. Thus, we use a 6-tuple notation $< c_m, d_m, c_p, d_p, s, e >$ to identify a VC:

- c_m, d_m: characterizing MCR

- c_p, d_p: characterizing PCR

- s: the station ID of the source

- e: the station ID of the destination

We can see that each CBR VC has its own deadline (d_m), or *local* cycle length. From a global view, the system has a *global* cycle length L which is defined before. To be more accurate, assume d_m^i is the local cycle length of VC_i and Δ is the set of all d_m^i's, then L is calculated as follows:

$$L = lcm(\Delta), where$$

$$\Delta \equiv \{\ d_m^i \mid d_m^i \text{ is the deadline of } VC_i\text{'s MCR}\}$$

For example, if $\Delta = \{5, 6, 13\}$, then $L = lcm(5, 6, 13) = 390$. Without losing the generality, we can view the PCR of a CBR VC as its MCR so that for either a CBR VC or ABR VC, we can use its MCR to determine how many time slots should be allocated. To further distinguish a CBR VC from an ABR VC, the PCR of a CBR VC is set as $< -1, -1 >$. For example, Fig. 2 shows a valid schedule for two CBR VC's with $W = 3$ and $\delta = 1$. The MCR of VC_1 and VC_2 are $< 3,\ 8 >$ and $< 3,\ 4 >$ respectively. This means that for VC_1, three slots should be allocated within any eight time slots and for VC_2, three slots should be allocated within any four time slots. In this example, within the 8-slot cycle, the first slot on λ_1, the third slot on λ_2, and the sixth slot on λ_3 are assigned to VC_1; on the other hand, the first three slots on λ_3 and the three consecutive slots starting from the fifth slot on λ_1 are allocated to VC_2.

Note that for considering the tuning latency, any two consecutive transmissions on different wavelengths are separated by at least one slot ($\delta = 1$). The deadline requirement of CBR traffic can also be examined by a sliding window. The notation w_i^j stands for sliding window j of VC_i. The size of each window w_i^j is set to the deadline of VC_i (e.g., the size of w_1^1 is 8-slot). No matter how we move the window frame, it can always cover as many slots as the first item (c_m) in the 6-tuple notation. Because a cell arriving in any slot has to be sent out before the deadline and the maximum number of cells may arrive in any d_m slots is c_m, it is necessary to assign c_m slots for any window frame with size of d_m slots. Assume $< c_m^i, d_m^i >$ is the MCR of VC_i. For illustration, for each of the windows w_1^1 and w_1^2 ($d_m^1 = 8$), the number of allocated slots is three ($c_m^1 = 3$), and for each of the windows w_2^1, w_2^2, and w_2^3 ($d_m^2 = 4$), the number of allocated slots is also three ($c_m^2 = 3$). Nevertheless, this requirement is not necessary in ABR traffic. For demonstration, consider another example shown in Fig. 3, where an

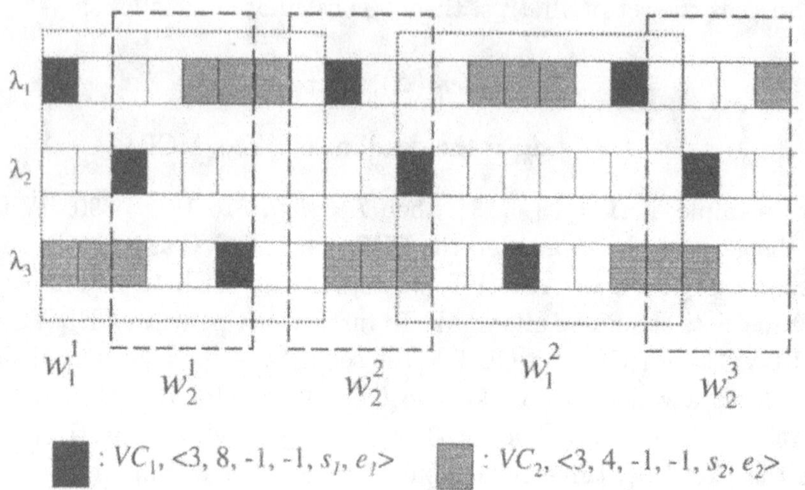

(a) Two CBR VC's

$$D = \begin{bmatrix} 1 & 0 & 0 & 0 & 2 & 2 & 2 & 0 \\ 0 & 0 & 1 & 0 & 0 & 0 & 0 & 0 \\ 2 & 2 & 2 & 0 & 0 & 1 & 0 & 0 \end{bmatrix} \Biggr\} \ W{=}3$$

$$\underbrace{}_{L=8}$$

(b) The corresponding slot-allocation matrix D

Fig. 2: A valid schedule example wih two CBR VC's.

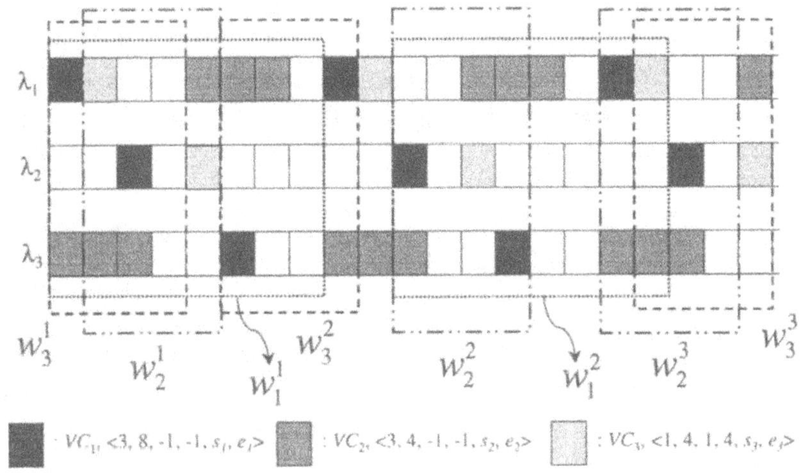

(a) Two CBR VC's and one ABR VC

$$D=\begin{bmatrix} 1 & 3 & 0 & 0 & 2 & 2 & 2 & 0 \\ 0 & 0 & 1 & 0 & 3 & 0 & 0 & 0 \\ 2 & 2 & 2 & 0 & 0 & 1 & 0 & 0 \end{bmatrix} \Bigg\} W=3$$

$$\underbrace{}_{L=8}$$

(b) The corresponding slot-allocation matrix D

Fig. 3: A valid schedule example wih two CBR VC's and one ABR VC.

ABR VC (VC_3) with MCR = PCR = $< 1, 4 >$ is added into the schedule shown in Fig. 2. In this case, the second slot on λ_1 and the fifth slot on λ_2 are assigned to VC_3 for every eight slots. Note that in this assignment not all window frames with a size of four ($d_m^3 = 4$) cover one slot for VC_3. For example, no slot is assigned to VC_3 in w_3^2 and two slots are assigned to w_3^3. However the assignment guarantees that 2 slots ($\frac{1}{4} \times 8 = 2$) are arranged to VC_3 in every cycle.

To simplify the problem, a *normalization* scheme is applied to MCR $< c_m, d_m >$ to get a normalized MCR $< c_m', d_m'>$. For QoS guarantee, rate reduction is not allowable after the normalization (specialization in [22]). Both the relation between d_m and d_m' and the relation between c_m and c_m' are shown as follows [22]:

$$d_m' = 2^{\lfloor \log_2 d_m \rfloor}, \; c_m' = c_m$$

We can see that the deadline d_m' is more restricted than d_m and c_m' remains the same as c_m, so the normalized data rate will not be lower than the original one. Recall the previous example ($\Delta = \{5, 6, 13\}$), if we use Δ' to represent the set of normalized local cycle lengths, we can derive $\Delta' = \{4, 4, 8\}$. Therefore, we can simplify L to be some integer that is power of two ($L = 2^i$, $i \in \mathbf{N}$); in this way, L is equal to the maximum element in Δ' ($L = max(\Delta') = 8$). Apparently, L is much shorter if normalization scheme is applied and so the related databases discussed in the following section will consume less space. As [22] addressed, the base can be any integer; however it will not affect what we propose in this paper, hence the base is set to 2 for simplicity.

As to ABR VC's, the normalization operation is slightly different. It has been mentioned earlier that there is no deadline constraint for ABR VC's; if the above approach is applied to ABR VC's, the allocated bandwidth will be more than what those ABR VC's actually requested. Let us use a case to explain this phenomenon. Note that the same normalization operation is applied to both MCR and PCR, hence in the following case, for simplicity, only MCR will be specified. Consider an ABR VC with MCR $< c_m, d_m >$ $= < 5, 27 >$. If the above normalization approach is employed, we have the normalized MCR $< c_m', d_m' > = < 5, 16 >$. Actually, finer granularity obtains a more accurate normalization but the deadline constraint will be violated. Fortunately, as deadline restriction is meaningless to ABR VC's, a more precise normalization can be employed to save bandwidth. So first we can set $d'_m = 32$, then calculate c'_m as follows:

$$\frac{c'_m}{32} \geq \frac{5}{27} \;\Rightarrow\; c'_m \geq 5.9 \;\Rightarrow\; c'_m = 6$$

Finally, we derive the normalized rate $< c'_m, \; d'_m > \; = \; < 6, \; 32 >$. It is apparent that a lot of bandwidth is saved in this way. The proposed transformation for ABR traffic can be described by the following mathematical expressions:

$$d'_m = 2^{\lceil \log_2 d_m \rceil}$$

$$\frac{c'_m}{d'_m} \geq \frac{c_m}{d_m} \;\Rightarrow\; c'_m = \lceil \frac{c_m}{d_m} \times 2^{\lceil \log_2 d_m \rceil} \rceil$$

2.2 Problem Definition

Based on assumptions described in Section 2.1, the scheduling problem can be defined as follows:

Given N stations, W available wavelengths for data transmission, L-slot global cycle and a $W \times L$ slot-allocation matrix D; each station is equipped with a pair of tunable transmitter and tunable receiver and each transceiver needs δ slots for tuning from λ_i to λ_j, $i \neq j$.

(1) *For a setup request $r_s = < c_m, \; d_m, \; c_p, \; d_p, \; s, \; e >$, find a new feasible slot-allocation matrix D_{new} with a new global cycle length L_{new} such that r_s is arranged into D_{new} subject to that all the QoS of accepted VC's in D can not be affected.*

(2) *For a termination request $r_t = < vc\_id >$, find a new feasible slot-allocation matrix D_{new} with a new global cycle length L_{new}, such that the bandwidth of the released VC becomes available.*

3 Proposed Slot Allocation Algorithm

Before we present our slot allocation algorithm[13], some data structures and system parameters used in the algorithm are introduced first.

A database, named VC Table, is used to keep the information of each VC. This includes the VC's ID, source ID, destination ID, c'_m, d'_m, c'_p, and d'_p. Each entry in the table is formatted as follows:

$$< vc\_id, \; s, \; e, \; c'_m, \; d'_m, \; c'_p, \; d'_p >$$

A $W \times L$ matrix D, named slot-allocation matrix, is used to represent current traffic schedule. Each entry $d_{ij} = 0$ in D indicates the j-th slot of λ_i is free. Otherwise, it specifies the ID of a specific VC which occupies the slot. (e.g. Fig. 3(b))

Since each station is equipped with only a pair of tunable transceivers, for each L-slot time, at most L slots can be allocated to a station for transmission (reception). To satisfy this constraint, it is necessary to record the bandwidth of allocated VC's for each transmitter (called the transmitter utilization, and denoted as ρ_i^t), and that for each receiver (called the receiver utilization, and denoted as ρ_i^r). We should have $\rho_i^t \leq 1$ and $\rho_i^r \leq 1$. Thus,

$$\rho_i^t = \sum \left(\frac{c_m^{j\,\prime}}{d_m^{j\,\prime}} \,\bigg|\, \frac{c_m^{j\,\prime}}{d_m^{j\,\prime}} \text{ is the normalized MCR of } VC_j \text{ emitting from } i \right)$$

$$\rho_i^r = \sum \left(\frac{c_m^{j\,\prime}}{d_m^{j\,\prime}} \,\bigg|\, \frac{c_m^{j\,\prime}}{d_m^{j\,\prime}} \text{ is the normalized MCR of } VC_j \text{ destining to } i \right)$$

The skeletons of the proposed connection setup algorithm and release algorithm are described in the section 3.1 and section 3.2.

3.1   Connection Setup Algorithm

The proposed connection setup algorithm consists of four steps to accommodate a new connection request: normalization, affordability check, available slot scan, and slot assignment as shown in Fig. 4. Steps of connection setup are explained as follows:

Step 1. Normalization

For a CBR request with $c_p = d_p = -1$, the MCR $< c_m,\ d_m >$ is normalized to $< c_m',\ d_m' >$ as proposed in [22].

For an ABR request, the MCR $< c_m,\ d_m >$ is normalized to $< c_m',\ d_m' >$ as described in Section 2.1.2.

Step 2. Affordability Check

This step checks whether the adding of normalized rate $\frac{c_m'}{d_m'}$ into the ρ_s^t and ρ_e^r violates the constraints of $\rho_s^t \leq 1$ and $\rho_e^r \leq 1$. Note that $\rho_s^t > 1$ ($\rho_e^r > 1$) means that the transmitter of station s (the receiver of station e) is unable to support this request and therefore, it should be rejected.

Step 3. Available Slot Scan

Even passing Step 2, we still need to make sure that we have enough free slots for this assignment. First, the new cycle length L', $L' = max(L,\ d_m')$ is calculated and a new slot-allocation matrix D' is created with $D' = D$

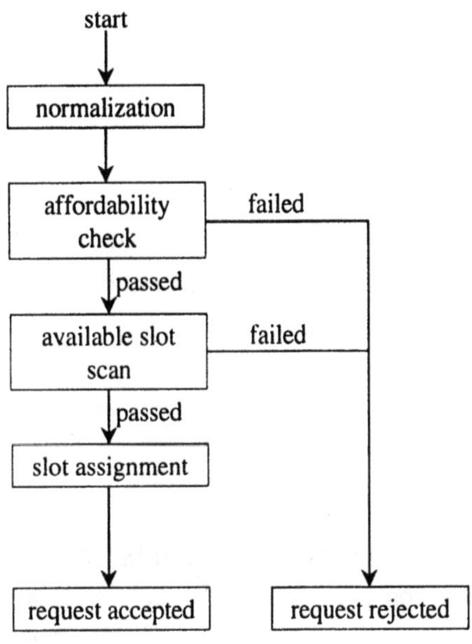

Fig. 4: Flowchart of Connection Setup Algorithm.

initially. If $L' > L$, we append D to the right of D' by $(L'/L) - 1$ times. We then build a $W \times L'$ matrix A, called *available slot matrix*, for available slot scanning,which is defined as follows

$$A = [a_{ij}]_{W \times L'}, \; a_{ij} \in \{0, 1\} \text{ such that}$$

$$a_{ij} = \begin{cases} 0 & \text{if either transmitter or receiver is busy at slot } j \text{ on } \lambda_i \\ 1 & \text{if both transmitter or receiver are free at slot } j \text{ on } \lambda_i \end{cases}$$

A is initialized with $a_{ij} = 1$. For any VC assigned to d'_{ij} in D', we set $a_{ij} = 0$ and for a VC assigned to d'_{ij} in D' whose source is s or destination is e, we set

$$a_{xj} = 0 \text{ for all } x, 1 \leq x \leq W \text{ (constraints (1) and (2))}$$

$$a_{x(j \pm y)} = 0 \text{ for all } x \neq i, 0 < y \leq \delta \text{ (constraints (3) and (4))}$$

Before explaining in detail how A is exploited to scan available slots, a decomposition method, named *grouping* is introduced to horizontally partition A in every d'_m slots. Each derived $W \times d'_m$ matrix is called a *group* and we have $\dfrac{L'}{d'_m}$ groups. The k-th group is denoted as G_k, where

$$G_k = \{ \ g_{ij}^k \ | \ g_{ij}^k = a_{i(j+(k-1) \times d'_m)}, \ 1 \leq i \leq W, \ \text{and} \ 1 \leq j \leq d'_m \}$$

Since each group represents a deadline d'_m , the same c'_m slots should be allocated at each group. To further evaluate whether for each offset within d'_m slots, at least one wavelength on this slot position is available, a *candidate vector* (denoted as κ) is designed, where

$$\kappa[j] = \begin{cases} 0 & \exists k \forall i, \ g_{ij}^k = 0 \\[2ex] 1 & \forall k \exists i, \ g_{ij}^k = 1 \end{cases} , \ 1 \leq k \leq \frac{L'}{d'_m}, \ 1 \leq i \leq W, \ 1 \leq j \leq d'_m$$

Let $|\kappa|$ denote the number of ones in κ. If $|\kappa| < c'_m$, it reveals a fact that less than c'_m slot positions pass affordability check and meanwhile satisfy latency constraints (i.e. constraints (3) and (4)) within every group G_k, $1 \leq k \leq \dfrac{L'}{d'_m}$. As a result, the request should be rejected. On the contrary, if $|\kappa| \geq c'_m$, the request is accepted.

For illustration, consider the slot-allocation matrix D shown in Fig. 5(a), where only VC_1 is scheduled with λ_1 at slot 3 and λ_2 at slot 6. For an incoming $VC_2 = \ < 1, \ 16, \ -1, \ -1, \ s_1, \ e_2 >$, we have $L' = max(8, 16) = 16$ and a new slot-allocation matrix D' is created with 16-slot cycle (Fig. 5(b)). The available matrix A is initialized to all 1's. First we set $a_{1,3} = a_{2,3} = a_{1,11} = a_{2,11} = 0$ and $a_{1,6} = a_{2,6} = a_{1,14} = a_{2,14} = 0$ according to constraints (1) and (2). Then we set $a_{2,1} = a_{2,2} = a_{2,4} = a_{2,5} = 0$, $a_{2,9} = a_{2,10} = a_{2,12} = a_{2,13} = 0$, $a_{1,4} = a_{1,5} = a_{1,7} = a_{1,8} = 0$, and $a_{1,12} = a_{1,13} = a_{1,15} = a_{1,16} = 0$ according to constraints (3) and (4). Since $L' = d'_m$, there is only one group G_1 for the requested VC_2 which is equal to A. The candidate vector κ can then be built with all entries equal to 1 except entries 3, 4, 5, 6, 11, 12, 13, and 14 where both wavelengths λ_1 and λ_2 are not available. For this case, $|\kappa| = 8 > c'_m = 1$,the request VC_2 is accepted.

If the incoming $VC_2 = \ < 1, \ 4, \ -1, \ -1, \ s_1, \ e_2 >$ (Fig. 5(c)), then L' remains 8 and the slot allocation matrix D' is the same as D. Again, the available matrrix A is initialized to 1. According to constraints (1) and (2), we set $a_{1,3} = a_{2,3} = 0$ and $a_{1,6} = a_{2,6} = 0$. Similarly, according to constraints (3) and (4), we set $a_{2,1} = a_{2,2} = a_{2,4} = a_{2,5} = 0$ and $a_{1,4} = a_{1,5} = a_{1,7} = a_{1,8} = 0$. Owing to $\dfrac{L'}{d'_m} = 2$, we have two groups for VC_2

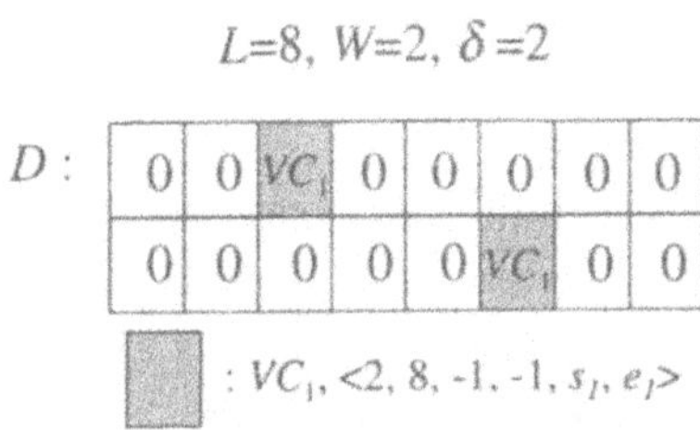

(a) Original Slot-Allocated Matrix D

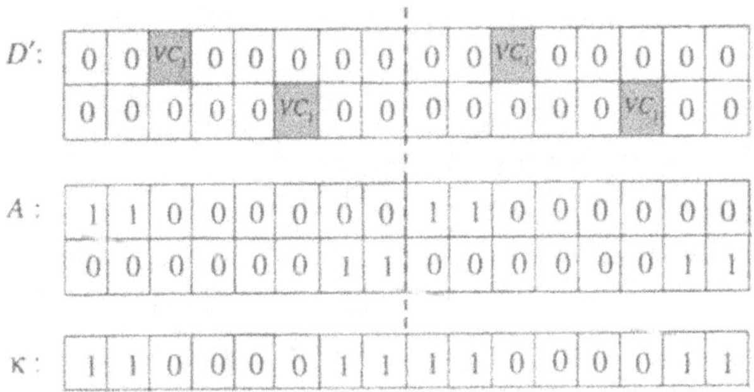

(b) $VC_2 = <1, 16, -1, -1, s_1, e_2>$

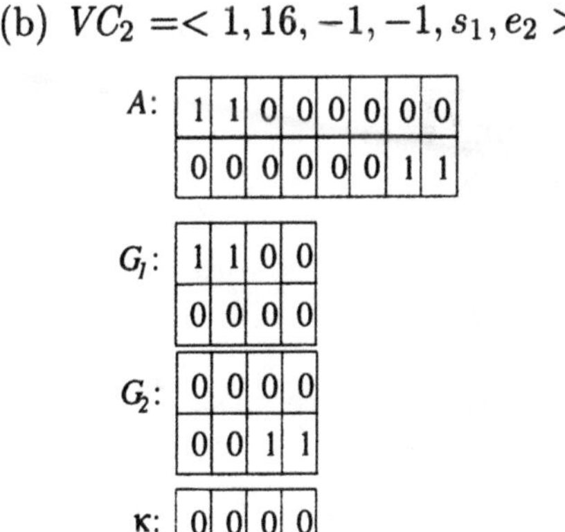

(c) $VC_2 = <1, 4, -1, -1, s_1, e_2>$

Fig. 5: An Example of Available Slot Scan.

and A is divided into G_1 and G_2. After calculation, all entries in κ are zero because both wavelengths λ_1 and λ_2 are unavailable at the last two slots in G_1 and the first two slots in G_2. For this case, $|\kappa| = 0 < c_m' = 1$ and VC_2 is rejected.

Step 4. Slot Assignment

If the request passes through Step 3, we can make sure that the bandwidth it demands can be granted. We pick up c_m' indices at which κ is equal to one and assign free slots at these indices in each group. Finally, we add the VC to VC table, update the information to D', and set $D_{new} = D'$, $L_{new} = L'$.

There are two slot assignment policies: row-major or *horizontal policy*, and column-major or *vertical policy*. With horizontal policy, free slots on higher-indexed wavelengths will not be considered only if there are not enough slots on lower-indexed wavelengths. On the other hand, with vertical policy, higher-indexed slots are not considered unless lower-indexed slots on all wavelengths are busy.

For example, there is only one group with $L = 8$, $W = 2$, and the current assignment for VC_1 whose $< c_m', d_m' >=< 4, 8 >$ is shown in Fig. 6(a). Assume that there is an request VC_2 with $< c_m', d_m' >=< 4, 8 >$, the assignments with two different policies are shown in Fig. 6(b). We observe that if there is another incoming request VC_3 with $< c_m', d_m' >=< 5, 8 >$, there is enough bandwidth only with horizontal policy applied. In our simulation, we take both policies to inspect the difference in blocking rates performance.

The detailed algorithm is shown as follows:

Algorithm Connection_Setup;

Input: A connection request $< c_m, d_m, c_p, d_p, s, e >$, a $W \times L$ current slot-allocation matrix D, a VC table T and transmitter/receiver utilization ρ_i^t and ρ_i^r for all station i.

Output: Acceptance or rejection to the request. If accept, output also contains updated global cycle length L_{new}, $W \times L_{new}$ slot-allocation matrix D_{new}, VC table T_{new}, and transmitter/receiver utilization $\rho_{new}{}_i^t$ and $\rho_{new}{}_i^r$

/ 1. Normalization Step */*
Step 1.1 Normalize the $< c_m, d_m >$ to $< c_m', d_m' >$.
         if $(c_p = d_p = -1)$ **then** /* *CBR traffic* */
            $c_m' = c_m;\ d_m' = 2^{\lfloor \log_2 d_m \rfloor}$;

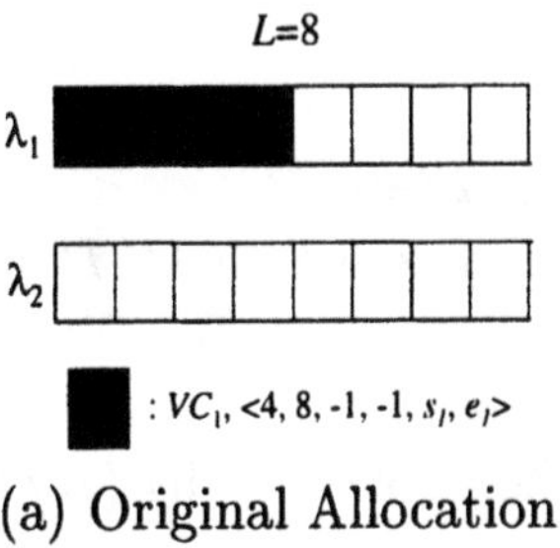

(a) Original Allocation

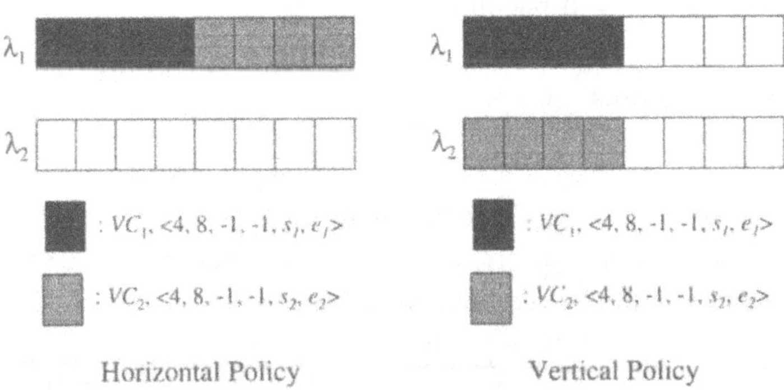

(b) Allocation using two policies

Fig. 6: Slot Allocation with Horizontal and Vertical Policies.

      else /* *ABR traffic* */

$$c_m' = \left\lceil \frac{c_m}{d_m} \times 2^{\lceil \log_2 d_m \rceil} \right\rceil; \quad d_m' = 2^{\lceil \log_2 d_m \rceil};$$

/* *2. Affordability Check Step* */

Step 2.1 Check the affordability of transmitter of s and receiver of e.

    if $(\rho_s^t + \frac{c_m'}{d_m'}) > 1$ **or** $(\rho_e^r + \frac{c_m'}{d_m'}) > 1$ **then**

      return failed;

/* *3. Available Slot Scan Step* */

Step 3.1 Calculate the new cycle length L', $L' = \textbf{max}\ (L,\ d_m')$.

Step 3.2 Create slot-allocation matrix D', $D' = D$ initially.

    if $(L' > L)$

      append D to the right of D' by $(L'/L) - 1$ times.

Step 3.3 Create available slot matrix A

    Build a $W \times L'$ matrix A with each entry initialized to 1.

    for (a virtual connection assigned to d_{ij}' in D'

    with source s or destination e)

      set $a_{xj} = 0$ for all x, $1 \le x \le W$

      set $a_{x(j\pm y)} = 0$ for all $x \ne i$, $0 \le y \le \delta$.

Step 3.4 Create groups

    Build $W \times d_m'$ matrix G_n, $1 \le n \le \dfrac{L'}{d_m'}$.

    The k-th group G_k, is composed of g_{ij}^k,

    where $g_{ij}^k = a_{i(j+(k-1)\times d_m')}$, $1 \le i \le W$, and $1 \le j \le d_m'$.

Step 3.5 Setup candidate vector κ, where

$$\kappa[j] = \begin{cases} 0 & \exists k \forall i,\ g_{ij}^k = 0 \\ \\ 1 & \forall k \exists i,\ g_{ij}^k = 1 \end{cases},\ 1 \le k \le \frac{L'}{d_m'},\ 1 \le i \le W,\ 1 \le j \le d_m'$$

Step 3.6 Scan available slots

    If $\sum\limits_{\forall i} \kappa[i] < c_m'$ then return failed

/* *4. Slot Allocation Step* */

Step 4.1 Slot Allocation

      if (Horizontal allocation scheme is adopted)

        assign vc_id to free slots in D' in row-major basis for c_m' times.

      if (vertical allocation scheme is adopted)

        assign vc_id to free slots in D' in column-major basis for c_m' times.

Step 4.2 Update Database and Return

add the *vc_id* to VC table T to form new table T_{new};

$$\rho_{new}\,{}^t_s = \rho^t_s + \frac{c'_m}{d'_m};$$

$$\rho_{new}\,{}^r_e = \rho^r_e + \frac{c'_m}{d'_m};$$

$$L_{new} = L';$$

$$D_{new} = D';$$

return success;

3.2 Connection Release Algorithm

After being active for the holding time, the established connection with *vc_id* is going to be released. We remove it from the VC table and for the entries in D whose value is equal to the *vc_id*, we mark them as 0. The cycle length L is updated by calculating the LCM of all the deadlines in the VC table. If L is smaller than the original value, we reduce the column size of D to L.

Algorithm Connection_Release;

Input: A $W \times L$ slot-allocation matrix D, a VC table T, and the releasing connection with *vc_id*.

ouput: Updated global cycle length L_{new}, $W \times L_{new}$ slot-allocation matrix D_{new}, and VC table T_{new}.

Step 1. In order to form the new VC Table Tnew,
Remove the entry whose VC identifier is equal to *vc_id* from T.

Step 2. Update the slot-allocation matrix D
$\forall i,\ j,$ **if** $(D[i,\ j] = vc\_id)$ **then** $D[i,\ j] = 0$.

Step 3. Calculate L_{new}
$\Delta \equiv \{\ d_m^{\,i}\ |\forall\ VC_i \in$ the set of existing VC's, $d_m^{\,i}$ is the deadline of VC_i's MCR$\}$
$L_{new} = lcm(\Delta)$

Step 4. Build D_{new} by reducing the column size of D to L_{new}
Create a $W \times L_{new}$ matrix D_{new}, where
$D_{new}[i,j] - D[i,j],$ for $1 \le i \le W$ and $1 \le j \le L_{new}$

---

4   Numerical Results

4.1   Simulation Model and Assumptions

To evaluate the performance of the proposed algorithm, we build a simulation model with 20-node WDM single-hop star-coupled network to simulate the results. The following are the detailed assumptions in the simulation:

- The total simulation time is 10,000 time units.

- The generated traffic pattern is in a uniform distribution.

- W is assumed to be 2, 4, 8, 16, 32, or 64.

- The tuning latency δ ranges from 0 to 10.

- The arrival of requests in the system is generated in a Poisson distribution with a mean of γ.

- All connection requests demand CBR traffic unless stated.

- There is no second chance for a rejected request; in other words, a rejected request will not enter the request queue again.

- The connection holding time is assumed to follow an exponential distribution with a mean of 10 time units.

- Both horizontal and vertical policies are observed for the impact on blocking rates.

4.2   Simulation Results

4.2.1   The Impact of Number of Wavelengths *(W)* and Slot-Assignment Policies

Fig. 7 shows the blocking rate under different value of W and slot assignment policies. The blocking rate decreases with the increasing of W when $W \leq 8$. The blocking rate remains the same after $W = 8$. Theoretically, the blocking rate decreases as W increases. It holds since the limitation of number of transceivers at each station. When the transceivers of each station are nearly fully utilized, the performance also achieves its extremity and the blocking rate will not be improved even W is larger.

We notice that the blocking rate with horizontal policy is lower than that with vertical policy under the same configuration. Note that horizontal

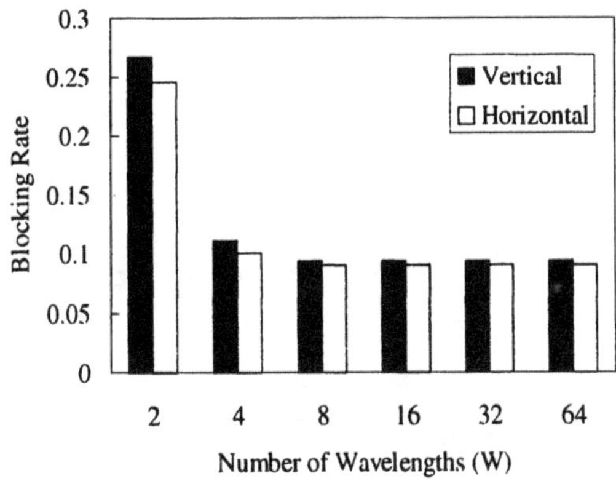

Fig. 7: Blocking Rate under different Number of Wavelengths W ($\gamma = 10, \delta = 2$).

policy tends to leave more unused wavelengths for higher QoS demands while vertical policy exhausts lower-indexed slots and leaves fewer free slots at each wavelength. As the result, the horizontal policy offers a lower blocking rate. However, as the number of wavelengths increases, the vertical policy generates a less number of exhausted slots and the blocking rates of the two policies are closer to each other.

4.2.2 The Impact of ABR Traffic Ratio

With the normalization scheme proposed in [22], the allocated bandwidth will exceed the required bandwidth for ABR VC's. However, with our approach, the bandwidth allocation is more accurate. The blocking rates produced by these two schemes are evaluated individually with both horizontal and vertical allocation policies. Fig. 8 indicates that the proposed scheme derives better improvement on blocking rate with heavier ABR traffic load. At least 10% improvement is achieved as the ABR traffic ratio higher than 50%.

Fig. 9 presents the relation between ABR traffic ratio and average allocated bandwidth per request. With our ABR normalization scheme, less bandwidth is required and thus improves the blocking rate. Notice that although the blocking rate of horizontal policy is smaller than that of vertical policy, the allocated bandwidth per request with horizontal policy is still slightly higher than that of vertical policy. This indicates that horizontal policy provides better bandwidth management and offers less blocking rate

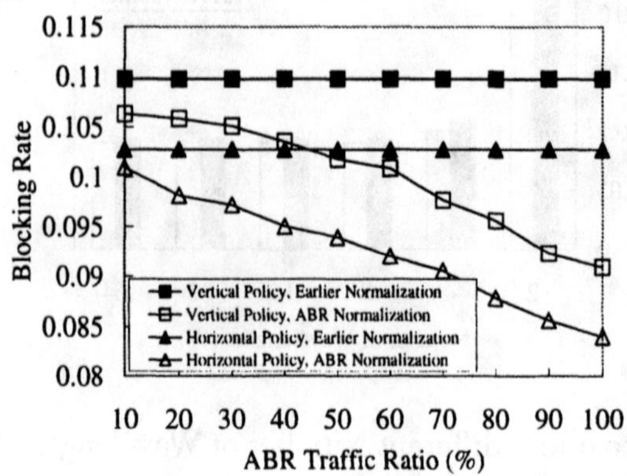

Fig. 8: Blocking Rate under different ABR Traffic Ratios ($\gamma = 10, \delta = 2, W = 4$).

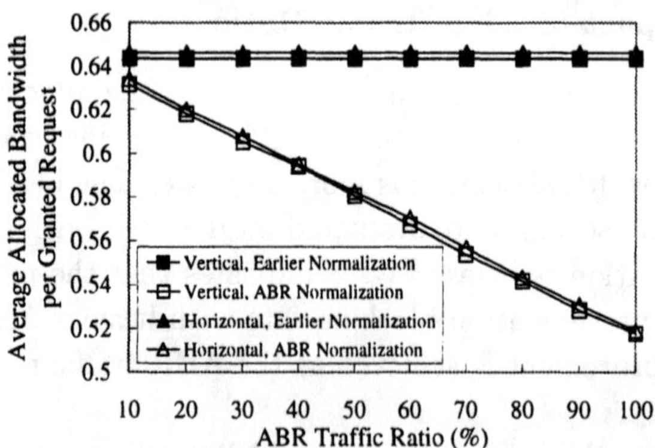

Fig. 9: Average Allocated bandwidth per granted request under different ABR Traffic Ratios ($\gamma = 10, \delta = 2, W = 4$).

than that of vertical policy.

4.2.3   The Impact of Tuning Latency

Fig. 10 depicts relationship between tuning latencies and blocking rate under different number of wavelengths. It is interesting to see that with vertical policy, higher tuning latency derives higher blocking rate. However, the impact of tuning latency for horizontal policy is not so obvious. Another observation indicates that the tuning latency has only limited impact to both policies with $W = 2$. This is because with a smaller W, the blocking rate will reach its extremity soon even under smaller tuning latency.

4.2.4   The Impact of Arrival Rate

Fig . 11 illustrates how arrival rate affects the blocking rate using two different slot allocation policies under $W = 2$, 4, and 8. As previous explanation, we ignore the cases with $W = 16$, 32, and 64. It is obvious that the blocking rate is higher when arrival rate increases and the blocking rate with horizontal policy is lower than vertical policy. However, the blocking rate is closer between two policies as W increases. We can still notice that the arrival rate has limited impact to the blocking rate difference between horizontal policy and vertical policy.

5   Conclusions

We proposed a dynamic scheduling algorithm under single-hop WDM/TDM networks with QoS support and tuning latency consideration. Both CBR traffic and ABR traffic are supported and a new normalization mechanism for ABR traffic is provided. In addition, two slot allocation schemes: horizontal policy and vertical policy are introduced. The simulation results indicated that the proposed ABR normalization scheme requires less bandwidth resources and hence efficiently improves the blocking rate. We also conclude that the horizontal policy is superior to the vertical policy. The horizontal policy offers a better bandwidth management while still maintaining lower blocking rate. Furthermore, the performance of blocking rate produced by this policy is nearly not affected by tuning latency.

               Nen-Fu Huang, Te-Lung Liu, and Ching-Fang Hsu

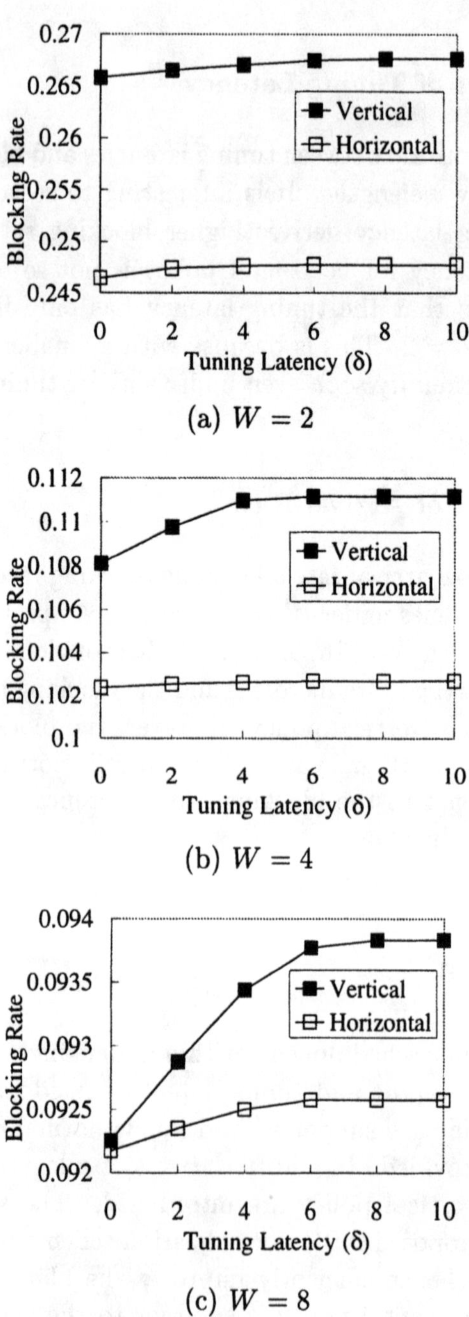

(a) $W = 2$

(b) $W = 4$

(c) $W = 8$

Fig. 10: Blocking Rate under different Tuning Latencies ($\gamma = 10$).

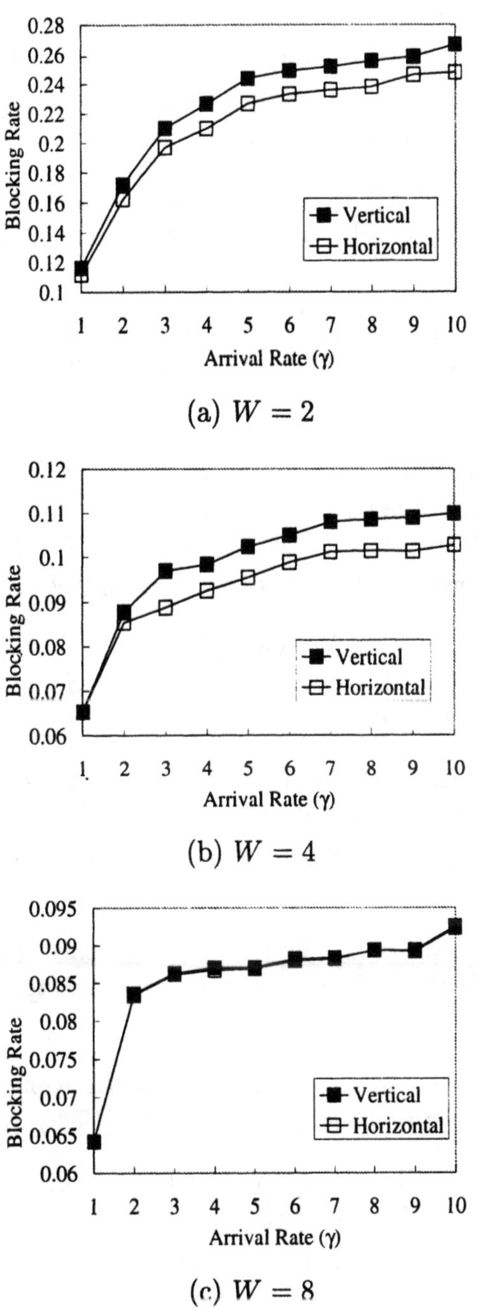

(a) $W = 2$

(b) $W = 4$

(c) $W = 8$

Fig. 11: Blocking Rate under different Arrival Rates ($\delta = 2$).

References

[1] Murat Azizoglu, Richard A. Barry, and Ahmed Mokhtar, Impact on Tuning Delay on the Performance of Bandwidth-Limited Optical Broadcast Networks with Uniform Traffic, *IEEE Journal on Selected Areas in Communications* (June 1996) pp. 935-944.

[2] Ilia Baldine, and George N. Rouskas, Dynamically Load Balancing in Broadcast WDM Networks with Tuning Latencies, *IEEE INFOCOM* (1998), pp. 78-85.

[3] A. Birman and A. Kershenbaum, Routing and Wavelength Assignment Methods in Single-Hop All-Optical Networks with Blocking, *IEEE INFOCOM* (1995), pp. 431-438.

[4] Michael S. Borella, and Biswanath Mukherjee, Efficient Scheduling of Nonuniform Packet Traffic in a WDM/TDM Local Lightwave Network with Arbitrary Transceiver Tuning Latencies, *IEEE Journal on Selected Areas in Communications* (June 1996) pp. 923-934.

[5] Charles A. Brackett, Dense Wavelength Division Multiplexing Networks: Principles and Applications, *IEEE Journal on Selected Areas in Communications* (August 1990), pp. 948-964.

[6] Feng Cao, David H. C. Du, and A. Pavan, Topological Embedding into WDM Optical Passive Star Networks with Tunable Transmitters of Limited Tuning Range, *IEEE Transactionsl on Computers* (December 1998), pp. 1404-1413.

[7] Abel Dasylva and R. Srikant, Optimal WDM Schedules for Optical Star Networks, *IEEE/ACM Transactions on Networking,* (June 1999), pp. 446-456.

[8] Aura Ganz and Yao Gao, A Time-Wavelength Assignment Algorithm for a WDM Star Network, *IEEE INFOCOM* (1992), pp. 2144-2150.

[9] Aura Ganz, and Yao Gao, Time-Wavelength Assignment Algorithms for High Performance WDM Star Based Systems, *IEEE Transactions on Communications* (February/March/April 1994), pp. 1827-1836.

[10] I. M. I. Habbab, M. Kavehrad, and C.-E. W. Sundberg, Protocols for Very High Speed Optical Fiber Local Area Networks Using a Passive Star Topology, *IEEE Journal of Lightwave Technology* (December 1987), pp. 1782-1794.

[11] N. F. Huang and H. I. Liu, On the Isochronous and Asynchronous Traffic Scheduling Algorithm for Single-Star WDM Networks, *IEEE ICC* (1996) pp. 1766-1770.

[12] N. F. Huang and H. I. Liu, An Isochronous and Asynchronous Traffic Scheduling Algorithm for Dual-Star WDM Networks, *IEEE Journal of Lightwave Technology* (March 1996), pp. 273-287.

[13] N. F. Huang, T. L. Liu, and C. F. Hsu, QoS Supported Dynamic Traffic Scheduling in WDM/TDM Network with Arbitrary Tuning Latencies, *IEEE GLOBECOM* (2000), to appear.

[14] A. C. Kam, K-Y. Siu, A. Barry, and E. A. Swanson, Toward Best-Effort Services over WDM Networks with Fair Access and Minimum Bandwidth Guarantees, *IEEE Journal on Selected Areas in Communications* (September 1998) pp. 1024-1039.

[15] Bo Li, and Yang Qin, Traffic Scheduling with Per VC QoS Guarantee in WDM Networks, *IEEE GLOBECOM* (1998) pp. 339-344.

[16] N. Medhravari, Performance and Protocol Improvements for Very High Speed Optical Fiber Local Area Networks Using a Passive Star Topology, *IEEE Journal of Lightwave Technology* (April 1990), pp. 520-530.

[17] B. Mukherjee, WDM-Based Local Lightwave Networks Part I: Single-Hop Systems, *IEEE Network* (May 1992), pp. 12-27.

[18] B. Mukherjee, WDM-Based Local Lightwave Networks Part II: Multi-hop Systems, *IEEE Network* (July 1992), pp. 20-32.

[19] Gerard R. Pieris, and Galen H. Sasaki, Scheduling Transmissions in WDM Broadcast-and-Select Networks, *IEEE/ACM Transactions on Networking* (April 1994), pp. 105-110.

[20] George N. Rouskas, and Vijay Sivaraman, On the Design of Optimal TDM Schedules for Broadcast WDM Networks with Arbitrary Transceiver Tuning Latencies, *IEEE INFOCOM* (1996), pp. 1217-1224.

[21] George N. Rouskas, and Vijay Sivaraman, Packet Scheduling in Broadcast Networks with Arbitrary Tansceiver Tuning Latencies, *IEEE/ACM Transactions on Networking* (June 1997) pp. 359-370.

[22] Bin Wang, Chou-Ju Hou, Ching-Chih Han, On Dynamically Establishing and Terminating Isochronous Message Streams in WDMA-based Local Area Lightwave Networks, *IEEE INFOCOM* (1997) pp. 1263-1271.

[23] Chiung-Shien Wu, Link-sharing Method for ABR/UBR Services in ATM Networks, *Computer Communications* (1998), pp. 1131-1142.

OPTICAL NETWORKS - RECENT ADVANCES
L. Ruan and D.-Z. Du (Eds.) pp. 75 - 97
©2001 Kluwer Academic Publishers

Optimal Placement of Wavelength Converters in WDM Networks for Parallel and Distributed Computing Systems

X.-H. Jia
Department of Computer Science
City University of Hong Kong, Kowloon, Hong Kong (SAR of China)
E-mail: jia@cs.cityu.edu.hk

D.-Z. Du
Department of Computer Science and Engineering
University of Minnesota, Minneapolis, MN 55455, USA
E-mail: dzd@cs.umn.edu

X.-D. Hu
Institute of Applied Mathematics
Chinese Academy of Sciences, Beijing 100080, P. R. China
E-mail: xdhu@public.bta.net.cn

H.-J. Huang
Department of Computer Science
City University of Hong Kong, Kowloon, Hong Kong (SAR of China)
E-mail: hejiao@cs.cityu.edu.hk

D.-Y. Li
Department of Computer Science
City University of Hong Kong, Kowloon, Hong Kong (SAR of China)
E-mail: dyli@cs.cityu.edu.hk

Contents

1   Introduction

Advances in electro-optic technologies have made optical networks a promising choice to meet the increasing demands for higher bandwidth and lower communication latency of high-performance computing and communication applications. *Wavelength division multiplexing* (WDM) [5, 13] is basically frequency division multiplexing in the optical frequency domain, where on a single optical fiber there are multiple communication channels at different wavelengths.

There are basically two types of architectures of WDM optical networks: single-hop systems and multihop systems. In *single-hop* systems [15], each pair of communication nodes has a logical channel configured and the same wavelength should be used through out the route of the channel. There is no wavelength conversion in the intermediate nodes in the route of a channel. Although it is simple and has high speed in data transmission, single-hop architecture is not suitable to large sized systems due to the limited number of available wavelengths and the limited switching capacity of channels going through a node. As a result, multihop architecture becomes necessary for large sized networks. In *multihop* systems [12], the channel of a pair of nodes can consist of several path segments, each uses a wavelength. Wavelength conversion is thus needed at a node where the input and output fibers of the channel use different wavelengths. Wavelength conversion can be done either electronically or optically. In electronical method, optical signal is first converted into electronic signal and then switched (routed) electronically. Finally, the electronical signal is converted back to optical signal and transmitted optically. This *optical-electronical-optical* conversion is very time consuming compared with optical speed of data transmission. In all-optical method, hardware devices called optical wavelength converters are used for wavelength conversions. In either cases, the number of converters (the nodes which are capable of wavelength conversions) should be minimized in a system due to the delay of electronical conversions or the hardware cost of optical converters.

Wavelengths are a kind of scarce resource in WDM networks. The wavelength conflict law governing WDM networks states that wavelengths used in different channels over a fiber link must be mutually exclusive. That is, two channels sharing a physical fiber must use different wavelengths over the shared fiber. It is obvious that the number of wavelengths required is at least equal to the maximal number of channels over a fiber, because each channel over the fiber requires a different wavelength. However, it is not known about how many wavelengths are precisely sufficient, for a network whose load is given. We define the network load as the maximal number of channels over a fiber in the network. It has been pointed out [19] that by using wavelength converters in a network the number of wavelengths needed can be made equal to the network load, which is called *load-wavelength assignable*. A simple example is to make every network node capable of wavelength conversions, which is clearly a case of load-wavelength assignable. However, it is too expensive and also not necessary to place a wavelength converter at every network node.

In this paper we consider the problem of *optimal placement of converter* (OPC) in multihop WDM networks: Given a WDM network, place the minimal number of wavelength converters on the network so that the load-wavelength assignability can be achieved. Our work focuses on a class of network topologies for parallel and distributed computing system, which includes trees, rings, meshes, and hyper-cubes. A suite of efficient algorithms are proposed to produce optimal solutions to this problem on networks.

The study of OPC problem has great implications to network and system design. Firstly, by load-wavelength assignability, the number of wavelengths needed in a system is made minimal, because the low bound of the number of wavelengths is equal to the network load. Secondly, with load-wavelength assignability, the system maximally utilizes the network bandwidth without causing wavelength conflicts, so long as the network load is kept less or equal to the number of wavelengths employed. Thirdly, by using the minimal number of converters to achieve the load-wavelength assignability, it reduces the hardware cost of building a network, and it also makes the data communication faster because of less wavelength conversions.

2   Technical Preliminaries

2.1   Network Model

The WDM network under consideration is modeled as a connected graph $G(V, E)$, where V is the vertex-set representing the set of nodes in the network and E is the edge-set representing physical fiber links between nodes in the network. Each link carries two oppositely-directed fibers, for data transmissions in the two directions of the link.

In this paper we consider two types of communication channels: duplex (bidirectional) and unidirectional. In a duplex channel, data can be transmitted in both directions of the channel. The wavelength conflict rule for duplex channels is that channels over the same link must use different wavelengths on the link. In a unidirectional channel, data can be transmitted only in one direction from the source to the destination. The wavelength conflict rule for unidirectional channels is that channels over the same link and in the same direction must use different wavelengths. That is, two unidirectional channels over the same link but in opposite directions can use the same wavelength.

We assume *full conversion* at every converter. That is, each converter can provide conversions of all possible permutations of wavelengths. Fig. 1 shows a simple example of channel establishment on a WDM network of ring. Two wavelengths are available in each link, w_1 and w_2. Three channels, C_1, C_2, and C_3, are set up in this network of four nodes. C_2 uses wavelength w_1. As C_1 and C_3 share link (2,3) and link (1,2) with C_2, respectively, and at the same time they share links (3,0) and (0,1). As a result, C_1 must use two different wavelengths, w_1 on link (3,0) and w_2 on links (0,1) and (1,2); C_3 uses w_1 on link (0,1) and w_2 on links (2,3) and (3,0). Wavelength conversion needs to be done at node 0 (where a converter is equipped).

Our problem is to, for a given network, locate a minimal set of nodes so that the load-wavelength assignability can be achieved if this set of nodes are made capable of wavelength conversions. Formally, we define this set as following:

Definition A set of nodes S, $S \subseteq V$, *is said to guarantee load-wavelength*

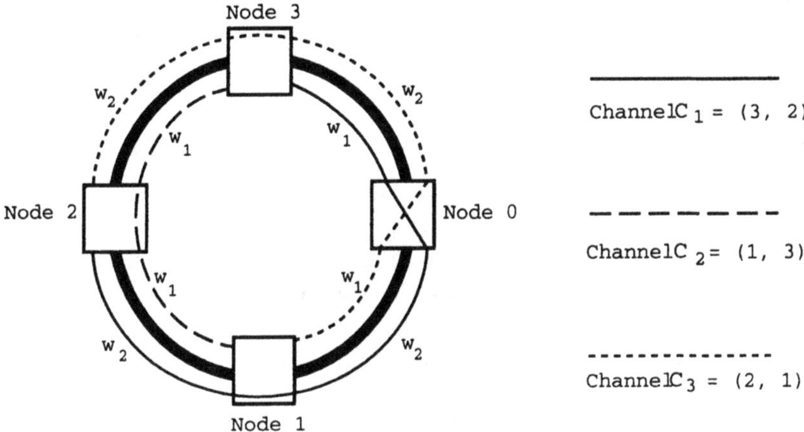

Figure 1: Channel establishment in a ring network equipped with a wavelength converter.

assignability if, by configuring each node in S a wavelength converter, the number of wavelengths needed to set up all channels in the system is equal to the network load.

2.2 Related Work

In single-hop systems, minimizing the number of wavelengths in use can be done at two steps: routing and wavelength assignment. Since the number of wavelengths needed in a system is at least equal to the network load, an important method to reduce the number of wavelengths is to route channels in a load balancing fashion [1, 19], which aims to minimize the network load. This problem has been proved to be NP-hard even for some simple network topologies, such as ring [17]. A widely used approach [7] is to formalize the problem to a type of integer linear program and then find an approximate solution. After the routing is done (i.e., the network load is determined), assigning wavelengths properly to channels (by applying coloring algorithms) can also reduce the number wavelengths needed in a system. Some other research work is to study the minimal number of wavelengths needed when the network load is given [3, 6, 10, 14]. Due to the complexity of the problem, this kind of study is only limited to the networks of trees or rings. Some best results obtained so far are summarized at Table 1. By "αL wavelength being sufficient" we mean that any set of channels with network load L can be assigned by using at most αL wavelengths, and by "αL being necessary" we mean that assigning some set of channels with network load L requires at least αL wavelengths.

	Unidirectional channels	Duplex channels
Tree	$\frac{5}{3}L$ wavelengths are sufficient [3] $\frac{5}{4}L$ wavelengths are necessary [3]	$\frac{3}{2}L$ wavelengths are sufficient [14] $\frac{3}{2}L$ wavelengths are necessary [14]
Ring	$(2L-1)$ wavelengths are sufficient [19] $(2L-1)$ wavelength are necessary [19]	

Table 1. The number of wavelengths required for tree and ring networks.

In multihop systems, the major research topic is the study of configuring a physical network topology (where edges are fiber links) into a logical topology where edges are lightpaths [8, 13, 15]. A lightpath is a logical path in which only one wavelength is used, i.e., there is no wavelength conversion in the intermediate nodes of a lightpath. An important goal of the design of logical topology is to minimize the number of wavelengths in the system, which is similar to the design of single-hop systems (a lightpath is a single-hop channel). Then for a channel setup, it routes the channel through lightpaths. A channel consists of one or more lightpaths, with wavelength conversions at nodes where two lightpaths joint. Therefore, converters are required at the nodes where a lightpath originates or terminates.

Another major research area of multihop systems is the study of introducing wavelength converters and placing the converters to reduce the number of wavelengths needed [9, 16, 18, 19, 20]. In [20], a network with limited wavelength conversion is studied and the performance is analyzed. The work in [18] studies *sparse wavelength conversion*, where networks are comprised of a mix of nodes having full or no wavelength conversion. Some heuristics for wavelength assignment in the networks with wavelength conversions are proposed in [9, 18, 20], and their performance analysis are based upon probabilistic models and techniques (i.g., blocking probabilities of setting up channels). The recent work in [16] discusses how to improve traffic-carry capacity by providing limited wavelength conversion capability in a network. The discussion includes ring, star, tree and general topology networks.

Our work is to find the minimal number of converters and their placement in a network to achieve load-wavelength assignability. It is proved in [19] that to achieve load-wavelength assignment for unidirectional channels, no converter is needed in star networks and only one converter is required in ring networks. Inspired by [19], we study this problem in a class of networks including trees, meshes, and hyper-cubes. We propose some efficient algorithms that can produce the optimal placements in these networks for unidirectional and duplex channels.

2.3   The Methodology

Since load-wavelength assignment in networks of stars and paths has been well studied [19], our approach is to decompose a general topology network into edge-disjoint simple subgraphs, such as paths or spiders. A *spider* is a tree which consists of several paths (called *legs*) with one end of each of these paths incident to a common node (called *body*). Paths and stars are two special cases of spiders.

The decomposition of a graph is done by a *splitting operation* described as follows. Given a graph $G(V, E)$ and a subset $S \subseteq V$, generate a new graph $G_S(V, E)$ by splitting each node $s \in S$ into $\delta(s)$ nodes, where $\delta(s)$ is the degree of s in $G(V, E)$. Each edge (s, t) in $G(V, E)$ becomes edge (s', t) in $G_S(V, E)$, where s' is a newly generated node from splitting s in $G(V, E)$. Fig. 2 shows an example of splitting operation on S which consists of three nodes (in black).

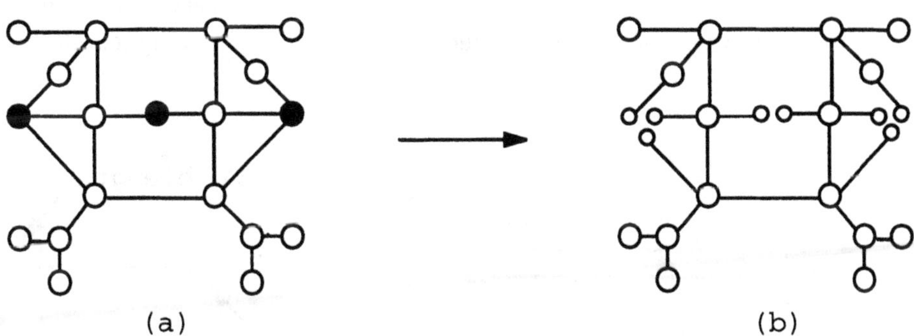

Figure 2: (a) The original graph $G(V, E)$. (b) The obtained graph $G_S(V, E)$ from splitting nodes in S.

By studying the reduced graph $G_S(V, E)$, we divide the OPC problem of general topology networks into the cases of special networks, such as paths and spiders. We will develop the necessary and sufficient conditions of OPC problem for duplex channels and unidirectional channels, respectively.

Based on these conditions we can design efficient algorithms that can find the optimal solutions for the networks with special topologies including tree, mesh, and hyper-cube.

3 OPC Problem for Duplex Channels

3.1 The Basic Theory

The following two theorems give two sets of necessary and sufficient conditions for load-wavelength assignability.

Theorem 1 *Given graph $G(V, E)$, subset $S \subseteq V$ guarantees load-wavelength assignability if and only if every connected component of $G_S(V, E)$ is a path.*

Proof.　"If": Given any set of routed channels with network load L in $G(V, E)$. Now consider them in $G_S(V, E)$. These channels which go through nodes in S are broken into several pieces. In each connected component, which is a path in $G_S(V, E)$, the channels which go through the entire or part of the path can be assigned wavelengths in the following way (see Fig. 3):

Step 1: order these channels from left to right according to the starting points in the path. Ties are broken arbitrarily.

Step 2: assign each of these channels the wavelength which is free and has the least value. A wavelength is reclaimed to be free once its assigned channel terminates.

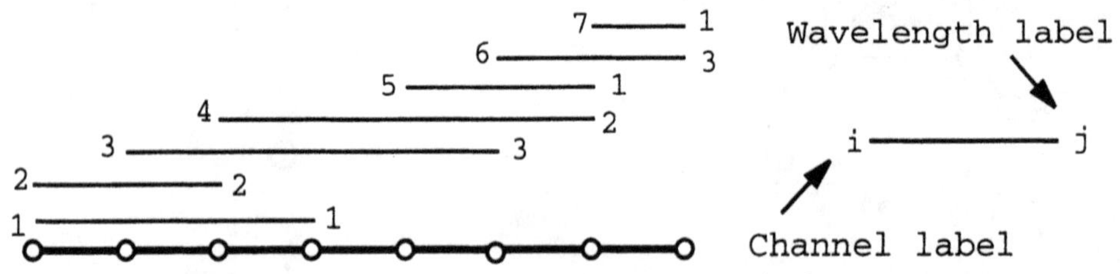

Figure 3: Wavelength assignment for channels in a path.

Since the network load is L, it is obvious that Step 2 uses at most L wavelengths to assign wavelengths to channels without causing wavelength conflict. The wavelength assignment in each connected component of $G_S(V, E)$, can be done independently, and there will be no wavelength conflict between the channels in different components. This is because each node in S has full wavelength conversion capability.

"Only if": we prove it by contradiction. Assume that one of the connected components in $G_S(V, E)$ is not a path. There are only two possibilities of this component: 1) at least one node in the component has degree greater than two, or 2) the component is a ring. We now analyze the two cases below.

(a)   (b)

Figure 4: (a) Three channels on a star. (b) Three channels on a ring.

Case 1. There exists a node adjacent to (at least) three other nodes in the component. Consider three channels in Fig. 4 (a). Apparently, three wavelengths are needed even though the network load is two, because each of three channels shares a link with the other two. This contradicts the load-wavelength assignability.

Case 2. The component is a ring. Consider three channels in Fig. 4 (b). Again, three wavelengths are needed, but the network load is two, because each of these three channels shares a link with the other two. This results in the same contradiction. □

Theorem 2 *For graph $G(V, E)$, let S be the set of nodes whose degrees are greater than or equal to three. If S is not an empty set, then S is both sufficient and necessary for guaranteeing load-wavelength assignability.*

Proof. "Sufficiency": Since S contains all nodes whose degrees are greater than or equal to three in $G(V, E)$, every connected component in $G_S(V, E)$ is a path. Thus it follows from Theorem 1 that S guarantees load-wavelength assignability.

"Necessity": If subset S' guarantees load-wavelength assignability, it must include all nodes that have degree greater than two; otherwise one of the components in $G_{S'}(V, E)$ contains a node whose degree is greater than two, which is not a path. According to Theorem 1, S' is not able to guarantee load-wavelength assignability.   □

3.2   OPC Problem on the Networks of Special Topologies

In this section, we discuss the OPC problem in the networks for parallel and distributed computing systems, which include trees, rings, meshes, and hyper-cubes.

Corollary 1 *To guarantee load-wavelength assignability on trees of n nodes,* $\lfloor (n-2)/2 \rfloor$ *converters are sufficient for all trees and necessary for some trees.*

Proof. "Sufficiency": Let n_i be the number of nodes with degree i, for $i = 1, 2$, and let n_{3+} be the number of nodes with degree at least three. Clearly, $n = n_1 + n_2 + n_{3+}$ and the total degree of n nodes is at least $n_1 + 2n_2 + 3n_{3+}$. In addition, a tree of n nodes has $n - 1$ edges and the total degree of n nodes equals $2(n - 1)$. Therefore, $n - 1 \geq (n_1 + 2n_2 + 3n_{3+})/2$. This implies, $n_{3+} \leq (n - n_2 - 2)/2 \leq (n - 2)/2$. According to Theorem 2, we know that $(n - 2)/2$ converters are sufficient.

"Necessity": For the tree that has a path of $(n + 2)/2$ nodes and each of $(n - 2)/2$ intermediate nodes in the path is adjacent to a degree-one node, it is obvious that $(n - 2)/2$ converters are necessary to guarantee load-wavelength assignability.   □

Corollary 2 *To guarantee load-wavelength assignability on a ring, one converter is both sufficient and necessary.*

Proof. Let S be a singleton including one node. Clearly, $G_S(V, E)$ is a path. Hence according to Theorem 1 a set containing one node in V is both sufficient and necessary for guaranteeing load-wavelength assignability.   □

Corollary 3 *To guarantee load-wavelength assignability on meshes of n nodes,* $(n - 4)$ *converters are both sufficient and necessary.*

Proof. Except four degree-two nodes in the four corners all other nodes in a mesh have degree three or four. The conclusion follows immediately from Theorem 2.   □

Corollary 4 *To guarantee load-wavelength assignability on* $(n \times n)$-*torus or* n-*hypercube, for* $n \geq 3$, n^2 *or* 2^n *converters are both sufficient and necessary.*

Proof. Note that all nodes in $(n \times n)$-torus or n-hypercube have degree at least three, for $n \geq 3$. The conclusion then follows from Theorem 2.   □

4 OPC Problem for Unidirectional Channels

4.1 The Basic Theory

Different from duplex channels, the wavelength conflict rule for unidirectional channels requires that channels over the same link and in the same direction are assigned different wavelengths. The following two theorems give a set of necessary and sufficient conditions that a set guarantees load-wavelength assignability for unidirectional channels.

Theorem 3 *Given graph $G(V,E)$, subset $S \subseteq V$ guarantees load-wavelength assignability if and only if each connected component in $G_S(V,E)$ is a spider.*

Proof.   "If": Given any set of routed channels with network load L in $G(V,E)$. Now consider them on $G_S(V,E)$. For a connected component in $G_S(V,E)$, which is a spider, channels intersecting with this spider are of the two cases (see Fig. 5(a)): 1) crossing the spider body from one leg to another; 2) traversing within a leg. We consider wavelength assignment for two cases below.

Case 1.   Channels go through from one leg to another. We introduce a bipartite multigraph $B(V_1 \cup V_2, E_{12})$ to represent these channels in the spider, where V_1 and V_2, $V_1 = V_2$, are the set of legs (i.e., each leg of the spider is represented as a node in V_1 and V_2). Each edge $e_{ij} \in E_{12}$ if there is a channel which goes through legs i and j for $i \neq j$. Note there may exist multi-edges between two nodes, because there are multiple channels crossing from leg i to leg j. See Fig. 5(b). Assigning wavelengths to the channels without causing wavelength conflict becomes the problem of coloring the edges of graph $B(V_1 \cup V_2, E_{12})$ under the condition that two adjacent edges must be in different colors. Since there are at most L channels going from leg i to leg j for $i \neq j$, multigraph $B(V_1 \cup V_2, E_{12})$ has the maximum degree L. Thus, $B(V_1 \cup V_2, E_{12})$ can be colored by using at most L colors (for details please refer to [2]).

Case 2.   Channels traverse within a leg. See Fig. 5(c). Note that each channel belonging to Case 1 is cut into two parts, one in a leg heading the body and the other in another leg leaving the body. Using the same method as discussed in the proof of Lemma 1 for assigning wavelengths to channels in a path, the channels heading the body of the spider are ordered from left to right (as shown in Fig. 5(c)) by the spider body (ties are broken arbitrarily) or their destinations, while the channels leaving the spider body are also ordered from left to right by the spider body or their sources. The channels which have been assigned wavelengths in the Case 1 retain the

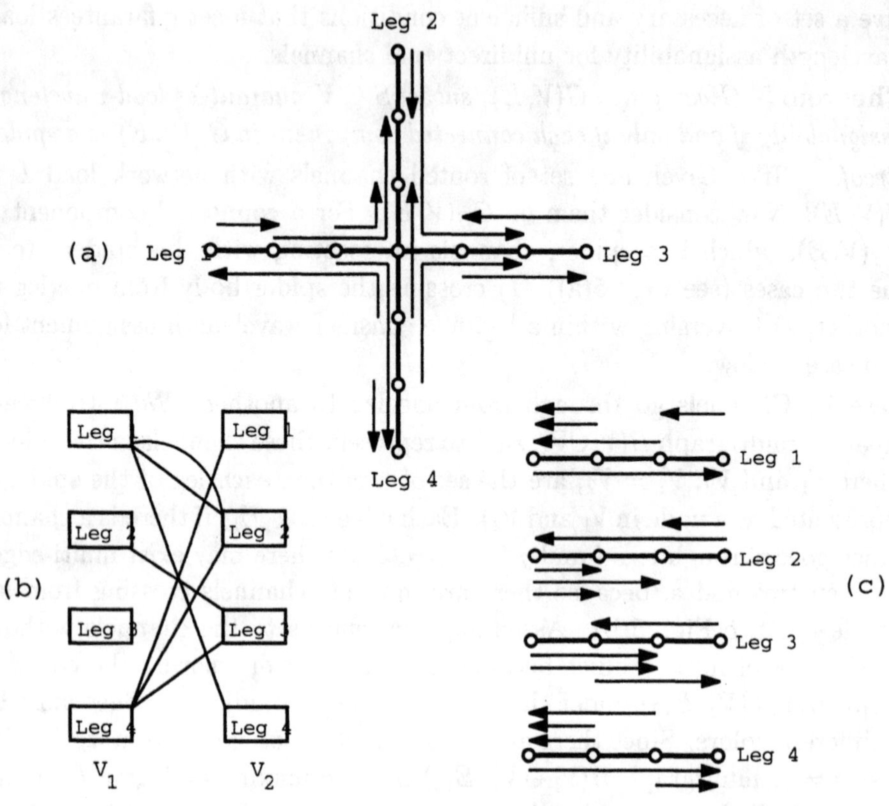

Figure 5: (a) Wavelength assignment in a spider. (b) Assigning wavelengths to channels crossing the body. (c) Assigning wavelengths to channels within legs.

same wavelengths, and the rest of the channels in this leg will be assigned the wavelengths which are free. Thus, all channels (both in Case 1 and Case 2) can be assigned with at most L wavelengths.

For each of those channels that traverse different connected components in $G_S(V, E)$, it may be assigned different wavelengths and this will not cause wavelength conflict, because each node in S has full wavelength conversion capability.

"Only if": we prove it by contradiction. Assume that one of the connected components of $G_S(V, E)$ is not a spider. There are only two possibilities of the component: 1) there are two (or more) nodes having degree greater than two; or 2) it is a ring. We consider them separately as below.

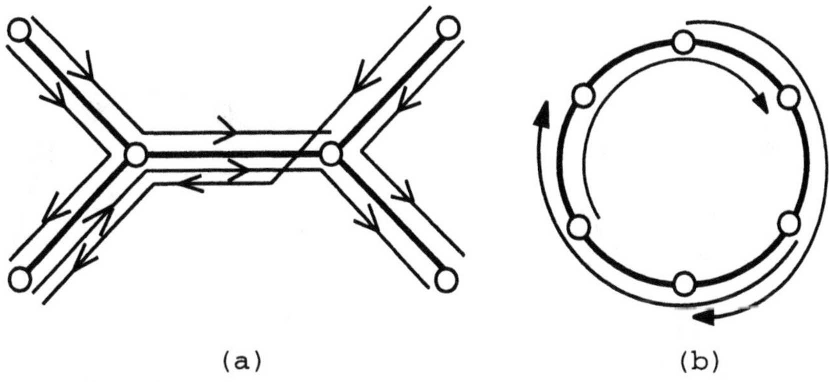

(a)   (b)

Figure 6: (a) Five channels on a tree. (b) Six channels on a ring.

Case 1. There are two nodes having degree greater than two. We construct a case in Fig. 6 (a), where the network load is two. It is not difficult to find that three wavelengths are needed for all the channels in this case. This contradicts load-wavelength assignability.

Case 2. The component is a ring. Consider a case we constructed in Fig. 6 (b), where the network load is two. Again, three wavelengths are needed in this case. This produces the same contradiction.   □

Theorem 4 *If every node in $G(V, E)$ has degree greater than two, then $S \subseteq V$ guarantees load-wavelength assignability if and only if S is a vertex cover of $G(V, E)$.*

Proof. "If": It is obvious to see that every connected component of $G_S(V, E)$ is a spider (actually a star). The conclusion follows directly from Theorem 3.

"Only if": Assume, by contradiction, that there are two nodes u and v in $G(V, E)$ such that edge $(u, v) \in E$ is not incident to any node in S. Then the connected component in $G_S(V, E)$ which contains edge (u, v) is not a spider, because it has two nodes (u and v) with degree greater than two. This contradicts Theorem 3. □

When $G(V, E)$ has degree-one or degree-two nodes, Theorem 4 does not hold any more. Taking the example of a tree network in Fig. 7 (a), it is obvious that one converter installed at the center (black node) can guarantee load-wavelength assignability; However, the minimal vertex cover set has eight nodes (in grey). In another example of a ring network, obviously, one converter at (any) one node (in black) can guarantee load-wavelength assignability; However the minimal node cover set contains half number of nodes (in grey) in the network.

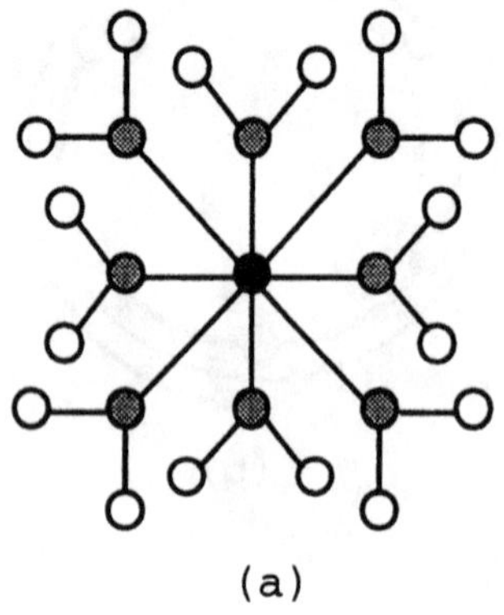
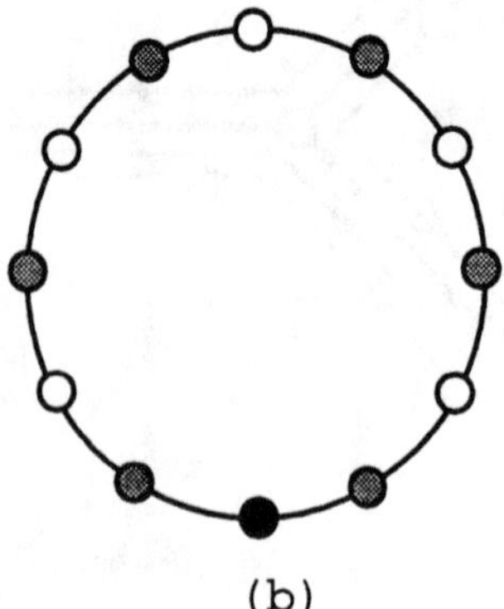

(a)   (b)

Figure 7: Two simple examples.

Theorem 5 *If graph $G(V, E)$ has a node with degree greater than two, then there exists a minimal sized subset of V that guarantees load-wavelength assignability and every node in the set has degree greater than two in $G(V, E)$.*
Proof. Let $v \in V$ be the node with degree greater than two, and S be a minimal sized subset of V that guarantees load-wavelength assignability. By Theorem 3 each connected component of $G_S(V, E)$ is a spider. Now assume there is a node $s \in S$ with degree less than or equal to two. Let u be the closest node to s in a path between v and s, and u has degree greater than two.
Case 1. $u \in S$. Let $S' = S \backslash \{s\}$. It is obvious that each connected component of $G_{S'}(V, E)$ is still a spider, because s is a node of degree less than or equal

to two. See Fig. 8 (a).

Case 2. $u \notin S$. Let $S' = S \cup \{u\} \setminus \{s\}$. It is easy to see that each connected component of $G_{S'}(V, E)$ is still a spider, because substituting s with u only makes one component of $G_S(V, E)$ have a longer leg and another split into several paths. See Fig. 8 (b).

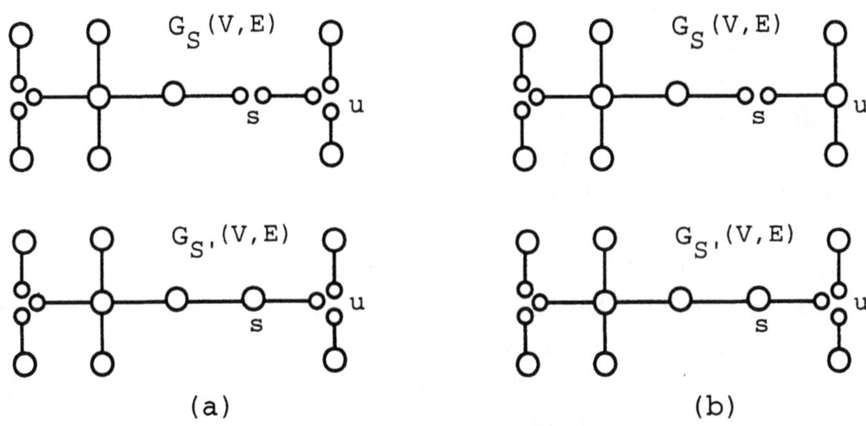

Figure 8: (a) Case 1. (b) Case 2.

In either cases, S' can guarantee load-wavelength assignability because of Theorem 3. In Case 1, S' is a proper subset of S, this contradicts to the minimal size of S. In Case 2, the substitution operation can be repeated until all nodes in S having degree less than or equal to two are replaced by the nodes with degree greater than two. The desired subset can thus be obtained. □

4.2 OPC Problem on Networks of Special Topologies

Again, in this section we consider our solutions for the networks of trees, meshes, torus, and hyper-cubes. Since ring network has been well studied in [19], it will not be discussed here.

For tree networks, if no node has degree greater than two, then a tree becomes a path. According to Theorem 3 no converter is necessary for load-wavelength assignability. So we only consider the case that there exists at least one node with degree greater than two. Notice that Theorem 4 can not be applied for trees, as any trees have nodes of degree one. According to Theorem 5, we do not need to consider degree-one or degree-two nodes for placing converters. We use the following steps to find the minimal set that guarantees load-wavelength assignability.

1) remove every degree-two node by substituting two edges incident to the node with one edge incident to two endpoints (other than the degree-two node).

2) remove every degree-one node by condensing it to the node adjacent to it.

3) find the minimal vertex cover of the resultant tree.

Fig. 9 illustrates this process. At the first step, three degree-two nodes (grey nodes in Fig. 9 (a)) are removed. At the second step, eleven degree-one nodes (grey nodes in Fig. 9 (b)) in the reduced network are condensed to five nodes that they are incident to, respectively (Fig. 9(c)). Finally a vertex cover of two nodes (black nodes in Fig. 9 (d)) is found. Notice that the final tree, reduced from by removing degree-one and degree-two nodes in the original tree, may have some degree-one or degree-two nodes, which are condensed nodes. The vertex cover of the finally reduced tree can be proved to be the optimal solution to OPC problem of the original tree. The proposed method is formally presented at below, and its validity is proved in the following theorem.

Input   A tree $G(V, E)$

Output   A subset $C \subseteq V$ guaranteeing load-wavelength assignability

Step 1   Remove degree-two nodes from $G(V, E)$

     $V' := V, \ E' := E.$

     $V_2 := \{v \mid v \text{ is adjacent to exactly two nodes } v_1 \text{ and } v_2\},$

     $V' := V' \setminus V_2.$

     while $V_2 \neq \emptyset$ **do begin**

         choose $v \in V_2,$

         $E' := E' \cup \{(v_1, v_2)\} \setminus \{(v, v_1), (v, v_2)\},$

         $V_2 := V_2 \setminus \{v\}.$

     end-while

Step 2   Remove degree-one nodes from $G(V', E')$

     $V_1 := \{u \mid u \text{ is adjacent to exactly one node } u_1\},$

     $V' := V' \setminus V_1.$

     while $V_1 \neq \emptyset$ **do begin**

         choose $u \in V_1,$

         $E' := E' \setminus \{(u, u_1)\},$

         $V_1 := V_1 \setminus \{u\}.$

     end-while

Step 3 Generate a minimal vertex cover of $G(V', E')$
$\quad C := \emptyset$
$\quad \overline{V} := V', \overline{E} := E'.$
$\quad$**while** $\overline{E} \neq \emptyset$ **do begin** (Loop ends until all edges in E' are covered)
$\quad\quad V_1 := \{v \in \overline{V} \mid v \text{ has degree one in } G(\overline{V}, \overline{E})\},$
$\quad\quad C := C \cup \{u \in \overline{V} \setminus V_1 \mid u \text{ is adjacent to some } v \in V_1\},$
$\quad\quad \overline{V} := \overline{V} \setminus \{v \in \overline{V} \mid v \in C \text{ or } v \text{ is adjacent to some } u \in C\},$
$\quad\quad \overline{E} := \{(u,v) \in \overline{E} \mid u \text{ and } v \text{ are belong to } \overline{V}\}.$
end-while
return C.

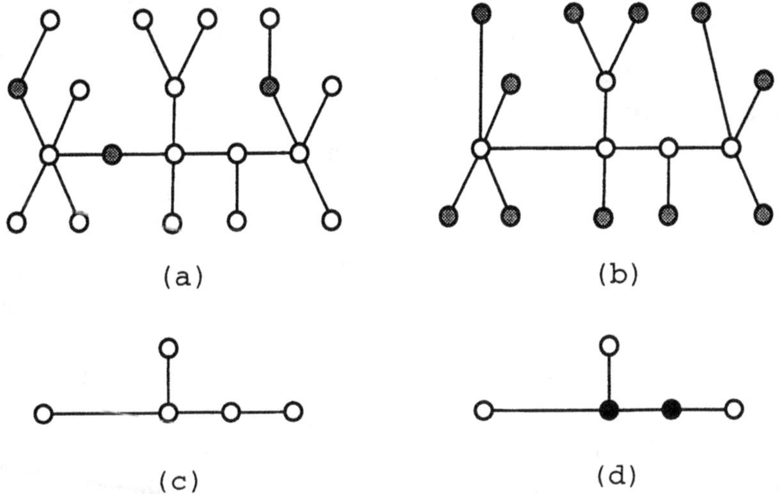

(a)

(b)

(c)

(d)

Figure 9: Illustration of the algorithm: (a) remove degree-two nodes, (b) remove degree-one nodes, (c, d) find a vertex cover of the induced tree.

Theorem 6 *The proposed algorithm can find the optimal solution of OPC problem in a network of tree in time of $O(|E| + |V|)$.*
Proof. Let $G(V', E')$ be the final graph obtained after Step 2 and C be the returned vertex cover of $G(V', E')$. According to Theorem 5, we know that there exists a solution of OPC problem in $G(V, E)$ that is a subset of V'.

In the following, we prove that a subset S of V' guarantees load-wavelength assignability in $G(V, E)$ if and only if it is a vertex cover of $G(V', E')$.
"If": We prove it by contradiction. Assume that $S \subset V'$ is a vertex cover of $G(V', E')$ but it does not guarantee load-wavelength assignability in $G(V, E)$. According to Theorem 3, there exists a connected component of $G_S(V, E)$

which is not a spider. This means that there exist two nodes x and y which are not in S and have degree greater than two in $G_S(V, E)$. Note there is a path in $G_S(V, E)$ between x and y and no node in the path is included in S. Moreover, nodes x and y are included in V', and removing degree-one and degree-two nodes from $G(V, E)$ does not destroy the connectivity between them. Thus x and y remain connected by the path in $G(V', E')$, which implies that all edges in the path are not covered by any node in S. This contradicts that S is a vertex cover of $G(V', E')$.

"Only if": We prove it again by contradiction. Assume that there exists a subset $S \subset V'$ that guarantees load-wavelength assignability in $G(V, E)$ but it is not a vertex cover. Then there are two nodes $x \in V'$ and $y \in V'$ such that $(x, y) \in E'$ but $x \notin S$ and $y \notin S$. Note x and y have degree greater than two in $G(V, E)$; otherwise they are removed away at Step 1 or Step 2. Hence the component of $G_S(V, E)$, which contains (x, y), is not a spider. This contradicts Theorem 4.

Now we prove that C, produced by our algorithm above, is the minimal vertex cover of $G(V', E')$. In the while-loop at Step 3, the nodes that are adjacent to degree-one nodes in current $\overline{V}$ are included in C. They can cover the edges incident to them. Thus these edges are removed from $\overline{E}$. In such way when $\overline{E}$ becomes empty, all edges in E' can be covered by some nodes in C. This means that C is a vertex cover of $G(V', E')$. Then, we prove C is the minimal. Assume C^* is a minimal vertex cover of $G(V', E')$. If C^* includes some degree-one nodes in $V_1 \subset \overline{V}$, then we can replace them with the nodes that are adjacent to them. It is obvious that such modified C^* remains a minimal vertex cover of $G(V', E')$. It is not difficult to see that the finally modified C^* is exactly the set of C, because this is the exact way how C is produced. Therefore C is minimal.

From above discussion we deduce that C is the optimal solution of OPC problem in $G(V, E)$.

Furthermore, the algorithm can finish in time of $O(|E| + |V|)$, this is because the key operation in all three steps is to check the degree of each node in the current graph and include or exclude the node and its incident edges in some sets. $\square$

Corollary 5 *To guarantee load-wavelength assignability on trees of n nodes, $\lfloor (n-2)/4 \rfloor$ converters are sufficient for all trees and necessary for some trees.*
Proof. "Sufficiency": For a tree of n nodes, let n_3 be the number of nodes with degree at least three. When the algorithm runs with the tree, the final graph of $G(V', E')$ includes n_3 nodes and $n_3 - 1$ edges. Because each node can cover at least two edges, the minimal vertex cover has size at most

$\lceil (n_3 - 1)/2 \rceil$, which can be proved to be less than or equal to $\lfloor (n-2)/4 \rfloor$.

"Necessity": It can be verified that $\lfloor (n-2)/4 \rfloor$ converters are necessary for trees that consists of a path and each node is adjacent to exactly three nodes (see Fig. 10).   □

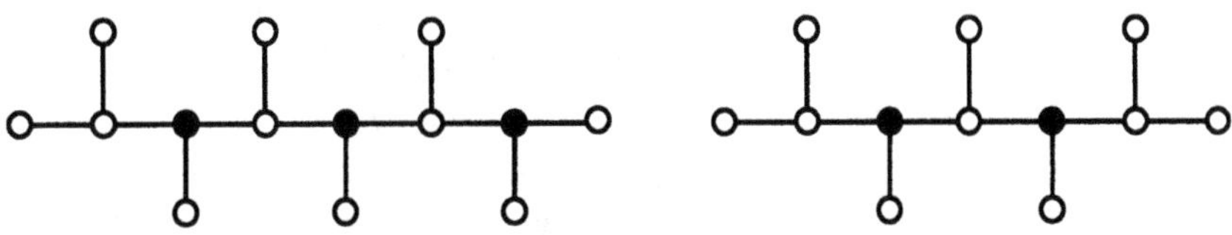

Figure 10: Tree of n nodes needs the most number of converters.

Corollary 6 *To guarantee load-wavelength assignability on $(n \times n)$-mesh, $2\lceil n/2 \rceil \lfloor n/2 \rfloor$ converters are both sufficient and necessary.*

Proof. For $(n \times n)$-mesh, according to Theorem 5 we can remove each of four degree-two nodes at the corners by substituting two edges incident to the node with one edge linking the two endpoints directly. By Theorem 4, the minimal vertex cover of modified mesh is both sufficient and necessary for load-wavelength assignability. It is easy to see that the minimal vertex cover is the set that includes every other node in each row and column (extra node must be added in even rows and columns for odd n), and thus the size of the cover is $n^2/2$ for even n and $(n^2 - 1)/2$ for odd n. See Fig. 11. Converters are optimally placed at black nodes.   □

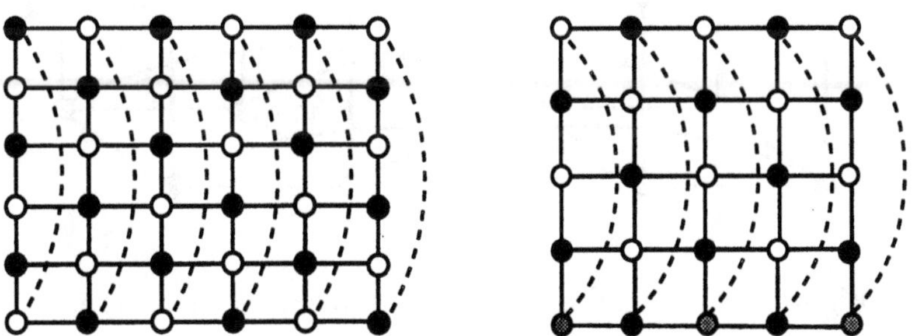

Figure 11: Optimal placement of converters on meshes and torus.

Corollary 7 *To guarantee load-wavelength assignability on $(n \times n)$-torus, $n\lceil n/2 \rceil$ converters are both sufficient and necessary.*

Proof. According to Theorem 4, the minimal vertex cover of $(n \times n)$-torus is both sufficient and necessary for load-wavelength assignability. It is easy to see that the minimal vertex cover is the set that includes every other node in each row and column (extra node must be added for odd n), and thus the size of the cover is $n^2/2$ for even n and $n(n+1)/2$ for odd n. See Fig. 11. Converters are optimally placed at black (and grey) nodes.   □

Corollary 8 *To guarantee load-wavelength assignability on n-cube for $n \geq 3$, 2^{n-1} converters are both sufficient and necessary.*

Proof. According to Theorem 4, the minimal vertex cover of n-cube is both sufficient and necessary for load-wavelength assignability. Construct two subsets of the vertex-set of n-cube $V(n) = \{i = (i_1, i_2, \cdots, i_n) \mid i_j = 0 \text{ or } 1, j = 1, 2, \cdots, n\}$ as follows: $C_1(2) = \{(0,0), (1,1)\}$ and $C_2(2) = \{(0,1), (1,0)\}$. For $n \geq 3$,

$$
\begin{aligned}
C_1(n) &= \{i, \text{ for } i \in C_1(n-1)\} \cup \{i + 2^{n-1}, \text{ for } i \in C_2(n-1)\}, \\
C_2(n) &= \{i, \text{ for } i \in C_2(n-1)\} \cup \{i + 2^{n-1}, \text{ for } i \in C_1(n-1)\}.
\end{aligned}
$$

It is not difficult to verify (by mathematical induction) that for $n \geq 3$, $C_1(n)$ and $C_2(n)$ make an equal partition of 2^n nodes in $V(n)$, and they are two minimal vertex covers of n-cube. See Fig. 12, where $C_1(n)$ consists of black nodes while $C_2(n)$ white nodes.   □

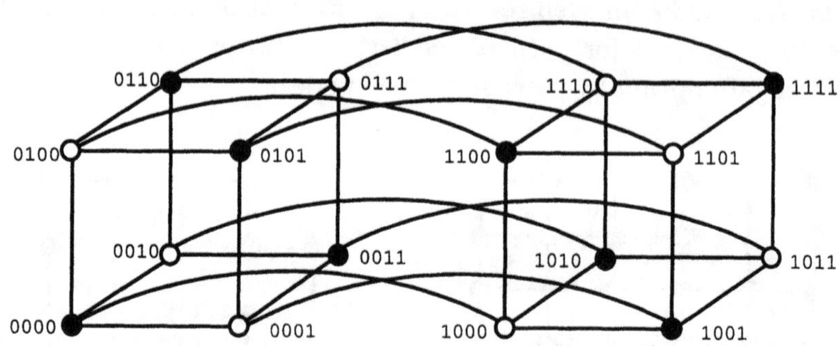

Figure 12: Optimal placement of converters on n-cube.

5   Conclusions

We have found the necessary and sufficient conditions that a subset of nodes guarantees load-wavelength assignability on general WDM networks. From these conditions we obtained the optimal placement of converters on a class of WDM networks of special topologies including trees, rings, meshes, torus, and hyper-cubes. The proposed theorems and methodology can also be applied to other interconnecting networks for parallel computing networks, such as Banyan networks and Shuffle networks.

The results obtained from this paper are summarized in Table 2.

Networks	Unidirectional Channels	Duplex Channels
Star	1	0
tree of n nodes	$\lfloor (n-2)/4 \rfloor$	$\lfloor (n-2)/2 \rfloor$
ring of n nodes	1	1
$(n \times n)$-mesh	$2\lceil n/2 \rceil \lfloor n/2 \rfloor$	$n^2 - 4$
$(n \times n)$-torus	$n\lceil n/2 \rceil$	n^2
n-cube	2^{n-1}	2^n

Table 2. Number of converters needed for load-wavelength assignability.

Acknowledgement

This work is supported in part by Research Grants of City University of Hong Kong under grant No. 7000677, 7000927, 7001035, and by CERG Research Grants of Hong Kong under grant No. 9040442; and National 973 Information Technology and High-Performance Software Program of China under grant No. G1998030402.

References

[1] D. Banerjee and B. Mukherjee, A practical approach for routing and wavelength assignment in large wavelength-routed optical networks, *IEEE J. Select. Areas Commun.*, 14 (5) (1996), pp. 903-908.

[2] J. A. Bondy and U. S. R. Murty, Graph Theory with Applications, American Elsevier Publishing Co., Inc., New York, 1976.

[3] T. Erlebach, K. Jansen, C. Kaklamanis, and P. Persiano, An optimal greedy algorithm in directed tree networks, *DIMACS Series in Discrete Mathematics and Theoretical Computer Science*, 40 (1998), pp. 117-129.

[4] M. R. Garey and D. S. Johnson, Comuters and Intractability: A Guide to the Theory of NP-Completeness, W. H. Freeman, San Francisco, CA, 1979.

[5] P. E. Green, Fiber-Optic Networks, Prentice-Hall, Cambridge, MA, 1992.

[6] E. J. Harder and H. A. Choi, Gossiping in WDM all-optical square mesh networks, *DIMACS Series in Discrete Mathematics and Theoretical Computer Science*, 48 (1998), pp. 75-84.

[7] P. Klein, S. Plotkin, C. Stein, and É. Tardos, Faster approximation algorithms for the unit capacity concurrent flow problem with applications to routing and finding sparse cuts, *SIAM J. Computing*, 23 (3) (1994), pp. 466-487.

[8] J. P. Labourdette and A. Acampora, Logically rearrangeable multihop lightwave networks, *IEEE Trans. Commun.*, 39 (8) (1991), pp. 1223-1230.

[9] K.-C. Lee and V. O. K. Li, A wavelength-convertible optical network, *J. Lightwave Technology*, 11 (5) (1993), pp. 962-970.

[10] M. Mihail, C. Kaklamanis, and S. Rao, Efficient access to optical bandwidth, *Proc. 36th Annual IEEE Symp. Foundations of Computer Science* (FOCS), 1995, pp. 548-557.

[11] B. Mukherjee, WDM-Based local lightwave networks Part I: single-hop system, *IEEE Network*, 6 (3) (1992), pp. 12-26.

[12] B. Mukherjee, WDM-based local lightwave networks Part II: multihop systems, *IEEE Network*, 6 (4) (1992), pp. 22-32.

[13] B. Mukherjee, D. Banerjee, S. Ramamurthy, and A. Mukherjee, Some principles for designing a wide-area WDM optical network, *IEEE/ACM Trans. Networking*, 4 (5) (1996), pp. 684-695.

[14] P. Raghavan and E. Upfal, Efficient routing in all-optical networks, *Proc. 26th Annual ACM Symp. Theory of Computing* (STOC), (1994), pp. 134-143.

[15] R. Ramaswami and K. N. Sivarajan, Design of logical topologies for wavelength-routed optical networks, *IEEE J. Select. Areas Commun.*, 14 (5) (1996), pp. 840-851.

[16] R. Ramaswami and G. Sasaki, Multiwavelength optical networks with limited wavelength conversion, *IEEE/ACM Trans. Networking*, 6 (6) (1998), pp. 744-754.

[17] A. Schriver, P. Seymour, and P. Winkler, The ring loading problem, *SIAM J. Discrete Math.*, 11 (1) (1998), pp. 1-14.

[18] S. Subramaniam, M. Azizoglu, and A. K. Somani, Connectivity and sparse wavelength conversion in wavelength-routing networks, *Proc. 1st IEEE Conf. Computer Commun.* (INFOCOM), (1996), pp. 148-155.

[19] G. Wilfong and P. Winkler, Ring routing and wavelength translation, *Proc. 9th Annual ACM-SIAM Symp. Discrete Algorithms* (SODA), (1998), pp. 333-341.

[20] J. Yates, J. Lacey, D. Everitt, and M. Summerfield, Limited-range wavelength translation in all-optical networks, *Proc. 1st IEEE Conf. Computer Commun.* (INFOCOM), (1996), pp. 954-961.

ADVANCES IN OPTICAL NETWORKS
D.-Z. Du and L. Ruan (Eds.) pp. 99 - 122
©2000 Kluwer Academic Publishers

Lightpath Establishment in Wavelength-Routed WDM Optical Networks

Jason P. Jue
Center for Advanced Telecommunications Systems and Services
University of Texas at Dallas, Richardson, TX 75083
E-mail: jjue@utdallas.edu

Contents

References

1   Introduction

Wavelength-division multiplexing (WDM) technology has been improving steadily in recent years, with existing systems capable of providing huge amounts of bandwidth on a single fiber link. WDM systems are currently being deployed in long-distance telecommunication networks on a point-to-point basis, with the optical signals being converted back to electronics at each node. The electronic switching and processing costs at the nodes can potentially be very high, leading to severe performance bottlenecks and limiting the delivery of optical link bandwidth to the end users.

Emerging wavelength-routed WDM systems, which utilize photonic cross-connects (PXCs), are capable of switching data entirely in the optical domain. Configuring these optical devices across a network enables one to establish all-optical connections, or lightpaths [1], between source nodes and destination nodes (Fig. 1). Data carried on these lightpaths avoid electronic conversion and processing at intermediate nodes, thereby alleviating the electronic bottleneck.

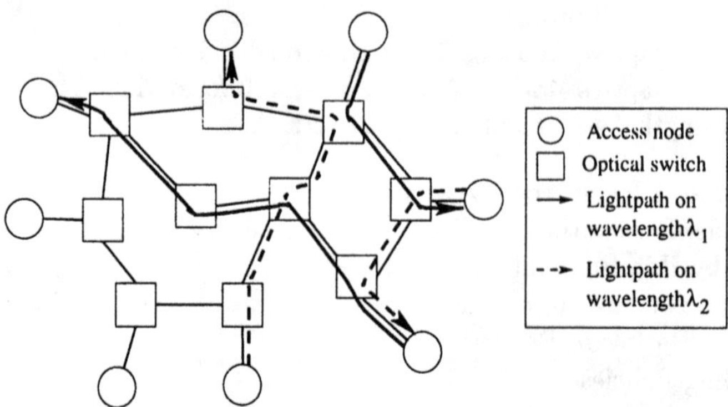

Figure 1: Wavelength-routed WDM network with lightpaths established between nodes.

One of the challenges involved in designing wavelength-routed networks is to develop efficient algorithms and protocols for establishing lightpaths in the optical network. The algorithms must be able to select routes and assign wavelengths to connections in a manner which efficiently utilize network resources and which maximizes the number of lightpaths that can be established. Signaling protocols for setting up lightpaths must effectively manage the distribution of control messages and network state information in order to establish a connection in a timely manner.

In this article, we discuss the various issues related to the control and management of lightpaths in a wavelength-routed optical network, and we review various approaches for handling these issues. Section 2 covers routing and wavelength assignment algorithms for WDM networks. Section 3 reviews various signaling protocols for reserving resources and setting up lightpaths in a wavelength-routed network. Section 4 presents various control and management models, as well as some standard protocols for lightpath establishment in an IP over WDM environment.

2   Routing and Wavelength Assignment

The problem of finding a route for a lightpath and assigning a wavelength to the lightpath is often referred to as the routing and wavelength assignment problem (RWA). The objective of the problem is to route lightpaths and assign wavelengths in a manner which minimizes the amount of network resources that are consumed, while at the same time ensuring that no two lightpaths share the same wavelength on the same fiber link. Furthermore, in the absence of wavelength conversion devices, the RWA problem operates under the constraint that a lightpath must occupy the same wavelength on each link in its route. This restriction is known as the wavelength-continuity constraint. The optimal formulation of the RWA problem is known to be NP-complete; therefore, heuristic solutions are often employed.

The RWA problem can be considered under two different traffic assumptions. The static RWA problem applies to the case in which the set of connections is known in advance, and a lightpath must be established for each connection. In the dynamic RWA problem, connections arrive to the network dynamically and remain for some amount of time before departing; thus, lightpaths must not only be established dynamically, but must also be taken down. In both the static and dynamic RWA problems, the routing component and the wavelength assignment component of the problem are often decoupled into two separate subproblems in order to make the problem more tractable. In the following sections, we discuss routing and wavelength assignment approaches for static and dynamic lightpath establishment. We assume that no wavelength conversion is available.

2.1   RWA for Static Lightpath Establishment

A number of studies have investigated the RWA problem for setting up a static set of lightpaths [1] [2] [3]. These studies formulate the problem using integer linear program (ILP) formulations, or rely on heuristic approaches in an attempt to minimize the number of wavelengths required to establish a given set of lightpaths. The ILP formulations are NP-complete and therefore may only be solved for very small systems. For larger systems, heuristic methods must be used.

In [3], the routing problem is formulated as an ILP in which the objective is to minimize the number of wavelengths required to establish a fixed set of lightpaths. The search space of the problem is reduced by restricting the set of links through which a lightpath for a given source-destination pair may traverse. The resulting ILP is then solved by relaxing the integer constraint, solving the resulting non-integer linear program, and then utilizing a randomized rounding approach on the result to obtain an integer solution.

Other heuristics consider only alternate shortest-hop paths between a source-destination pair, and choose one of the paths according to a predefined policy [4, 5]. In [4], a shortest-hop path is randomly chosen for each source-destination pair. Each source-destination pair is then considered individually, and the route for the pair of nodes is switched to an alternate shortest-hop path if doing so results in a reduction of load on the most heavily loaded link in the original shortest-path route. In [5], an approach similar to that in [4] is considered; however, the objective is to minimize the number of fibers in a multifiber network, and the set of alternate paths includes routes which may be longer than the shortest-hop routes.

The wavelength-assignment sub-problem of the RWA problem can itself be formulated as a graph coloring problem, which is also NP-complete. Greedy heuristics for the wavelength-assignment problem for a static set of lightpaths typically involve ordering the wavelengths, and assigning the same wavelength to as many lightpaths as possible before moving on to the next wavelength [1]. Also, the set of lightpaths may be ordered by length, such that wavelengths are assigned to longer lightpaths before wavelengths are assigned to shorter lightpaths.

2.2   RWA for Dynamic Lightpath Establishment

When lightpaths are established and taken down dynamically, routing and wavelength assignment decisions must be made as connection requests arrive to the network. It is possible that, for a given connection request, there may be insufficient network resources to set up a lightpath, in which case the connection request will be blocked. The connection may also be blocked if there is no common wavelength available on all of the links along the chosen route. Thus, the objective in the dynamic situation is to choose a route and a wavelength which maximizes the probability of setting up a given connection, while at the same time attempting to minimize the blocking for future connections. Similar to the case of static lightpaths, the dynamic RWA problem can also be decomposed into a routing subproblem and a corresponding wavelength assignment subproblem. Approaches to solving the routing subproblem can be categorized as being either fixed or adaptive, and as utilizing either global or local network state information.

2.2.1 Fixed Routing

In fixed routing, a single fixed route is predetermined for each source-destination pair. When a connection request arrives, the network will attempt to establish a lightpath along the fixed route. If no common wavelength is available on every link in the route, then the connection will be blocked. A fixed routing approach is simple to implement; however, it is very limited in terms of routing options and may lead to a high level of blocking. In order to minimize the blocking in fixed routing networks, the predetermined routes need to be selected in a manner which balances the load evenly across the network links. Fixed routing schemes do not require the maintenance of global network state information.

2.2.2 Adaptive Routing Based on Global Information

Adaptive routing approaches increase the likelihood of establishing a connection by taking into account network state information. For the case in which global information is available, routing decisions may be made with full information as to which wavelengths are available on each link in the network.

Centralized Versus Distributed Routing   Adaptive routing based on global information may be implemented in either a centralized or distributed manner. In a centralized algorithm, a single entity, such as a network manager, maintains complete network state information, and is responsible for finding routes and setting up lightpaths for connection requests. Since a centralized entity manages the entire network, there does not need to be a high degree of coordination among nodes; however, a centralized entity becomes a possible single point of failure. Furthermore, a centralized approach does not scale well, as the centralized entity would need to maintain a large database to manage all nodes, links, and connections in the network.

Alternate-Path Routing One approach to adaptive routing with global information is alternate-path routing. Alternate-path routing relies on a set of predetermined fixed routes between a source node and a destination node [6, 7, 8, 9]. When a connection request arrives, a single route is chosen from the set of predetermined routes, and a lightpath is established on this route. The criteria for route selection is typically based on either path length or path congestion. An example of a routing algorithm based on path length is the K-shortest paths algorithm [6], in which the first K shortest paths are maintained for each source-destination pair, and the paths are selected in order of length, from shortest to longest. A connection is first attempted on the shortest path. If resources are not available on this path, the next shortest path is attempted. A path selection policy based on path congestion examines the available resources on each of the alternate paths, and chooses the path on which the highest amount of resources are available. Choosing the shortest-path route consumes less network resources, but may lead to high loads on some of the links in the network, while choosing the path with the least congestion leads to longer paths, but distributes the load more evenly over the network.

Unconstrained Routing Another adaptive routing approach utilizing global information is unconstrained routing which considers all possible paths between a source node and a destination node. In order to choose an optimal route, a cost is assigned to each link in the network based on current network state information, such as wavelength availability on links. A least-cost routing algorithm is then executed to find the least-cost route [10, 11, 12]. Whenever a connection is established or taken down, the network state information is updated. Two examples of unconstrained routing approaches are link-state routing and distance-vector routing.

In a distributed link-state routing approach, each node in the network must maintain complete network state information [11]. Each node may then find a route for a connection request in a distributed manner. Whenever the state of the network changes, all of the nodes must be informed. Therefore, the establishment or removal of a lightpath in the network may result in the broadcast of update messages to all nodes in the network. The need to broadcast update messages may result in significant control overhead.

Furthermore, it is possible for a node to have outdated information, and for the node to make an incorrect routing decision based on this information.

A distance-vector approach to routing with global information is also possible [12]. This approach doesn't require that each node maintains complete link-state information at each node as in [11], but instead has each node maintain a routing table which indicates for each destination and on each wavelength, the next hop to the destination and the distance to the destination. The approach relies on a distributed Bellman-Ford algorithm to maintain the routing tables. Similar to [11], the scheme also requires nodes to update their routing table information whenever a connection is established or taken down. This update is accomplished by having each node send routing updates to their neighbors periodically or whenever the status of the node's outgoing links changes. Although each node maintains less information than in [11] and the updates are not broadcast to all nodes, the scheme may still suffer from a high degree of control overhead.

Although routing schemes based on global knowledge must deal with the task of maintaining a potentially large amount of state information which is changing constantly, these schemes often make the most optimal routing decisions if the state information is up to date. Thus, global-knowledge based schemes may be well suited for networks in which lightpaths are fairly static and do not change much with time.

2.2.3   Adaptive Routing Based on Local Information

While near-term emerging systems will be fairly static, with lightpaths being established for long periods of time, it is expected that, as network traffic continues to scale up and become more bursty in nature, a higher degree of multiplexing and flexibility will be required at the optical layer. Thus, lightpath establishment will become more dynamic in nature, with connection requests arriving at higher rates, and lightpaths being established for shorter time durations. In such situations, maintaining distributed global information may become infeasible. The alternative is to implement routing schemes which rely only on local information.

A number of adaptive routing schemes exist which rely on local information rather than global information. The advantage of using local information is that the nodes do not have to maintain a large amount of state information; however, routing decisions tend to be less optimal than in the case of global information. Two examples of local-information-based adaptive routing schemes are alternate routing with local information, and deflection routing.

Alternate-Path Routing with Local Information　While alternate-path routing schemes typically rely on global information, variations exist which utilize only local information. A least-congested alternate path routing scheme is investigated in [13]. In this scheme, the choice of a route is determined by the wavelength availability along the alternate paths. Two variations of the scheme are considered: the case in which wavelength availability information is known along the entire path, and the case in which only local information is available.

In the first approach, the decision making entity is aware of the wavelength availability information for all of the links in each of the alternate paths. In this case, the chosen route is that which has the greatest number of wavelengths which are available along all of the links in its path. For example, in Fig. 2, if we consider two alternate routes from source node A to destination node D, with available wavelengths as shown on each link, then two wavelengths (λ_1 and λ_3) are available along the entire length of route 1, while only one wavelength (λ_2) is available along the entire length of route 2; thus, route 1 will be chosen.

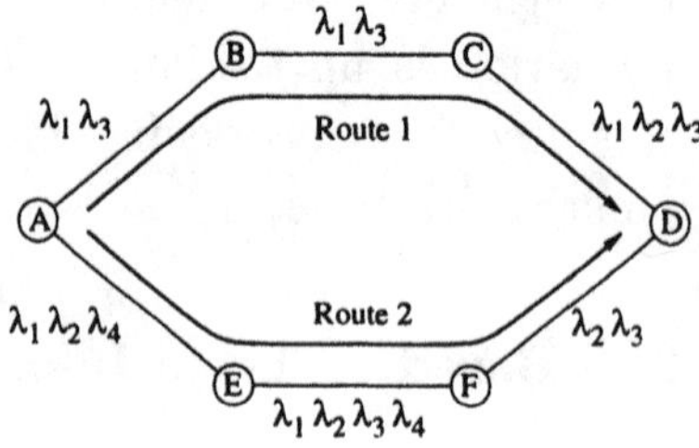

Figure 2: Alternate routing. Available wavelengths are shown on each link.

The limitation of basing the route selection decision on full path information is that the information may be difficult to maintain or difficult to obtain in a timely manner. Each node would be required to either maintain complete state information, or the information would need to be gathered in real time, as the lightpath is being established. The alternative, based on local information, is to gather wavelength availability information only along the first k hops of each path. The route is then chosen based on which path is the least congested along its first k hops. In Fig. 2, if $k = 2$, then route 2 would be chosen, since it has three wavelengths available on the first two links (λ_1, λ_2, and λ_4), while route 1 only has two wavelengths available on the first two links (λ_1 and λ_3). Although local information may provide a good estimate of the congestion along a path, it does not guarantee that any particular wavelength will be available along the entire path; thus, it is possible that after choosing a route, the connection will still be blocked due to lack of available wavelengths.

Deflection Routing   Another approach to adaptive routing with limited information is deflection routing, or alternate-link routing [14]. This routing scheme choses from alterate links on a hop-by-hop basis rather than choosing from alternate routes on an end-to-end basis. The routing is implemented by having each node maintain a routing table which indicates, for each destination, one or more alternate outgoing links to reach that destination. These alternate outgoing links may be ordered such that a connection request will preferentially choose certain links over other links as long as wavelength resources are available on those links. Other than a static routing table, each node will only maintain information regarding the status of wavelength usage on its own outgoing links. When choosing an outgoing link for routing, the decision can be determined on either a shortest-path or least-congested basis.

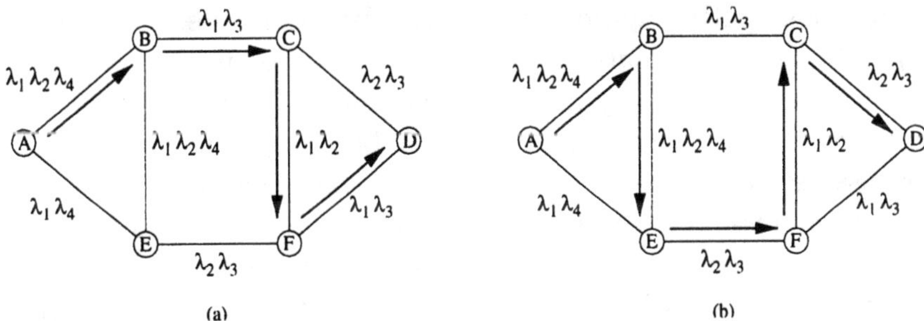

Figure 3: Deflection routing. Available wavelengths are shown on each link.

Under the shortest path criteria, the routing scheme will first attempt to choose the outgoing link which results in the shortest path to the destination. If there is no feasible wavelength available on the link, then the routing scheme will attempt to choose an alternate outgoing link which leads to the next shortest path to the destination. The routing scheme proceeds in this manner until the destination is reached or the connection is blocked. Figure 3(a) illustrates the deflection routing scheme for a connection request from node A to node D. The default shortest path in this example is along the path A→B→C→D. When the request reaches node C, it cannot continue over link CD, since no common wavelength is available on links AB, BC, and CD. The request is therefore deflected to node F, where it can continue to the destination along link FD. The wavelength selected for the lightpath will be λ_1. Note that, in the absence of any deflections, the default routing for any connection will be shortest-path routing. Also, once the routing for a lightpath is deflected at a node, the default routing from the point of deflection onward will again be shortest-path routing if no further deflections take place.

In a least-congested deflection routing approach, the routing scheme chooses, from among the alternate outgoing links, the link which has the largest number of feasible wavelengths. The set of feasible wavelengths consists of the set of wavelengths which are available on all of the previous hops as well as the next outgoing link. Least-congested deflection routing is illustrated in Fig. 3(b) for a connection from node A to node D. On the first hop, link AB is selected, since it has three available wavelengths, while link AE has only two available wavelengths. When the connection request arrives to node B, it will be routed to node E, since there are three feasible wavelengths (λ_1, λ_2, and λ_4) available on link BE, and there is only one feasible wavelength (λ_1) available on link BC. The least-congested deflection routing approach will generally result in longer paths than the shortest-path deflection routing approach; however, least-congested deflection will allow a lightpath to be routed around congested areas in the network, balancing the load more evenly across the network. The results in [14] show that a shortest-first policy results in lower blocking at low loads, while a least-congested policy results in lower blocking at higher loads.

A number of issues arise when implementing a deflection routing scheme. One such issue is the problem of looping, in which a connection request message returns to a node which has already been visited. Loop detection may be addressed by having each connection request message maintain a path vector containing a list of visited nodes. If a node receives a connection request message which indicates that the message has already visited this node, then the connection attempt will be blocked. An alternative to maintaining a path vector is to utilize a time-to-live field, which would prevent the connection request message from looping in the network indefinitely.

Another problem which may arise is that a connection request may be deflected a large number of times, leading to an unreasonably long route for the lightpath. Possible solutions to this problem include limiting the maximum length or number of hops in a lightpath, or limiting the number of deflections that a route can take. When a connection request message reaches its limit on the maximum number of hops or deflections, the connection attempt will be blocked. Further restrictions may also be placed on the selection of possible outgoing ports in order to prevent routes from heading back towards the source node.

2.3 Wavelength Assignment

In general, if there are multiple feasible wavelengths between a source node
and a destination node, then a wavelength assignment algorithm is required
to select a wavelength for a given lightpath. The wavelength selection may be
performed either after a route has been determined, or in parallel with finding a route. Since the same wavelength must be used on all links in a lightpath, it is important that wavelengths are chosen in a way which attempts
to reduce blocking for subsequent connections. A review of wavelength-
assignment approaches can be found in [15].

One example of a simple, but effective, wavelength-assignment heuristic
is first-fit. In first-fit, the wavelengths are indexed, and a lightpath will
attempt to select the wavelength with the lowest index before attempting
to select a wavelength with a higher index. By selecting wavelengths in this
manner, existing connections will be packed into a smaller number of total
wavelengths, leaving a larger number of wavelengths available for longer
lightpaths.

Another approach for choosing between different wavelengths is to simply
select one of the wavelengths at random. In general, first-fit will outperform
random wavelength assignment when full knowledge of the network state is
available [7]. However, if the wavelength selection is done in a distributed
manner, with only limited or outdated information, then random wavelength
assignment may outperform first-fit assignment. The reason for this behavior is that, in a first-fit approach, if multiple connections are attempting
to set up a lightpath simultaneously, then it may be more likely that they
will choose the same wavelength, leading to one or more connections being
blocked.

Other simple wavelength assignment heuristics include the most-used-
wavelength heuristic and the least-used-wavelength heuristic. In most-used
wavelength assignment, the wavelength which is the most used in the rest of
the network is selected. This approach attempts to provide maximum wavelength reuse in the network. The least-used approach attempts to spread
the load evenly across all wavelengths by selecting the wavelength which
is the least-used throughout the network. Both most-used and least-used
approaches require global knowledge.

A number of more advanced wavelength assignment heuristics which
rely on complete network state information have been proposed [16, 17].
It is assumed in these heuristics that the set of possible future lightpath
connections is known in advance. For a given connection, the heuristics
attempt to choose a wavelength which minimizes the number of lightpaths

in the set of future lightpaths that will be blocked by this connection. It is shown that these heuristics offer better performance than first-fit and random wavelength assignment.

3　Signaling and Resource Reservation

In order to set up a lightpath, a signaling protocol is required to exchange control information among nodes and to reserve resources along the path. In many cases, the signaling protocol is closely integrated with the routing and wavelength assignment protocols. Signaling and reservation protocols may be categorized based on whether the resources are reserved on each link in parallel, reserved on a hop-by-hop basis along the forward path, or reserved on a hop-by-hop basis along the reverse path. Protocols will also differ depending on whether global information is available or not.

3.1　Parallel Reservation

In [11], the control scheme reserves wavelengths on multiple links in parallel. The scheme, which is based on link-state routing, assumes that each node maintains global information on the network topology and on the current state of the network, including information regarding which wavelengths are being used on each link. Based on this global information, the node can calculate an optimal route to a destination on a given wavelength. The source node then attempts to reserve the desired wavelength on each link in the route by sending a separate control message to each node in the route. Each node that receives a reservation request message will attempt to reserve the specified wavelength, and will send either a positive or negative acknowledgement back to the source. If the source node receives positive acknowledgements from all of the nodes, it can establish the lightpath and begin communicating with the destination. The advantage of a parallel reservation scheme is that it shortens the lightpath establishment time by having nodes process reservation requests in parallel. The disadvantage is that it requires global knowledge, since both the path and the wavelength must be known ahead of time.

3.2　Hop-by-Hop Reservation

An alternative to parallel reservation is hop-by-hop reservation in which a control message is sent along the selected route one hop at a time. At each intermediate node, the control message is processed before being forwarded to the next node. When the control message reaches the destination, it is processed and sent back towards the source node. The actual reservation of link resources may be performed either while the control message is traveling in the forward direction towards the destination, or while the control message is traveling in the reverse direction back towards the source.

3.2.1   Forward Reservation

In forward reservation schemes, wavelength resources are reserved along the forward path to the destination on a hop-by-hop basis. The method of reserving wavelengths depends on whether or not global information is available to the source node. If the source node is maintaining complete state information, then it will be aware of which wavelengths are available on each link. Assuming that the state information is current, the source node may then send a connection setup message along the forward path, reserving the same available wavelength on each link in the path.

For the case in which a node only knows the status of its immediate links, the wavelength selection becomes more complicated, as the source node doesn't know which wavelength will be available along the entire path. The source node may utilize a conservative reservation approach, choosing a single wavelength and sending out a control message to the next node attempting to reserve this wavelength along the entire path; however, there is no guarantee that the selected wavelength will be available along every link in the path. If the wavelength is blocked, the source node may select a different wavelength and reattempt the connection. The limitation of this approach is that it may result in high setup times, since it may take several attempts before a node can establish a lightpath.

An alternate approach to maximizing the likelihood of establishing a lightpath in a forward reservation scheme is to use an aggressive reservation scheme which over-reserves resources [18]. Multiple wavelengths may be reserved on each link in the path, with the expectation that at least one wavelength will be available on all links in the path. In a greedy approach, all feasible wavelengths will be reserved at every link in the path. The source node will first reserve all available wavelengths on the desired outgoing link. A connection request message containing the wavelength reservation information is then sent to the next node along the path. At each intermediate node, the subset of wavelengths consisting of the intersection of the wavelengths reserved on the previous link and the wavelengths available on the next link will be reserved. For example, if S_n is the set of wavelengths available on the nth link, then the set of wavelengths reserved on the first

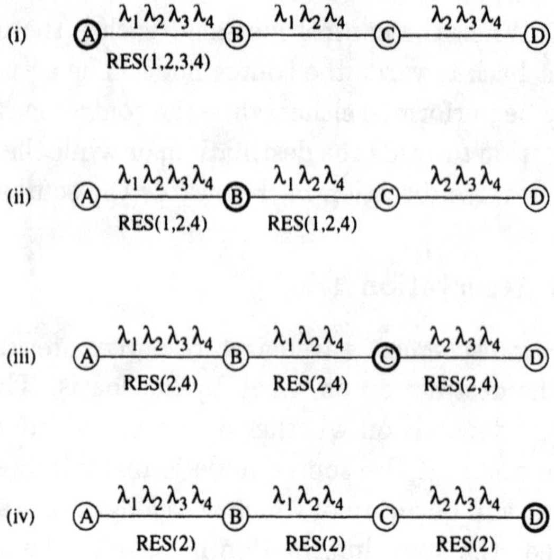

(i) (A) —$\lambda_1\lambda_2\lambda_3\lambda_4$— (B) —$\lambda_1\lambda_2\lambda_4$— (C) —$\lambda_2\lambda_3\lambda_4$— (D)
     RES(1,2,3,4)

(ii) (A) —$\lambda_1\lambda_2\lambda_3\lambda_4$— (B) —$\lambda_1\lambda_2\lambda_4$— (C) —$\lambda_2\lambda_3\lambda_4$— (D)
      RES(1,2,4)   RES(1,2,4)

(iii) (A) —$\lambda_1\lambda_2\lambda_3\lambda_4$— (B) —$\lambda_1\lambda_2\lambda_4$— (C) —$\lambda_2\lambda_3\lambda_4$— (D)
       RES(2,4)   RES(2,4)   RES(2,4)

(iv) (A) —$\lambda_1\lambda_2\lambda_3\lambda_4$— (B) —$\lambda_1\lambda_2\lambda_4$— (C) —$\lambda_2\lambda_3\lambda_4$— (D)
      RES(2)   RES(2)   RES(2)

Figure 4: Forward reservation. As the control message propagates from A to D, each node reserves the set of wavelengths which have been available on all links traversed by the control message.

link in the path will be S_1, the set of wavelengths reserved on the second link will be $S_1 \cap S_2$, the set of wavelengths reserved on the third link would be $S_1 \cap S_2 \cap S_3$, etc. When the connection request reaches the destination, one wavelength of the remaining set of wavelengths will be chosen, and all of the other wavelengths will be released. Figure 4 illustrates the forward reservation of wavelengths when establishing a lightpath from node A to node D.

One disadvantage of over-reserving resources is that, during the time that the resources are reserved, the reserved resources cannot be utilized by other users, even if these resources will never be used by the connection. In order to reduce the amount of time that an unused wavelength is reserved on a link, the wavelength may be released as soon as it is apparent that the wavelength is not viable for a given connection. For example, if wavelengths λ_1, λ_2, λ_3, and λ_4 are available on the first link, then all four of the wavelengths will be reserved on this link. However, if it is subsequently discovered that only λ_1, λ_2, and λ_4 are available on the second link, then not only will λ_1, λ_2, and λ_4 be reserved on the second link, but λ_3 will immediately be released on the first link.

Another approach to limiting the number of wavelengths that are reserved is to divide the wavelengths into groups. When reserving wavelengths on a link, a node will reserve only those wavelengths which belong to a specific group [19]. The choice of the group is made at the source node and is based on the number of available wavelengths in each group. The source node will find the group with the largest number of available wavelengths, and the node will reserve all of the available wavelengths in that group before sending the request on to the next node. The size of the group is a critical parameter. If the group is too large, then too many resources will be reserved, but if the group is too small, then the likelihood of establishing a lightpath will be smaller.

3.2.2   Backward Reservation

To prevent the over-reservation of resources altogether, reservations may be made after the control message has reached the destination and is headed back to the source. Such reservation schemes are referred to as backward reservation schemes [18]. By reserving wavelengths in the reverse direction, the reserved wavelengths are idle for less time than if the wavelengths are reserved in the forward direction. Another advantage is that the connection request message can gather wavelength usage information along the path in the forward direction. This information can then be used by the destination node to select an appropriate wavelength to reserve. Figure 5 illustrates the backward reservation scheme. It is shown in [18] that, in general, backward reservation schemes outperform forward reservation schemes for the case in which there is no wavelength conversion. One possible drawback of a backward reservation scheme is that if multiple connection are being set up simultaneously, it is possible that a wavelength that was available on a link in the forward direction will be taken by another connection request and will no longer be available when the reservation message traverses the link in the reverse direction.

3.3   Holding Policies

To improve the connection setup probability at the cost of higher setup times, it is possible to hold or buffer connection requests at intermediate nodes if wavelength resources are not immediately available [20, 21]. If an appropriate wavelength becomes available, the connection request will continue towards the destination. If, after waiting for some time, the appropriate resources do not become available, then the connection is blocked.

In [21], it is shown that a holding policy decreases the blocking probability without significantly increasing setup time. However, it is also shown in [20] that a holding policy reduces the network throughput compared to a

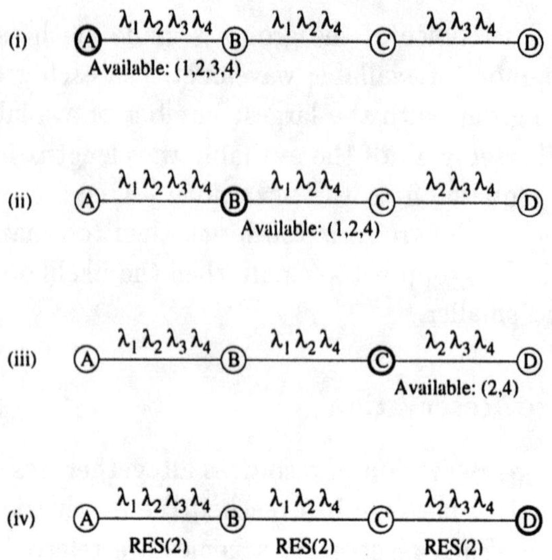

Figure 5: Backward reservation. As the control message propagates from A to D, it records the set of available wavelengths. Node D then chooses a wavelength and reserves the wavelength on each link by sending a reservation message back towards the source.

policy in which calls are blocked immediately if resources are not available.

4   IP over WDM Control and Management

With the rapid growth of the Internet, it is becoming apparent that IP traffic will soon become the dominant type of traffic in emerging networks; thus, there is much interest in optimizing the underlying optical network to handle IP traffic. When IP networks are implemented over optical WDM network, it is possible to modify existing IP protocols to handle the control and management associated with establishing and taking down lightpaths at the optical layer. The advantage of such an approach is that it reduces the need to develop new protocols for the optical layer, and it enables the control of both IP and optical layers through a single integrated control plane.

The relationship between the IP layer and the optical layer is defined by the control model. In an overlay model, the IP layer is a client to the optical layer, with the optical layer providing point-to-point lightpath services between IP routers. The control protocols for the IP layer are independent of the optical layer control protocols; however, it is possible for the optical layer protocols to be based on modified versions of IP protocols. The IP and optical layers interact through a user-to-network interface (UNI). The UNI defines specific services that the IP layer may request from the optical layer, such as lightpath establishment, lightpath removal, and lightpath modification.

Another control model is the peer model in which IP routers and optical crossconnects are peers with regard to control and management. Each optical node is treated as an IP addressable entity and runs the same control protocols as other IP routers. In this situation, the same protocol which is used to establish a virtual circuit connection between IP routers can also be used to establish a lightpath between two routers. One of the limitations of this approach is that the entire optical network must be made visible to all of the IP routers in the same domain; thus the approach is not very scalable.

A third control model is the integrated model in which the IP and optical layers are running separate instances of the same control protocols. In this case, optical crossconnects are still IP addressable; however, the optical network is viewed as a separate IP domain by the rest of the network. The integrated model is more scalable than the peer model, while at the same time, the IP nodes may still interact with the optical layer through existing IP protocols.

In each of the control models, the establishment of lightpaths in the optical network may be performed through a multi-protocol lambda switching (MPλS) control framework [22, 23], which is a variation of the multi-protocol label switching (MPLS) framework.

4.1 MPLS and MPλS

MPLS is a control framework which is currently being developed as a standard to enable fast switching in IP networks [24]. MPLS control mechanisms can be used to establish a label-switched path (LSP) between two non-neighboring IP routers. In an LSP, packets do not need to be processed at the IP layer at any of the intermediate IP routers; thus, the packets experience much lower delays. To route a packet over a LSP, labels are applied to packets, and these labels are used to switch the packets within the network. When a labeled packet arrives at an input port of a label-switched router, the label is used to index a table which specifies the outgoing port for the packet and the packet's new label. The old label is replaced by the new label, and the packet is sent to the appropriate output port.

The concept of MPLS can be extended to wavelength-routed optical networks as MPλS. In MPλS, rather than applying labels to packets, the wavelengths themselves act as labels and are used to route the packets through the optical network. A signal coming in on a given input port is directed to an appropriate output port based on the wavelength of the signal. Since a wavelength in MPλS is analogous to a label in MPLS, the process of swapping labels would correspond to the process of converting an optical signal from one wavelength to another. Thus, in the absence of wavelength conversion devices, it is not possible to swap "labels" in MPλS.

The MPλS control mechanism can be used to establish a lightpath in a similar way that MPLS establishes a label-switched path. The establishment of a lightpath in the MPλS framework requires a routing protocol and a signaling protocol. Neither of these protocols is specified by MPλS; however, it is possible to use standard IP-based protocols to provide routing and signaling functionality.

4.2   Routing

Routing in most IP networks is currently handled by the open shortest path first (OSPF) protocol [26]. OSPF is a link-state protocol in which the state of each link in the network is periodically broadcast to all nodes in the form of link-state advertisements (LSAs). The LSA will provide link information such as whether the link is up or down, and the amount of bandwidth on the link. In the context of wavelength-routed optical networks, the link-state information may indicate whether a wavelength is in use or not. Given the link-state information, each node will have full knowledge of the network state, and will use this information to construct the network topology and to route packets through the network. The routing itself is based on Dijkstra's shortest path algorithm. The limitations of OSPF are similar to the limitations of any other routing algorithm which utilizes global state information. The signaling overhead may potentially be very high when lightpaths and other connections are being established and taken down at a high rate. Also, incorrect routing decisions may result from outdated state information.

4.3   Signaling Protocols and Label Distribution

In MPLS and MPλS, signaling protocols are required to exchange control messages for setting up LSPs and lightpaths, to reserve network resources, and to distribute labels. Possible signaling protocols include RSVP (resource reservation protocol) [27] and CR-LDP (constraint-based label distribution protocol) [28]. Both protocols perform signaling on a hop-by-hop basis, with RSVP reserving resources in the backward direction (destination-initiated reservation), and CR-LDP reserving resources in the forward direction (source-initiated reservation). Routing can either be done based on OSPF, or an explicit route can be specified by the source node.

4.3.1 RSVP

In RSVP, signaling takes place between the source router and destination router. At the source router, a PATH message is created and sent to the destination node. The PATH message may be routed according to standard routing protocols, such as OSPF. Alternatively, an explicit route through the network may be specified if the source router is aware of the network topology. The PATH message may also contain information such as QoS requirements for the carried traffic and label requests for assigning labels at intermediate nodes. At an intermediate node, the request is recorded, and the PATH message is forwarded to the next node. If the message cannot be forwarded or if resources are not available, the path setup fails, and a message is sent back to the source router. At the destination node, an RESV message is generated to distribute labels, and is sent back to the previous node. The intermediate nodes reserve the appropriate resources, allocate new labels for the path, and send the RESV message back towards the source router.

4.3.2 CR-LDP

The CR-LDP protocol may also be used to provide signaling and to distribute labels. CR-LDP utilizes TCP sessions between nodes in order to provide a reliable distribution of control messages. At the ingress node, a LABEL_REQUEST message is created. The message indicates the route and the required traffic parameters for the route. Resources are reserved at the ingress node, and the LABEL_REQUEST is forwarded to the next node. At the intermediate node, resources are reserved, and the LABEL_REQUEST is forwarded. At the destination, resources are reserved and a label is assigned to the request. The destination node creates a LABEL_MAPPING message which contains the new label, and passes the message back towards the source node. Each intermediate node allocates a label and sets up its forwarding table before passing the LABEL_MAPPING to the previous node.

Both RSVP and CR-LDP may be used to reserve a single wavelength for a lightpath if the wavelength is known in advance. These protocols may also be modified to incorporate wavelength selection functionality into the reservation process.

5   Conclusion

As optical networks continue to develop and emerge, WDM systems will evolve from static point-to-point links to optically switched wavelength-routed networks. The establishment of lightpaths in such networks requires the implementation of control and management protocols to perform routing and wavelength assignment functions, as well as to exchange signaling information and to reserve resources. In this article, we have presented some of the routing, wavelength assignment, and signaling protocols for establishing lightpaths in a wavelength-routed network.

In routing and wavelength assignment algorithms for dynamic lightpaths, the goal is to minimize the number of blocked connections. The performance of these algorithms depends heavily on the amount of state information available to each node. If global information is known, then the routing and wavelength assignment decisions can be nearly optimal; however it is difficult to maintain complete up-to-date information in a very dynamic environment. When only limited information is known, routing and wavelength assignment decisions may be based on either shortest path or least congested criteria. By choosing routes or links which result in a shorter path, network resources are conserved, possibly leading to lower blocking for future connections. On the other hand, a least-congested approach to routing and wavelength assignment may result in the load being more evenly distributed across the network.

The effectiveness of a wavelength assignment policy depends on the amount of information available, as well as on the connection arrival rate. With global information, a first-fit policy will perform well. However, with only local information, a random policy may perform better, since lower indexed wavelengths are more likely to be used on other links throughout the network. Also, when connection requests are arriving at a high rate, using first-fit may result in higher blocking for simultaneous connection requests.

The performance of signaling protocols for reserving wavelengths along a lightpath will depend on whether or not global information is available, and whether or not multiple connection requests may be attempted simultaneously. For the case in which global information is available, reservations may be made either in parallel or on a hop-by-hop basis, with parallel reservations leading to lower connection establishment times. When only local information is available, wavelength selection may be combined with the reservation scheme. In both forward and backward reservation schemes, additional information is gathered along the path before deciding on a wavelength. The two schemes differ in how they are affected by simultaneous connection requests. In forward reservation, the over-reservation of wavelength resources leads to higher blocking for other connections, while in backward reservation, there is the possibility that a previously available link in the route will be taken by another connection request.

In emerging IP over WDM systems, the routing and signaling protocols for establishing lightpaths are likely to exist within an MPλS framework and are expected to be based on existing IP protocols such as OSPF, CR-LDP, and RSVP. However, it is important to note that the MPλS framework is not limited to any particular routing or signaling protocol, and that in certain situations, such as the case of rapidly arriving and departing lightpaths, alternative protocols may offer better performance.

References

[1] I. Chlamtac, A. Ganz, and G. Karmi, "Lightpath Communications: A Novel Approach to High Bandwidth Optical WANs", *IEEE Transactions on Communications* Vol.40 No.7 (1992) pp. 1171-1182.

[2] R. Ramaswami and K. Sivarajan, "Routing and Wavelength Assignment in All-Optical Networks," *IEEE/ACM Transactions on Networking* Vol.3 No.5 (1995) pp. 489-500.

[3] D. Banerjee and B. Mukherjee, "A Practical Approach for Routing and Wavelength Assignment in Large Wavelength-Routed Optical Networks," *IEEE Journal on Selected Areas in Communications* vol. 14 No.5 (1996) pp. 903-908.

[4] S. Baroni and P. Bayvel, "Wavelength Requirements in Arbitrary Connected Wavelength-Routed Optical Networks," *IEEE/OSA Journal of Lightwave Technology* Vol.15 No.2 (1997) pp. 242-251.

[5] M. Alanyali and E. Ayanoglu, "Provisioning Algorithms for WDM Optical Networks," *IEEE/ACM Transactions on Networking* Vol.7 No.5 (1999) pp. 767-778.

[6] K. Bala, T.E. Stern, and K. Bala, "Algorithms for Routing in a Linear Lightwave Network," *Proceedings, IEEE Infocom '91*, Miami, FL, Apr. 1991.

[7] A. Birman and A. Kershenbaum, "Routing and Wavelength Assignment Methods in Single-Hop All-Optical Networks with Blocking," *Proceedings, IEEE Infocom '95*, Boston, MA, Vol. 2, Apr. 1995, pp. 431-438.

[8] H. Harai, M. Murata, and H. Miyahara, "Performance of Alternate Routing Methods in All-Optical Switching Networks," *Proceedings, IEEE Infocom '97*, Kobe, Japan, Apr. 1997, pp. 516-524.

[9] S. Ramamurthy and B. Mukherjee, "Fixed-Alternate Routing and Wavelength Conversion in Wavelength-Routed Optical Networks," *Proceedings, IEEE Globecom '98*, Sydney, Australia, Nov. 1998, pp. 2295-2302.

[10] A. Mokhtar and M. Azizoglu, "Adaptive Wavelength Routing in All-Optical Networks," *IEEE/ACM Transactions on Networking* vol. 6 No.2 (1998) pp. 197-206.

[11] R. Ramaswami and A. Segall, "Distributed Network Control for Wavelength Routed Optical Networks," *Proceedings, IEEE Infocom '96*, San Francisco, CA, Mar. 1996, pp. 138-147.

[12] H. Zang, L. Sahasrabuddhe, J. P. Jue, S. Ramamurthy, and B. Mukherjee, "Connection Management for Wavelength-Routed WDM Networks," *Proceedings, IEEE Globecom '99*, Rio de Janeiro, Brazil, Dec. 1999, pp. 1428-1432.

[13] L. Li and A. K. Somani, "Dynamic Wavelength Routing Using Congestion and Neighborhood Information," *IEEE/ACM Transactions on Networking* Vol.7 No.5 (1999) pp. 779-786.

[14] J. P. Jue and G. Xiao, "An Adaptive Routing Algorithm with a Distributed Control Scheme for Wavelength-Routed Optical Networks," *Proceedings, Ninth International Conference on Computer Communications and Networks*, Las Vegas, NV, Oct. 2000.

[15] H. Zang, J. P. Jue, and B. Mukherjee, "Review of Routing and Wavelength Assignment Approaches for Wavelength-Routed Optical WDM Networks," *Optical Networks Magazine* Vol.1 No.1 (2000) pp. 47-60.

[16] S. Subramaniam and R. A. Barry, "Wavelength Assignment in Fixed Routing WDM Networks," *Proceedings, IEEE ICC '97*, Montreal, Canada, Vol.1, June 1997, pp. 406-410.

[17] X. Zhang and C. Qiao, "Wavelength Assignment for Dynamic Traffic in Multi-fiber WDM Networks," *Proceedings, 7th International Conference on Computer Communications and Networks*, Lafayette, LA, Oct. 1998, pp. 479-485.

[18] X. Yuan, R. Melhem, R. Gupta, Y. Mei, and C. Qiao, "Distributed Control Protocols for Wavelength Reservation and Their Performance Evaluation," *Photonic Network Communications*, vol. 1, No.3, 1999, pp. 207-218.

[19] A. G. Sotica, A. Sengupta, "On a Dynamic Wavelength Assignment Algorithm for Wavelength-Routed All-Optical Networks," *Proceedings, SPIE/IEEE/ACM OptiComm 2000*, Dallas, TX, Oct. 2000, pp. 211-222.

[20] Y. Mei and C. Qiao, "Efficient Distributed Control Protocols for WDM All-Optical Networks," *Proceedings, International Conference on Computer Communication and Networks*, Las Vegas, NV, Sept. 1997, pp. 150-153.

[21] A. Sengupta, S. Bandyopadhyay, A. R. Balla, and A. Jaekel, "Algorithms for Dynamic Routing in All-Optical Networks," *Photonic Network Communications* Vol.2 No.2 (2000) pp. 163-184.

[22] N. Ghani, "Lambda Labeling: A Framework for IP-Over-WDM," *Optical Networks Magazine* Vol.1 No.2 (2000) pp. 45-58.

[23] D. O. Awduche, "Multi-Protocol Lambda Switching: Combining MPLS Traffic Engineering Control with Optical Crossconnects," *IETF Draft*, draft-awduche-mpls-te-optical-01.txt, Nov. 1999.

[24] R. Callon, P. Doolan, N. Feldman, A. Fredette, G. Swallow, and V. Srinivasan, "A Framework for Multiprotocol Label Switching," *IETF Draft*, draft-ietf-mpls-framework-05.txt, Sept. 1999.

[25] D. Awduche, J. Malcolm, J. Agogbua, M. O'Dell, and J. McManus, "Requirements for Traffic Engineering Over MPLS," RFC 2702, Sept. 1999.

[26] W. R. Stevens, *TCP/IP Illustrated, Volume 1*, Addison-Wesley, 1994.

[27] R. Braden, L. Zhang, S. Berson, S. Herzog, and S. Jamin, "Resource ReSerVation Protocol (RSVP) – Version 1 Functional Specification," IETF RFC 2205, September 1997.

[28] B. Jamoussi et al., "Constraint-Based LSP Setup Using LDP," *IETF Draft*, draft-ietf-mpls-cr-ldp-04.txt, July 2000.

OPTICAL NETWORKS - RECENT ADVANCES
L. Ruan and D.-Z. Du (Eds.) pp. 123 - 149
©2001 Kluwer Academic Publishers

Multifiber WDM Networks

Ling Li
Axiowave Networks, Inc.
100 Nickerson Road, Marlborough, MA 01752
E-mail: `lingli@ieee.org`

Arun K. Somani
Department of Electrical and Computer Engineering
Iowa State University, Ames, IA 50011
E-mail: `Arun@iastate.edu`

Contents

1   Introduction

Computer and communication networks have changed the world dramatically in the 20th century, and will continue to do so in the future. Over the last two decades, optical fibers have revolutionized the communications industry. Researchers have been driven by a vision of accessing a larger fraction of the approximately 50-THz theoretical information bandwidth of single-mode fiber. With the advancement of optical technologies, a wide variety of optical components for building WDM networks have been developed, such as wide-band optical amplifiers (OAs), optical add/drop multiplexers (OADMs) and optical cross-connects (OXCs). A natural approach to utilize the fiber bandwidth efficiently is to partition the usable bandwidth into non-overlapping wavelength bands. Each wavelength, operating at several gigabits per second, is used at the electronic speed of the end-users. The use of wavelengths to route data is referred to as wavelength routing, and a network which employs this technique is known as a *wavelength-routed* network [1]. In such networks, each connection between a pair of nodes is assigned a path through the network and a wavelength on that path, such that connections whose paths share a common link in the network are assigned different wavelengths. The optical communication path between two nodes is called a *lightpath*. All-optical networks employing wavelength-division multiplexing and wavelength routing are a viable solution for future wide-area networks (WANs) and metropolitan-area networks (MANs). These wavelength-routed WDM networks offer the advantages of protocol transparency and simplified management and processing compared to routing in telecommunications systems using digital cross-connects [2].

The research issues in WDM networks can be broadly classified into two categories: network design and network operation. The network design problem is generally an optimization problem [3, 4]. The inputs to the problem are static traffic demands, a general network topology, and some specific requirements, e.g., network reliability and/or restoration time. The objective of the optimization problem could be minimizing the resources, including the number of wavelengths, the number of fibers, or the number of cross-connect ports, to meet the requirements. The outputs may include the network configuration, and the route and wavelength for each source-destination pair. The network design problem can be formulated as an integer liner programming (ILP) or mixed integer linear programming (MILP) problem. Since the number of variables and constraints can be very large in WDM networks, heuristics are usually used to find fast solutions.

After a network is built, one critical problem is how to operate the network such that the network performance is optimized under dynamic traffics. The traffic intensity of the dynamic traffic is usually known while the individual demands arrive and depart randomly. Since network resources are typically not sufficient enough to guarantee that every dynamic demand can be accommodated in the network, the average blocking probability for a given utilization is one of the metrics to measure the network performance. Some other metrics include control overhead and algorithm complexity.

In a wavelength-routed WDM network, the path of a signal is determined by the location of the signal transmitter, the wavelength on which it is transmitted, and the state of the network devices. In simple WDM networks, the same wavelength must be assigned to a connection on every link on a path if wavelength converters are not available at the switching nodes. Connection requests encounter higher blocking probability than they do in electronic-switched networks because of the wavelength continuity constraint. Routing and wavelength assignment algorithms (RWAA) are responsible for selecting a suitable route and a wavelength among the many possible choices for establishing a connection. Good routing and wavelength assignment algorithms are critically important in improving the performance of WDM networks.

The network performance can be improved by using wavelength converters at the switching nodes, which can convert data on one wavelength along a link into another wavelength at an intermediate node and forward it along the next link [14]. A WDM network without wavelength conversion is referred to as a *wavelength selective (WS)* network. The networks with wavelength conversion are called *wavelength interchanging (WI)* networks [5]. All-optical wavelength converters are being prototyped in research laboratories [6]. However, the techniques have not matured so far. Wavelength converters are likely to remain costly devices in the near future.

To reduce the cost of wavelength conversion in a network, sparse wavelength conversion and limited wavelength conversion techniques are proposed in [7, 8]. A network with only a few nodes having full conversion capability is called a network with sparse wavelength conversion. The results in [7] show placing converters on a fraction of nodes of a network is sufficient to ensure high network performance. A network with limited wavelength conversion is a network in which not every wavelength can be converted to every other wavelength in every switching node. This may be the result of having limited numbers of wavelength converters in the WDM network, or by using devices in which noise or other device limitations restrict the set of conversions which may be performed. Limited wavelength conversion has been shown to provide performance which is often close to that achieved with ideal full wavelength conversion in networks with tunable transmitters and receivers [8].

With the development of networks and optical technology, more and more researchers have realized that a single-fiber network may not have enough capacity to support the dramatically increasing bandwidth requirement. In fact, most of the optical networks, if not all, consist of multiple fibers on each link. A WDM network in which each link consists of multiple fibers, and each fiber carries information on multiple wavelengths is called a multifiber WDM network. The wavelength continuity constraint is relaxed in multifiber networks because a wavelength that cannot continue on one fiber can be switched to another fiber using optical cross-connect as long as the same wavelength is available on the other fibers on the outgoing link. Multifiber WDM networks provide an alternative solution to overcome the wavelength continuity constraint [9, 11]. Since an incoming wavelength can be switched to the same wavelength on any outgoing fibers using an optical cross-connect (OXC), multiple fibers in WDM networks have similar effects as the limited wavelength conversion.

A fundamental problem in multifiber WDM networks is how to design and operate multifiber networks such that the network performances are close to the corresponding WDM networks with full wavelength conversion at every node. The design and operation problems in multifiber networks are similar to those in single-fiber networks that have been discussed in the previous paragraphs. However, multifiber networks have their own characteristics as well. One important issue is the fiber and wavelength requirements in multifiber networks. This issue is considered in [15, 24] for arbitrarily connected networks as physical topologies. An off-line wavelength assignment problem in star and ring networks that deploy multiple fibers between nodes is studied in [22]. To overcome the limitation of conventional WDM rings in terms of network expansion capabilities, the number of nodes within the ring, and the number of optical paths (OPs) accommodated in the network, different architectures of multifiber WDM ring networks are proposed and studied in [20, 21]. The trade-off between link cost and node cost of conventional WDM ring networks is analyzed in [16]. The literature of designing multifiber WDM networks is surveyed and summarized in Section 2. In Section 3 we review the research work for the dynamic performance of multifiber networks.

Routing and wavelength assignment (RWA) algorithms still play a critical role in multifiber WDM networks with dynamic traffic inputs. The wavelength assignment algorithms (e.g., random, first-fit, most-used, and least-used) that are first proposed for single fiber networks can also be applicable in multifiber networks with or without modifications. A set of new RWA algorithms are proposed for multifiber WDM Networks, e.g., least-loaded (LL), minimum sum (MS) in [27], $M\sum$ in [28], and Relative Capacity Loss (RCL) in [9]. The performances of various routing algorithms have been studied, e.g., fixed path routing (FPR), alternate path routing (APR), and least congestion routing (LCR) [30, 35]. RWA algorithms that consider routing and wavelength assignment problems simultaneously are proposed in [27, 29]. New dynamic RWA algorithms based on layered-graph model are proposed in [33] for multifiber WDM networks. Details of these algorithms are discussed in Section 3. Simulation and analysis results show that using multiple fibers per link can improve the network performance significantly. A limited number of fibers, say F with W wavelengths each, is sufficient to guarantee high network performance that is comparable to a network with a single fiber with $F \times W$ wavelengths and full wavelength conversion at every node [11, 33]. We present our concluding remarks in Section 4.

2   Design of Multifiber Networks

Network design problem has been extensively studied in computer and communications networks. Network design is generally considered an optimization problem [12]. Sometimes the topology in network design problems is known and fixed. For given static traffic demands between node pairs in the network, and some specific requirements, e.g., network reliability and/or restoration time, the objective of the optimization problem is to minimize network cost, or the required resources, or to optimize network performance, e.g., minimize average network delay.

It is difficult to predict the traffic demands and statistical properties of the traffic that will be carried by future wavelength-routing WDM networks, but consideration of these factors is important for network design and analysis. Many design methodologies have been proposed in the literature for optical WDM networks with or without wavelength conversions. Possible objectives in the design of WDM networks include minimizing the number of wavelengths, the number of fibers, or the number of cross-connect ports. The outputs may include the network configuration, and the route and wavelength for each source-destination pair. There is no statistical modeling of the sources and there is not a blocking-probability consideration. The network design problem can be formulated as an integer liner programming (ILP) or mixed integer linear programming (MILP) problem. Since the number of variables and constraints can be very large in WDM networks, heuristics are usually used to find fast solutions.

Some of the existing work in the literature pursues minimization of the number of wavelengths for the WDM network design problem. On the other hand, in most practical cases, the number of wavelengths per fiber is determined by the technology and is fixed. If a single fiber does not suffice to carry existing traffic on a link, more fibers on that link need to be lit [15]. As such, minimization of the number of fibers, fiber miles, the size of cross-connects, etc., or a combination of the above, are more meaningful measures of the merit of a design. We review and summarize proposed solutions for multifiber network design in this section.

2.1   Multifiber WDM Ring Networks

The benefits of the ring topology are well displayed in the field of self-healing rings (SHR's) that employ electronic add-drop multiplexers (SONET ADM's) [17]. Some fault recovering schemes such as path switching protection or line switching protection can be implemented on ring networks [18]. Conventional WDM ring networks consist of one fiber on each link, and the single fiber carries multiple wavelengths. The difference from traditional SONET ring networks is that it uses spare wavelengths instead of protection fibers. The wavelength requirements, for the minimum number of wavelengths required for the ring, have been investigated to study the network performance of each ring architecture in [19].

To overcome the limitation of conventional WDM rings in terms of network expansion capabilities, the number of nodes within the ring, and the number of optical paths (OPs) accommodated in the network, different architectures of multifiber WDM ring networks are proposed and studied in [20, 21]. The algorithms proposed in [20] minimize the average number of fibers handled at the switching nodes. The required optical cross-connect (OXC) system scale at each node is obtained from the algorithms in both single-fiber and multifiber ring networks. Algorithms that consider failure restoration have also been proposed. Generic node architecture and several lightpath accommodation design algorithms are proposed in [21]. The total number of fiber ports, N, is used as performance criterion to compare the performance of different multifiber ring architectures. Here N includes both inter-office fiber ports $N1$, accommodating optical paths (OPs) from/to neighbor nodes, and the intra-office fiber ports $N2$, accommodating OPs terminated at the node. Several network architectures have been studied, including the $2 \times m$-fiber uni-directional ring (UR/2m), the $4 \times m$-fiber bi-directional ring (BR/4m), and the $2 \times m$-fiber bi-directional ring (BR/2m), where m represents the degree of the space division multiplexing. Multiple network design algorithms have been proposed, e.g., wavelength path (WP)-UR/2m, virtual wavelength path (VWP)[1]-BR/4m, and WP-BR/4m. The performances of the proposed algorithms are evaluated under various conditions. The effect of the ring architecture, e.g., uni-/bidirectional, single-/multi-multifiber rings, and different node connectivity patterns are demonstrated. The results show that a substantial difference does not exist between multifiber ring networks with or without wavelength conversion. The network performance of multifiber ring architectures, UR/2m or Br/4m, showed strong dependence on the path distribution pattern. Therefore, the choice of UR/2m or BR/4m is a key issue when applying optical paths to multifiber ring networks.

The trade-off between link cost and node cost of conventional WDM ring networks is analyzed in [16]. Conventional design approaches tend to emphasize a reduction in the number of wavelengths by utilizing wavelength reuse. The total cost of these wavelength-efficient networks could exceed that of the primitive fiber mesh networks without WDM due to node complexity. The results of the analysis lead to a new breed of multifiber WDM ring networks with the most simple nodes and modest link efficiency. The general idea of the proposed architecture is that a cable of $N - 1$ unidirectional fibers is provided for interconnecting N nodes. Each fiber is terminated at its designated node, which is referred as the end node. All the wavelengths are dropped and terminated at the end node. Each fiber originates from the node next to its end node. The fiber traverses the other nodes which add signals with different wavelengths into the fiber via optical couplers. Thus, the insertion function is distributed along the fiber, while the drop function is concentrated in each end node. The paper also discusses the design and performance of the proposed architecture, and locates its most promising application areas with respect to the number of nodes and ring length.

An off-line wavelength assignment problem in star and ring networks that deploy multiple fibers between nodes is studied in [22]. The problem is formulated as follows: find the minimum number of wavelengths needed to accommodate a given set of connections (source-destination pairs) so that

---

[1]VWP refers to a path with a wavelength converter at every node on the path.

no two connections that share a link are assigned the same wavelength. The single-fiber version of the wavelength assignment problem has received considerable attention in the literature [23]. The paper [22] considers wavelength assignment in multifiber networks. Both dual-fiber networks (two fibers on each link) as well as the general case of k fibers per link are considered. For dual-fiber ring networks, the paper gives an upper bound of $3/2L-1$ for the total number of wavelengths required (counting wavelengths that are reused in the second fiber), where L is the network load. It is shown that this upper bound is tight when the number of network nodes is large enough. This bound is a sharper per-fiber bound $(3L/4)F$ than the equivalent per-fiber bound (L) for two superimposed single-fiber rings. The upper bound is also extended to general k fiber ring networks, resulting in at most $(k+1)/KL - 1$ required wavelengths counting over all k fibers. It is also shown that this upper bound is tight. For arbitrary dual-fiber undirected networks where connection path-lengths are of length at most two, the paper shows that the wavelength assignment problem is optimally solvable (in contrast to the single-fiber version, which is NP-complete). A few results are also shown for general topologies with fibers that relate the number of required wavelengths to that of an equivalent single-fiber network.

2.2   Arbitrary Multifiber Networks

Resilient multifiber wavelength-routed optical networks (WRONs) are studied in [10]. The paper investigates the influence on the network performance of the maximum number of wavelengths per fiber W, restoration strategies, node functionality, and physical topology. Fiber requirements are analyzed for numerous network topologies both with and without link failure restoration, considering different optical cross-connect (OXC) configurations and terminal functionalities. An integer linear program (ILP) formulation is presented for the exact solution of the routing and wavelength allocation (RWA) problem, with minimal total number of fibers, $F_T/(W)$. Lower bounds on $F_T/(W)$ are discussed, and heuristic algorithms proposed. Three restoration strategies are considered:

1. *Edge-Disjoint Path Restoration*: Each node-pair is assigned an active lightpath and an edge-disjoint restoration lightpath, the latter being used to restore any failure in the active lightpath. Since both source and destination nodes are involved, this is necessarily an end-to-end restoration process. The restoration capacity can be shared among edge-disjoint active lightpaths to restore non-concurrent link failures.

2. *Path Restoration:* In the case of a link failure, for each interrupted lightpath, any physical path from source to destination which does not include the failed link may be considered for restoration. Therefore, the restoration lightpath may be different according to which link in the active lightpath has failed.

3. *Link Restoration:* When a link fails, restoration is achieved by re-routing the interrupted lightpaths around, to avoid the failed link whilst maintaining the reminding of the paths.

The three restoration schemes are compared in terms of capacity requirement in the paper. Different network topologies are analyzed, to evaluate the influence of physical connectivity and network size on the restoration capacity. Simulation results show that path restoration in wavelength-convertible networks is always equal or very close to the lower bound of the required number of fibers, implying that path restoration enables near-optimal restoration, even by reallocating only the interrupted lightpaths. A few more fibers are required for Edge-Disjoint Path Restoration, whereas a significant difference was observed for link restoration. The number of required fibers for the three restoration schemes in WDM networks without wavelength conversion is very close to that in wavelength-convertible networks, particularly with path restoration.

Wavelength and fiber requirements are also considered for arbitrarily connected networks as physical topologies in [24]. By analyzing a large number of randomly generated networks, bounds on the network wavelength requirements are evaluated as a function of the physical connectivity. The advantages achievable by multifiber connections and the consequence of single link failure restoration are then assessed for several existing or planned network topologies. The results show that the wavelength requirement strongly depends on the physical connectivity, while it is almost independent of the network size. Moreover it is shown that WRON's can provide high transport capability with a modest number of wavelengths and that wavelength translation does not lead to a reduction of the number of required wavelengths.

Network provisioning algorithms for both primary networks that do not account for restoration and restorable networks are studied in [15]. Provisioning of a network refers to assigning network resources to a static traffic demand. Efficient provisioning is essential in minimizing the investment made on the network required to accommodate a given demand. It is assumed that there is a fixed set of wavelengths available on each fiber in [15]. The objective of network design is to minimize the number of fibers required for a set of given traffic demands. The problem can be formulated as an integer linear program (ILP), however the computational complexity of the ILP is prohibitive for a network whose size is not trivial. Therefore efficient heuristic solution methods are proposed in [15]. The first algorithm for primary networks without restoration is an iterative algorithm which is based on greedy decisions by the connections to decrease a certain metric whose minimum value corresponds to an optimal assignment. Numerical results show that the computational complexity per iteration of the algorithm is low. On a mesh-like topology, the algorithm generates efficient solutions in a reasonable number of iterations. The second algorithm studies the design of fault-tolerant networks, and is obtained by an adaptation of the first method. The method entails coordinated planning of several failure scenarios. Numerical results show that it provides more efficient designs than those obtained by considering the failures independently.

We have surveyed and summarized the literature of network design algorithms in multifiber networks. In the next section, we study dynamic performance of various routing and wavelength assignment algorithms in multifiber WDM networks.

3 Dynamic Performance of Multifiber Networks

In the previous section, we have reviewed the literature on the design of multifiber networks. In this section we review and summarize the routing and wavelength assignment algorithms and the performance analysis of dynamic traffic in multifiber networks. The difference from the network design problem is that in this section we assume that the network traffic arrives dynamically and randomly. Each arriving connection request has to be handled individually without the knowledge of any future arrivals. The objective of dynamic traffic processing is to find a path that can satisfy the connection requirement, with the consideration of optimizing the network performance in the long run. Since network resources are typically not sufficient enough to guarantee that every dynamic demand can be accommodated in the network, the average blocking probability for a given utilization is one of the metrics to measure the network performance.

3.1 Wavelength Assignment Algorithms in Multifiber Networks

The wavelength assignment problem is a unique problem in WDM networks. Unlike electrical circuit-switched networks, the same wavelength has to be free on all of the links of a path in order to establish a connection in all-optical WDM networks without wavelength conversion. If full-range wavelength converters are available at every node, wavelength assignment is a trivial problem. However, the technology of all-optical wavelength conversion is not mature yet. Wavelength converters are likely to be costly devices in the near future. Therefore, good wavelength assignment algorithms along with routing algorithms are critically important in improving the network performance and reducing the network cost.

The wavelength assignment algorithms in the literature can be broadly classified into two categories: the algorithms proposed for single-fiber networks and those for multifiber networks. The following algorithms are first proposed for single-fiber networks:

1. *Random (R)*: The random wavelength assignment algorithm chooses one of the available wavelengths randomly with a uniform distribution to establish a connection.

2. *First-Fit (FF)*: Assume that the wavelengths are arbitrarily ordered, e.g., $\lambda_1, \lambda_2, \ldots, \lambda_W$, where W is the maximum number of wavelengths per fiber. The first-fit algorithm checks the status of the wavelengths sequentially and chooses the first available wavelength to establish a connection.

3. *Most-Used (MU)*: The free wavelength that is used on the most number of links in the network is chosen to establish a connection.

4. *Least-Used (LU)*: The free wavelength that is used on the least number of links in the network is chosen to establish a connection.

The random wavelength assignment algorithm is usually assumed by analytical models because of its simplicity. However, the used wavelengths are randomly distributed and mixed up with free wavelengths in the network. It may be hard for a connection request to find a wavelength free on consecutive links from a source to a destination node if random wavelength assignment is used. The least-used algorithm attempts to route a connection on the least utilized wavelength in order to achieve a near-uniform distribution of the load over the wavelength set. The results in [29] show that both the random and the least-used algorithms distribute the load evenly over the wavelengths. The first-fit and the most-used methods attempt to pack the connections together to use fewer wavelengths, and leave more wavelengths consecutively free. Simulation results in [27, 29, 37] show that the blocking probability of the random and least-used wavelength assignment algorithms are higher than that of the first-fit and most-used algorithms. The random assignment algorithm has a performance close but better than the least-used algorithm. The blocking probability of the first-fit algorithm with fixed-path routing is considerably lower than that of the random wavelength assignment algorithm. However the most-used algorithm performs slightly better than the first-fit algorithm [28].

As we have discussed in Section 1, the wavelength continuity constraint is relaxed in multifiber networks. This is because a wavelength that cannot continue on one fiber can be switched to another fiber using an optical cross-connect as long as the same wavelength is available on the other fibers on the outgoing link. Several wavelength assignment algorithms have been proposed in the literature for multifiber WDM networks:

1. *Least-loaded (LL)*: the LL algorithm proposed in [27] selects the wavelength that has the largest residual capacity on the most loaded link along a path;

2. *Minimum sum (MS)*: the MS algorithm proposed in [27] chooses the wavelength that has the minimum average utilization.

   Both MS and LL algorithms select the most used wavelength when multiple wavelengths are tied; hence they reduce to the most-used rule in the single-fiber case.

3. $M\sum$: the $M\sum$ algorithm in [28] chooses the wavelength that leaves the network in a "good" state for future calls. The goodness of a state is measured by a new concept called the *value* of the network. The value function $V(\alpha)$ of the resulting state α after a call is established is restricted to be a function of path capacities, i.e., $V(\alpha) = g([C(\alpha, p) : p : P])$ where P is the set of all possible paths, and $C(\alpha, p)$ is the path capacity of p in an arbitrary state ϕ. Let $\alpha_{prime}(j)$ be the next network state if wavelength j is assigned to the connection. $M\sum$ chooses the

wavelength j that maximizes the quantity

$$\sum_{p \in P} C(\alpha_{prime}(j), p). \tag{1}$$

4. *Relative Capacity Loss (RCL)*: the relative capacity loss algorithm in [9] chooses the wavelength that minimizes the relative capacity loss. The relative capacity loss of path p on wavelength λ^*, denoted by $R_c(p, \lambda^*)$, is defined as

$$R_c(p, \lambda^*) = \frac{P_c(p, \lambda^*) - P_c'(p, \lambda^*)}{\sum_\lambda P_c(p, \lambda)}. \tag{2}$$

where $P_c(p, \lambda)$ is the wavelength path capacity of path p on wavelength λ [9].

Note here that the random, first-fit, most-used, and least-used algorithms are first proposed for single fiber network, but they can also be used in multifiber networks with or without modifications. The results in [27] show that the LL and MS algorithms performs better than the random, first-fit, and most-used algorithms in multifiber networks. The $M\sum$ algorithm in [28] performs considerably better than other algorithms except the RCL algorithm at the cost of increased computational complexity. The blocking probability of using the RCL algorithm is $5\% - 30\%$ better than the $M\sum$ depending on the traffic demands and network topology, but has the same worst case time complexity of $M\sum$ [9].

The computational complexity of $M\sum$ and RCL in the worst case is WN^3, where W is the number of wavelengths on each link and N is the number of nodes in the network. To reduce the computation cost, a distributed extension of RCL, called distributed relative capacity loss (DRCL), is proposed in [45]. In DRCL, the network routing is assumed to use the Bellman-Ford algorithm [12]. In Bellman-ford routing, each node exchanges routing tables with its neighbor nodes and updates its own routing table accordingly. An RCL table is introduced at each node in DRCL. The RCL tables are exchanged and updated in a similar way as the routing table. Upon the arrival of a connection request, the wavelength which yields the lowest total relative capacity loss is selected to establish the connection.

In the above discussions a connection establishment procedure is separated into two steps: select a route first (if not fixed-path routing), and then select a wavelength from the available free wavelengths. The routing and wavelength assignment algorithm can also be solved jointly as proposed in [27, 29, 33]. The route-wavelength pair that meets the specified criteria, i.e., maximizes the residual capacity over all wavelengths and considered paths is selected jointly. These joint routing and wavelength assignment algorithms outperform the disjoint approaches. However, the computational cost of these algorithms could be too high to be used practically.

3.2 Performance of Routing Algorithms in Multifiber Networks

Routing and wavelength assignment algorithms play a key role in improving the performance of WDM networks [1, 2]. We have reviewed the wavelength assignment algorithms in multifiber networks in the previous section. We study various routing algorithms and their performance in this section.

Network resources and congestion are two major concerns in dynamic network routing algorithms. One goal for routing algorithms is to minimize resource utilization so that more resources can be left for future connections. Because of the dynamic property of traffic arrivals, some network links may be over-utilized and some under-utilized, which may cause unnecessary connection blocking. Thus another goal to achieve is to balance the traffic load on each link of the network to avoid congestion. However, these two goals may not be achievable in some network topologies and under certain traffic conditions. In WDM networks, both routing and wavelength assignment algorithms have to put these two goals into consideration. The results in [32, 36] show that routing schemes have more impact on the performance of a network than wavelength assignment schemes in WDM networks. The routing schemes proposed in the literature can be classified into two categories: static routing and dynamic routing. In static routing, a path or a set of paths to establish a connection between a source-destination (SD) pair is pre-determined without considering dynamic network status. Two typical static routing algorithms are fixed path routing and alternate path routing, which are defined later. In dynamic routing, a path to establish a connection is dynamically determined according to up-to-date network status. A typical example of dynamic routing is least-congested routing. These three routing algorithms are defined as follows:

1. *Fixed Path Routing:* In the fixed routing, one fixed path is pre-selected for each source-destination (SD) pair, e.g., the shortest path. Any connection request for the SD pair is attempted to set up on the fixed path.

If several wavelengths are free on the path, a specified wavelength assignment algorithm is used to select a wavelength for the connection. The connection is blocked if no wavelength is free on the path.

2. *Alternate Path Routing:* A set of fixed paths are pre-selected for each SD pair and ordered according to specified criteria, e.g., the path length. An arrival connection request is first attempted to set up on the first path. If the first path has no free wavelength, the request is overflowed to the second path. If none of the pre-selected paths has free wavelength to accommodate the connection, the request is blocked.

3. *Least-congestion routing:* In the above two routing algorithms, the path/paths are pre-selected for node pairs without considering the network status. In least-congestion routing, the least-congested path among all paths between the SD pair is selected to establish the connection.

The fixed path routing is a simple and straightforward routing approach. It is easy to implement. However, since only one path is provided for each SD pair, the connections encounters high blocking probability when the traffic load at each node is not very low. Several paths are provided in the alternate path routing. Connections have more chances to be accommodated on one of the paths than the fixed path routing. Minimizing the usage of network resources is the goal of both these two routing algorithms. Since the paths are pre-determined without considering the current network status, congestion avoidance is hard to achieve for these static routing algorithms. On the contrary, the goal of the least-congestion routing is to distribute traffic load evenly in the network so that network congestion can be avoided. However, the least-congested path may not be the shortest path. In some cases, it may be longer than the shortest path. Therefore the least-congestion routing may consume more network resources than the shortest path and alternate path routing algorithms. Another disadvantage of using dynamic routing is its computing complexity. Since the statuses of all the paths between a SD pair have to be compared to determine the least-congested path, it requires more computing power than the static routing.

To take advantage of both static and dynamic routing, a new routing algorithm, called fixed-paths least-congestion routing, is proposed in [32]. In the fixed-paths least-congestion routing, a set of paths are provided for each SD pair. Upon the arrival of a connection request, the least-congested path among the given paths is selected to establish the connection. To minimize network resource usage, the first K shortest paths are selected as path candidates. Since one of the K paths is dynamically selected, network congestion can be effectively avoided. A key parameter here is K, the number of considered shortest paths. If K is large, the fixed-paths least-congestion routing performs similar to the least-congestion routing. If $K = 1$, the fixed-paths least-congestion routing becomes fixed shortest path routing. The analysis and simulation results in [32] show that the number of shortest paths that need to be considered depends on many factors, such as network topologies and traffic patterns. However, in most of the cases considered in [32], e.g., general mesh topologies, a small number of K, e.g., 2 or 3, is enough to guarantee high network performance.

New dynamic RWA algorithms based on layered-graph model are proposed in [33] for multifiber WDM networks. In the layered-graph model, each node in a given graph G is replicated W times in a newly constructed layered graph LG, here W is the number of wavelengths per fiber. Each bi-directional link in G is replicated into W pairs of uni-directional links in LG. In this layered-graph model, the RWA problem can be reduced to finding the shortest path from source to destination in LG. Two RWA algorithms, PACK Strategy and SPREAD Strategy, are proposed in [33]. As the names suggest, PACK Strategy attempts to *pack* the existing calls on as fewer edges and fewer layers as possible. The calls are routed on the most utilized wavelength and link first, in order to maximize the utilization of available wavelengths with more network resources reserved for the calls arriving in the future. The SPREAD Strategy can *spread* the existing calls over all the edges and layers as even as possible by using any shortest path algorithm. In order to achieve a near-uniform distribution of the load over the wavelength set and link set, the calls are routed on the least utilized wavelength and link first. The performances of PACK and SPREAD strategies are compared to fixed routing with first-fit wavelength assignment heuristic (FR/FF), fixed routing with maxsum heuristic (FR/MS), alternate routing with random heuristic (AR2), alternate routing with relative capacity loss heuristic (AR/RCL) and fixed-path least-congestion routing (FPLC) in [33]. Simulation results show that using multiple fibers per link can improve the network performance significantly. Fixed routing and alternate routing algorithms perform poorly even with full wavelength conversion when compared to RWA algorithms that consider routing and wavelength assignment sub-problems simultaneously (e.g. SPREAD Strategy). The SPREAD algorithm performs much better than FR/FF, and outperforms AR2 and FPLC, in regular 2×3 mesh, 5×5 mesh-torus, irregular CERNET-Like, and NSF T1 backbone networks.

The performance of these routing algorithms has been extensively studied in the literature. In [27, 30, 35, 36, 37, 38, 39, 40] the performance of the shortest path (SP) routing and the alternate shortest paths (ASP) routing methods are investigated through approximate analysis and simulation. Dynamic routing approaches are more efficient than static routing methods [27, 31, 32, 35, 36, 38, 41, 42]. In [35], simulation results show that the dynamic routing method can significantly improve the network performance compared to the SP and the ASP. Efficient algorithms for wavelength rerouting in multifiber WDM ring networks are studied in [34].

Several approximate analytical methods for computing the blocking probabilities of networks have been proposed in the literature. In [25], a model to compute the approximate blocking probability with Poisson traffic input is presented. However, the model is inappropriate for networks with sparse topologies because it does not consider the correlation among the use of wavelengths between successive links of a path. This model is improved in [26] with the consideration of this dependence. A Markov chain based reduced load model with state-dependent arrival rate is presented in [37, 39]. A more accurate model in [7] accounts for link load correlation. Both of the analytical methods, i.e., the link load correlation model [7] and the approximate reduced load model [37, 39], are used to compute the blocking performance of dynamic fixed-path least-congestion routing algorithm in [43]. The results show that the model considering link load correlation is simpler and more accurate than the reduced load model.

There has also been considerable interest to analyze the blocking performance of multifiber WDM networks. The independent wavelength load model in [25] is extended to multifiber networks in [5]. The results of this model are not numerically accurate for Poisson traffic because of the assumption that the load on one wavelength is independent of those on the other wavelengths on a link. The link load independence model proposed in [37] is extended to multifiber networks in [28]. However, this independent model is not accurate [28]. It overestimates the blocking performance for $F = 1$ and underestimates it for $F > 1$ in a mesh-torus network. The blocking performance models for first-fit wavelength assignment in [27, 29] are also proposed to be applicable in multifiber networks. However, both of these models require intensive computation due to their iterative procedure to solve the Erlang fixed-point equation.

The performance of multifiber WDM networks with different routing and wavelength assignment algorithms is studied in [11, 30, 31]. A new analytical model, multifiber link-load correlation (MLLC) model, is presented in [30] to compute the blocking performance of multifiber networks with fixed-path routing. Comparing this model to the link independence model in [28], the MLLC model is a more accurate and general model that is applicable to not only regular networks but also irregular networks.

In the MLLC model, it is assumed that traffic arrivals at each node follow Poisson distribution with arrival rate λ. The holding time of each connection is exponentially distributed with mean $1/\mu$. A single path is pre-selected for each SD pair, and a wavelength assigned to a connection is randomly selected with uniform distribution from the set of free wavelengths on that path. Let F be the number of fibers/link, and W be the number of wavelengths/fiber. F and W are assumed to be the same on all links and fibers, respectively. It is also assumed that an incoming request on one channel can be switched to any output port using an OXC as long as the output port has the same wavelength free regardless of which fiber it is on. If the wavelength is not free on all of the F fibers, the request is blocked on this wavelength. No wavelength converter is available at any node.

Define a Light Channel (LC) as a wavelength on a fiber on a link. A lightpath (LP) is a connection between a SD pair using the same wavelength on all the links of a path. Note that a lightpath consists of several LCs on successive links. However, the LCs on a path may or may not be on the same fiber. Let a wavelength trunk (WT) λ_i be a collection of the LCs/LPs using λ_i on all the fibers. We define a WT "free" on a <u>link</u> if the wavelength is free on at least one of the fibers on the link. A WT is "busy" on the link otherwise. A WT is "free" on a <u>path</u> if that WT is free on all of the links constituting the path. A WT is "busy" on the path otherwise.

An ordinary view of multifiber networks is that each link consists of F fibers and each fiber consists of W wavelengths. An alternate view of the multifiber network is that each link consists of W wavelength trunks, and each wavelength trunk consists of F fibers. This is a simplified view because connections will be established on wavelength trunks and wavelength continuity constraint is not required in wavelength trunks.

To improve the accuracy of the analytical model, link load correlation between two adjacent links is considered in the MLLC model. Assuming that the traffic load on the lth hop of a path is dependent only on the load on the $(l-1)$th hop, the blocking probability on a l-hop path can be computed recursively by viewing the first $l-1$ hops as the first hop and the lth hop as the second hop of a two-hop path. Therefore the problem is converted to the analysis of a two-hop path. The difficulty in computing the blocking performance of a two-hop path results from the continuing calls from the first hop to the second hop. To simplify the computation, the W wavelength trunks on the two-hop path are divided into different groups as shown in Figure 1. Each wavelength trunk consists of F fibers. A filled slot in the figure indicates that the wavelength trunk is busy, that is, it is fully occupied on the link. An unfilled slot indicates that the wavelength trunk is free, that is, the wavelength trunk may be partially occupied or free on every fiber. The conditional distribution of continuing calls is computed in each group. A closed form expression of the free wavelength trunk distribution can then be obtained on the two-hop path. For additional details the reader should refer to [30].

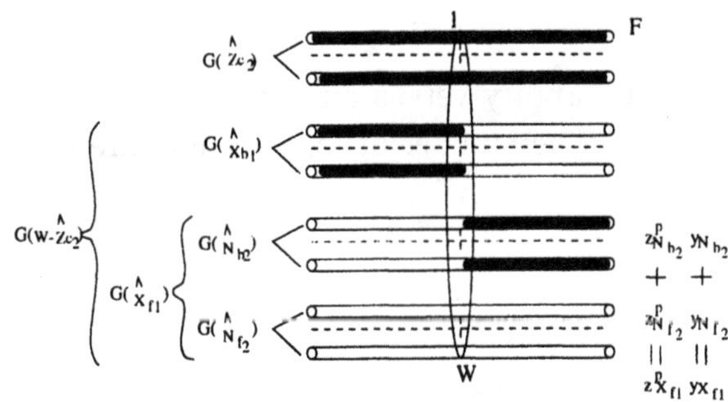

Figure 1: The wavelength trunks on a two-hop path.

Analytical results obtained from this model are illustrated in Figure 2, Figure 3, and Figure 4. Three typical network topologies were studied: a 10-node unidirectional ring network, a 5×5 mesh-torus network, and the 14-node NSFnet (Figures of the network topologies are available in [46]). For comparison, the analytical results obtained from the independence model in [28] are also plotted. It is seen that the independence model severely overestimated the blocking probability when the number of fibers per link, F, is small, and underestimated the blocking when F is large. It is observed that 6 fibers per link in the ring, and 4 fibers per link in the mesh-torus and NSFnet, are sufficient to provide similar performance as that in full-wavelength-convertible networks (F=24, W=1).

The effects of multiple fibers in WDM networks with the alternate-path routing (APR) is studied in [11]. The paper attempts to answer how many

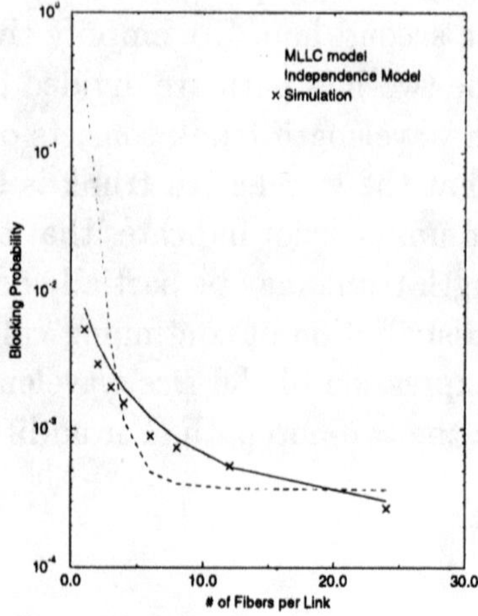

Figure 2: Blocking probability versus number of fibers in a 10-node unidirectional ring network. The number of LCs per link are fixed at 24.

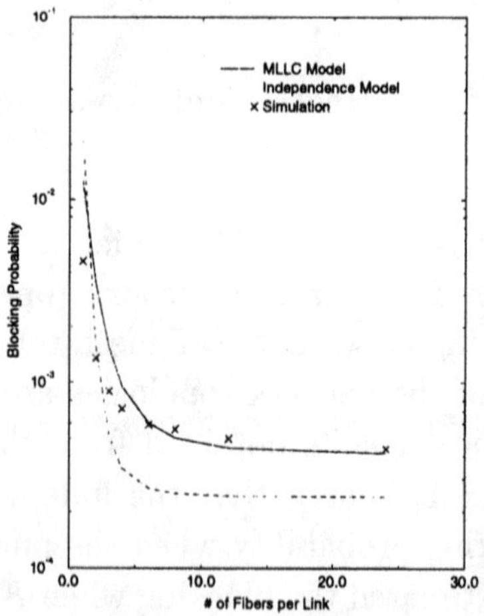

Figure 3: Blocking probability versus number of fibers in a 5×5 mesh-torus network. The number of LCs per link are fixed at 24.

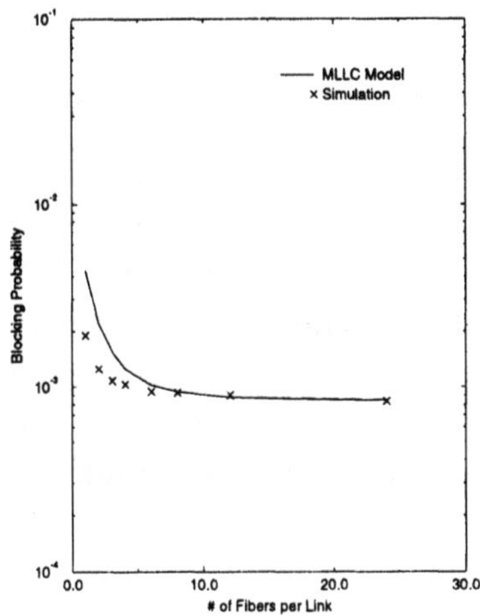

Figure 4: Blocking probability versus number of fibers in the NSFnet. The number of LCs per link are fixed at 24.

fibers per link are required to guarantee high performance in a WDM network with the APR. The MLLC model in [30] is used and extended to analyze the performance of WDM networks with the APR. It is observed that the number of fibers required to guarantee high performance is slightly higher in the APR than in the FPR. However, a limited number of fibers is still sufficient to guarantee that the blocking performance of a multifiber WDM network is similar to the blocking performance of a full-wavelength-convertible network.

As shown in [32], dynamic routing algorithms can significantly improve the network performance compared to the fixed-path routing and alternate-path routing. The blocking performance of multifiber WDM networks with fixed-paths least-congestion (M-FPLC) routing is studied in [31]. Two routing algorithms are studied in the paper: wavelength-trunk (WT)-based FPLC and light-path (LP)-based FPLC. The multifiber link-load correlation model developed in [11, 30] is extended to analyze the performance of multifiber WDM networks with WT-based FPLC routing. It is observed that the number of fibers required to provide high performance in a multifiber network with dynamic routing is higher than those multifiber networks with fixed-path routing and alternate-path routing. However, multifiber networks with dynamic routing can still achieve similar blocking performance to full-wavelength-convertible networks with limited number of fibers.

3.3   Multifiber Linear Lightwave Networks

Linear lightwave networks (LLNs) are optical networks in which network nodes perform only linear operations on optical signals: power splitting, combining, and non-regenerative amplification [44]. A multi-fiber linear lightwave network (M-LLN) architecture is considered in [13] for telecommunications where switching exchanges are normally connected by multi-fiber cables. The authors proposed the use of the shortest path (SP) to accommodate new arriving requests. If several shortest paths exist between a source-destination pair, the following linear path (LP) allocation schemes are considered:

1. shortest path minimum utilization (SP-MIN), which selects the path that has the least number of occupied wavelengths. The idea of this scheme is to distribute the traffic over all of the shortest paths and minimize the traffic on each path;

2. shortest path maximum utilization (SP-MAX), which selects the path that has the largest number of occupied wavelengths. This scheme can increase the utilization of existing paths, while avoiding the use of free fiber-links in the network.

3. Shortest Path (SP), which selects one of the shortest paths randomly.

Simulation results of a randomly-generated 30-node network show that SP-MAX achieves the best performance in terms of blocking probability under most traffic loads. The SP-MIN gives a slightly better performance than the SP-MAX under very heavy traffic. The paper also discusses in detail the implementation issues of multifiber linear lightwave networks. It is shown that M-LLNs can be implemented with commercially available components.

4   Conclusions

In this chapter, we have provided an overview of the research work for multifiber WDM networks. Both static network design approaches and dynamic routing and wavelength assignment algorithms are reviewed and summarized. Since most of the current optical networks are based on ring topologies, various multifiber ring architectures have been proposed and studied for scalability, reliability, and network cost [16, 19, 21]. Theoretical results on the number of fibers and wavelengths required in multifiber ring and tree networks to carry certain traffic is shown in [22]. The design of resilient multifiber networks is studied in [10]. Three restoration schemes are compared in terms of capacity requirement in the paper. Different network topologies are analyzed, to evaluate the influence of physical connectivity and network size on the restoration capacity.

Routing and wavelength assignment algorithms play key roles in dynamic wavelength routing networks. In addition to the wavelength assignment algorithms that are proposed for single-fiber networks but can also be extended to multifiber networks, various other wavelength assignment algorithms have been proposed for multifiber networks [9, 27, 28, 45] to further improve the network performance. Both static routing and dynamic routing algorithms have been studied in multifiber WDM networks [27, 29, 30]. The results show that the benefit of wavelength conversion in multifiber networks is very limited. A small number of fibers on each link are sufficient to guarantee high network performance in various network topologies [11, 30, 31]

References

[1] B. Mukherjee, *Optical Communication Networks*, Mc-Graw-Hill, NY, NY, 1997.

[2] R. Ramaswami and K.N. Sivarajan, *Optical Networks: A Practical Perspective*, Morgan-Kaufman, San Francisco, CA, 1998.

[3] B. Mukherjee, D. Banerjee, S. Ramamurthy, A. Mukherjee, "Some Principles for Designing A Wide-Area WDM Optical Network", *IEEE Journal on Selected Areas in Communications*, Vol. 4, No. 5, pp. 684-696, Oct. 1996.

[4] R. Ramaswami, K. N. Sivarajan, "Design of Logical Topologies for Wavelength-Routed Optical Networks", *IEEE Journal on Selected Areas in Communications*, Vol. 14, No. 5, pp. 840-851, Oct. 1996.

[5] G. Jeong and E. Aynoglu, "Comparison of Wavelength-Interchanging and Wavelength-Selective Cross-Connects in Multiwavelength All-Optical Networks", *Proc. IEEE INFOCOM '96*, pp. 156-163, 1996.

[6] B. S. Glance, J. M. Wiesenfeld, U. Koren, and R. W. Wilson, "New Advances in Optical Components Needed for FDM Optical Networks", *Journal of Lightwave Technology*, Vol. 11, No. 5/6, pp. 882-890, May/June 1993.

[7] S. Subramaniam, M. Azizoğlu, and A. K. Somani, "All-Optical Networks with Sparse Wavelength Conversion", *IEEE/ACM Transactions on Networking*, Vol.4, pp. 544-557, Aug. 1996.

[8] J. Yates, J. Lacey, D. Everitt, and M. Summerfield, "Limited-Range Wavelength Translation in All-Optical Networks", *Proc. IEEE INFO-COM '96*, pp. 954-961, March 1996.

[9] X. Zhang and C. Qiao, "Wavelength Assignment for Dynamic Traffic in Multi-fiber WDM Networks", *Proc. ICCCN '98*, pp. 479-485, 1998.

[10] S. Baroni, P. Bayvel, R. Gibbens and S. K. Korotky, "Analysis and Design of Resilient Multifiber Wavelength-Routed Optical Transport Networks", *Journal of Lightwave Technology*, Vol. 17, No. 5, pp. 743-758, May 1999.

[11] L. Li and A. K. Somani, "Fiber Requirement in Multifiber WDM Networks with Alternate-Path Routing", *Proc. ICCCN '99*, Boston, MA, 1999.

[12] D. Bertsekas and R. Gallager, *Data Networks*, Prentice Hall, Englewood Cliffs, NJ, 1992.

[13] P. C. Wong and H. H. Chan, "Multi-Fiber Linear Lightwave Networks; Design and Implementation Issues", *IEICE Transactions on Communications*, Vol. E77-B, No. 8, pp. 1040-1047, Aug. 1994.

[14] N. Wauters and P. Demeester, "Wavelength Translation in Optical Multi-Wavelength Multi-Fiber Transport Networks," *International Journal of Opto-Electronics*, Vol. 11, No. 1, pp. 5370, Jan./Feb. 1997.

[15] M. Alanyali and E. Ayanoglu, "Provisioning Algorithms for WDM Optical Networks," *IEEE/ACM Transactions on Networking*, vol. 7, no. 5, Oct. 1999, pp. 767-778.

[16] H. Obara, H. Masuda, K. Suzuki, and K. Aida, "Multifiber Wavelength-Division Multiplexed Ring Network Architecture for Tera-bit/s Throughput", *Proc. IEEE ICC '98*, Vol. 2, pp. 921-925, June 1998.

[17] T. H. Wu, *Fiber Network Service Survivability*, Artech House, Norwood, MA, 1992.

[18] T. H. Wu and R. C. Lau, "A Class of Self-Healing Ring Architectures for SONET Network Applications," *Proc. GLOBECOM '90*, pp. 444-451, 1990.

[19] L. Wuttisittikulkij and M. J. O'Mahony, "Multiwavelength Self-Healing Ring Transparent Networks," *Proc. GLOBECOM '95*, pp. 45-49, 1995.

[20] N. Nagatsu, A. Watanabe, S. Okamoto, K. Sato, "Optical Path Cross-Connect System Scale Evaluation Using Path Accommodation Design for Restricted Wavelength Multiplexing", *IEEE Journal on Selected Areas in Communications*, Vol.14, No. 5, pp. 893-902, June 1996.

[21] N. Nagatsu, A. Watanabe, S. Okamoto, K. Sato, "Architectural Analysis of Multiple Fiber Ring Networks Employing Optical Paths", *Journal of Lightwave Technology*, Vol.15, No. 10, pp. 1794-804, Oct. 1997.

[22] G. Li and R. Simha, "On the Wavelength Assignment Problem in Multifiber WDM Star and Ring Networks", *IEEE INFOCOM '2000*, March, 2000.

[23] R. Ramaswami and K. N. Sivarajan. "Routing and Wavelength Assignment in All-Optical Networks", *IEEE/ACM Transactions on Networking*, Vol. 3, No. 5, pp. 489-500, Oct. 1995.

[24] S. Baroni, P. Bayvel, "Wavelength Requirements in Arbitrarily Connected Wavelength-Routed Optical Networks", *Journal of Lightwave Technology*, Vol. 15, No. 2, pp. 242-51, Feb. 1997.

[25] R. A. Barry and P. A. Humblet, "Models of Blocking Probability in All-Optical Networks with and without Wavelength Changers", *IEEE Journal on Selected Areas in Communications*, Vol. 14, pp. 858 - 867, June 1996.

[26] R. A. Barry and D. Marquis, "An Improved Model of Blocking Probability in All-Optical Networks", *LEOS 1995 Summer Topical Meeting*, pp. 43-44, Aug. 1995.

[27] E. Karasan and E. Ayanoglu, "Effects of Wavelength Routing and Selection Algorithms on Wavelength Conversion Gain in WDM Optical

Networks", *IEEE/ACM Transactions on Networking*, Vol. 6, No 2, pp. 186-196, April 1998.

[28] S. Subramaniam and R. Barry, "Dynamic Wavelength Assignment in Fixed Routing WDM Networks", *Proc. IEEE ICC '97*, pp. 406-410, Montreal, Canada, Nov. 1997.

[29] A. Mokhtar and M. Azizoğlu, "Adaptive Wavelength Routing in All-Optical Networks", *IEEE/ACM Transactions on Networking*, Vol. 6, No. 2, pp. 197-206, April 1998.

[30] L. Li and A. K. Somani, "A New Analytical Model for Multifiber WDM Networks", *Proc. GLOBECOM '99*, pp. 1007-1011, Rio de Janeiro, Brazil, December 1999.

[31] L. Li and A. K. Somani, "Blocking Performance Analysis of Fixed-Paths Least-Congestion Routing in Multifiber WDM Networks", *Proc. SPIE '99*, Boston, MA, 1999.

[32] L. Li and A. K. Somani, "Dynamic Wavelength Routing Using Congestion and Neighborhood Information", *IEEE/ACM Transactions on Networking*, pp. 779-786, Oct. 1999.

[33] S. Xu, L. M. Li, S. Wang, "Dynamic Routing and Assignment of Wavelength Algorithms in Multifiber Wavelength Division Multiplexing Networks", *IEEE Journal on Selected Areas in Communications*, Vol. 18, No. 10, pp. 2130 - 2137, Oct. 2000.

[34] G. Mohan, C. S. R. Murthy, "Efficient Algorithms for Wavelength Rerouting in WDM Multi-Fiber Unidirectional Ring Networks", *Computer Communications*, Vol.22, No.3, pp.232-43, Feb. 1999.

[35] H. Harai, M.Murata and H.Miyahara, "Performance of Alternate Routing Methods in All-Optical Switching Networks", *Proc. IEEE INFOCOM '97*, Vol. 2, pp. 517-525, April 1997.

[36] S. Ramamurthy and B. Mukherjee, "Fixed-Alternate Routing and Wavelength Conversion in Wavelength-Routed Optical Networks," *Proc. IEEE GLOBECOM '98*, Vol. 4, pp. 2295-2303, Sydney, Australia, Nov. 1998.

[37] M. Kovačević and A. Acampora, "Benefits of Wavelength Translation in All-Optical Clear-Channel Networks", *IEEE Journal on Selected Areas in Communications*, Vol. 14, pp. 868-880, June 1996.

[38] D. Banerjee and B. Mukherjee, "Practical Approaches for Routing and Wavelength Assignment in All-Optical Wavelength-Routed Networks," *IEEE Journal on Selected Areas in Communications*, Vol. 14, pp. 903-908, June 1996.

[39] A. Birman, "Computing Approximate Blocking Probabilities for a Class of All-Optical Networks", *IEEE Journal on Selected Areas in Communications*, Vol. 14, pp. 852-857, June 1996.

[40] A. Birman and A. Kershenbaum, "Routing and Wavelength Assignment Methods in Single-Hop All-Optical Networks with Blocking", *Proc. IEEE INFOCOM '95*, 1995.

[41] K. Chan, and T. P. Yum, "Analysis of Least Congested Path Routing in WDM Lightwave Networks", *Proc. IEEE INFOCOM '94*, Vol. 2, pp. 962-969, 1994.

[42] E. D. Lowe and D. K. Hunter, "Performance of Dynamic Path Optical Networks", *IEE Proceedings-Optoelectronics*, Vol. 144, No. 4, pp. 235-9, Aug. 1997.

[43] L. Li and A. K. Somani, "Dynamic Wavelength Routing Techniques and Their Performance Analyses", *OPTICAL WDM NETWORKS: PRINCIPLES AND PRACTICE* edited by K. M. Sivalingam and S. Subramaniam, pp. 247-275, Kluwer Academic Publishers, March 2000.

[44] Thomas E. Stern and Krishna Bala, *Multiwavelength Optical Networks: A Layered Approach*, Addison Wesley Longman, March 1999.

[45] H. Zang, J. Jue, B. Mukherjee, "A Review of Routing and Wavelength Assignment Approaches for Wavelength Routed Optical WDM Networks", *Optical Networks Magazine*, Vol. 1, No. 1, pp. 47-60, Jan. 2000.

[46] L. Li, "Dynamic Wavelength Routing in Multifiber WDM Networks," Ph.D. dissertation, Dept. Elect. Comput. Eng., Iowa State University, Ames, Iowa.

OPTICAL NETWORKS - RECENT ADVANCES
L. Ruan and D.-Z. Du (Eds.) pp. 151 - 185
©2001 Kluwer Academic Publishers

Recent Developments in Optical Multistage Networks

Yi Pan
Department of Computer Science
Georgia State University, Atlanta, GA 30303
E-mail: pan@cs.gsu.edu

Chunming Qiao
Department of Computer Science and Engineering
SUNY at Buffalo, Buffalo, NY 14260
E-mail: qiao@eng.buffalo.edu

Yuanyuan Yang
Department of Electrical and Computer Engineering
SUNY at Stony Brook, Stony Brook, NY 11794
E-mail: yang@ece.sunysb.edu

Jie Wu
Department of Computer Science and Engineering
Florida Atlantic University, Boca Raton, FL 33431
E-mail: jie@cse.fau.edu

Contents

1   Introduction

Fiber optic communication offers a combination of high speed, decreasing switching time, low error probability, and gigabit transmission capacity. It has been extensively used in wide-area networks and has received much attention in distributed processing community as well. Recent progress in electro-optic technologies has made optical communication a promising networking choice to meet the increasing demands for high channel bandwidth and low communication latency by high-performance computing/communication applications [10], [21]. The unique characteristics of optical interconnects have significant implications on architectural freedom and constraints as well as system configuration and complexity. It is expected that optical technology will drastically change the landscape of interconnection schemes.

Many different network topologies can be formed based on optical technology. The multistage interconnection network (MIN) is one of them and has been an important interconnecting scheme for communication and parallel computing systems. Depending on the number of stages, the number of switching elements (SEs) per stage, the capability of the SEs, and the interconnection patterns used between stages, a MIN can be either blocking such as a Banyan network [22], rearrangeably nonblocking such as a Benes network [22], or nonblocking such as a Clos network [22]. These classifications of blockingness have also been extended to multi-rate and/or multicast networks [23].

An optical MIN can be implemented with either free-space optics or guided wave technology [21]. In this chapter, we consider optical implementation with the guided wave technology. Two types of guided wave optical switching systems can be identified. The first is a hybrid (photonic) approach in which optical signals are switched, but both the switch control and routing decisions are carried out electronically at a speed that can be much lower than the bit rate of the optical signals being switched. The second approach is all-optical switching, which would potentially overcome the speed-mismatch problem associated with the hybrid approach.

Several switching methods can be used in an optical MIN. In a traditional electronic MIN, it is common to use packet switching, with which packets of data are processed at each stage by the logic associated with the SEs, and routed to the next stage towards its destination. Buffers are often used at each SE to hold packets in contention with other packets, and flow control mechanisms are used between stages to prevent buffers from overflowing. However, in hybrid optical MINs which use electronically controlled optical SEs, such as Lithium Niobate directional couplers, packet switching requires conversions between optical signals and electronic ones, which could be very costly. Compared with the switching speed of the SEs, which can be as fast as hundreds of picoseconds, the process of determining their settings based on the address information in each packet could become a significant overhead. For these reasons, circuit switching is usually preferred in optical MINs, with which a direct connection between the source and the destination is set up by a network controller (or controllers) before data are sent. Since neither packet processing nor buffering are needed at each SE, circuit switching can be implemented with SEs that are simpler and faster than those required for packet switching. Nevertheless, data transmission can still be packetized in order to utilize the network bandwidth better by letting multiple connections time-share an input/output port of a switch or a link between two switches. This essentially leads to Time Division Multiplexed (TDM) MINs proposed by Qiao and Melhem in [15]. Similarly, one may use Wavelength Division Multiplexed (WDM) MINs, in which a wavelength channel at a source may be switched either to the same wavelength channel at a destination independently of how other wavelength channels are switched, or through the use of wavelength converters, to a different wavelength channel at the destination.

In this chapter, we first describe unique characteristics and new problems of optical MINs and in particular, the crosstalk introduced by the SEs in the next section. We then discuss general approaches to avoid crosstalk via network dilation and especially the time domain dilation approach in Section 3. In Section 4, the ability of time domain dilated optical MINs to embed regular structures such as rings, meshes and trees, and to realize permutations is presented. In Section 5, an analytical model to evaluate the performance of optical MINs in establishing an arbitrary set of connections under individual switch control is described, followed by an optimal scheduling algorithm with a polynomial time complexity for optical MINs under stage control. In Section 6, a diagnostic process for detecting and locating crosstalk-faulty switching elements is described. Finally, we conclude our chapter by identifying new research topics in the area of optical MINs.

2   New Problems in Optical MINs

In an optical MIN, wide-band optical signals are switched under electronic control using directional couplers between Ti:LiNbO$_3$ waveguides on a planar LiNbO$_3$ crystal [12]. The basic SE is a directional coupler with two active inputs and two active outputs. Depending on the amount of voltage at the junction of the two waveguides which carry the two input signals, either of the two inputs can be coupled to either of the two outputs. Many architectures have been proposed to construct an $N \times N$ MIN using the 2×2 directional coupler as the basic component. These architectures are essentially similar to those of electronic MINs.

A good amount of research has been done on electronic MINs in the literature. Some of the analytical methods and results are also applicable to optical MINs. However, optical MINs hold their own challenges. For example, one problem is *path dependent loss*. In a large MIN, a substantial part of this path dependent loss is directly proportional to the number of couplers along an optical path, which is determined by the architecture used and the network size.

Another problem in optical MINs is *optical crosstalk*, which occurs when two signal channels interact with each other. There are two ways in which optical signals can interact in a planar switching network. The channels carrying the signals could cross each other in order to embed a particular topology. Alternatively, two paths sharing a SE will experience some undesired coupling from one path to another within a SE. For example, assume that the two inputs are y and z, respectively, the two outputs will have $ly + lxz$ and $lz + lxy$, respectively, where l is signal loss and x is signal crosstalk in a SE. Using the best device reported in the literature [21], $x = -35$ dB, and $l = 0.25$dB. For more practically available devices, it is more likely that $x = -20$ dB and $l = 1$dB [21]. Figure 1 shows an example of crosstalk in a SE. In the figure, the SE is set to straight (Figure 1(b)), the main signal is injected at the upper input as shown in Figure 1(a), and a crosstalk signal having a small fraction of the input signal power may be detected at the lower output. Hence, when a signal passes many SEs, the input signal will be distorted at the output due to the loss and crosstalk introduced on the path.

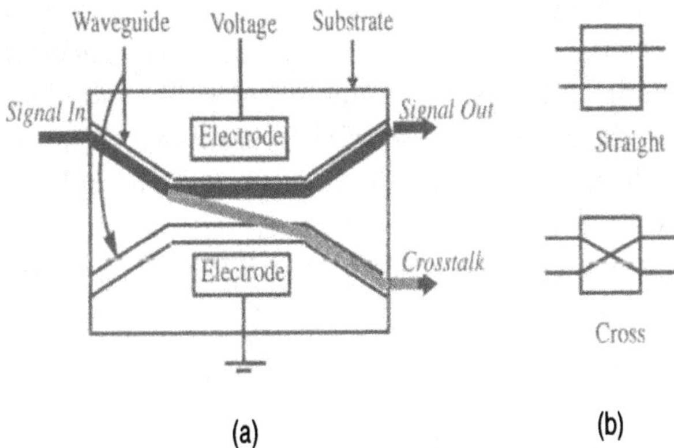

Figure 1: Crosstalk in an electro-optical switching element (SE).

Experimental results [12] show that it is possible to make the crosstalk from passive intersections of optical waveguides negligible by keeping the intersection angles above a certain minimum amount. Studies also indicate that the crosstalk problem is more severe than the path dependent loss problem with current optical technology [7], [21]. Thus, switch crosstalk is the most significant factor which reduces the signal-to-noise ratio and limits the size of a network. Fortunately, first order crosstalk can be eliminated by ensuring that a SE is not used by two input signals simultaneously. Once the major source of crosstalk disappears, crosstalk in an optical MIN will have very small effect on the signal-to-noise ratio and thus a large optical MIN can be built and effectively used in communication and parallel computing systems. In the following discussion, switch crosstalk will be simply referred to as *crosstalk*. This unique characteristic in an optical MIN causes new problems and leads to different design and analysis approaches from those used in its electronic counterpart. New routing strategies have to be designed to overcome the probems.

3   Crosstalk Avoidance in Optical MINs

If the size of an optical MIN is small, crosstalk will not present a severe problem. To reduce the negative effect of crosstalk optical MINs of large sizes, various approaches which apply the concept of *dilation* in either the space, time or wavelength domains have been proposed [5, 12, 19, 21]. With the space domain approach, extra SEs (and links) are used to ensure that at most one input and one output of every SE will be used at any given time. With the time domain approach, the same objective is achieved by treating crosstalk as a conflict, that is, two connections will be established at different times if they use the same SE (even if they use different inputs and outputs of the SE). With the wavelength domain approach, only the wavelength channels that are far apart from each others are assigned to the same SE. The following discussions will concentrate on the first two approaches.

3.1   Space and Time Domain Dilations

As an example of dilating a MIN in the space domain, Figure 2 shows a 2×2 dilated Benes network (DBN) proposed in [12]. The idea is to duplicate hardware to avoid crosstalk. Note that although the DBN has four inputs and outputs, only *half* of them will be available to its users. More specifically, only input and output numbered, in binary, 00 and 10 are used for establishing two connections. This way, the crosstalk signal generated from one connection will not interfere with the signal carried by the other connection. Its construction is recursive in nature. A 2×2 DBN, which is a basic building block, has four inputs and four outputs but can only be used for two connections. An $N \times N$ DBN has $2N$ inputs and outputs although only N of them are used for source and destination connections. The network can be constructed from two $N/2 \times N/2$ subnetworks by putting one on the top of another, and interconnecting the two subnetworks with a front and an end stages as depicted in Figure 3. The number of stages in an $N \times N$ DBN, $M(N)$, satisfies the recursively defined equation, $M(N) = M(N/2)+2$

where $M(2) = 2$. By solving the equation, we have $M(N) = 2\log N$, which is referred to simply as M in Figure 3.

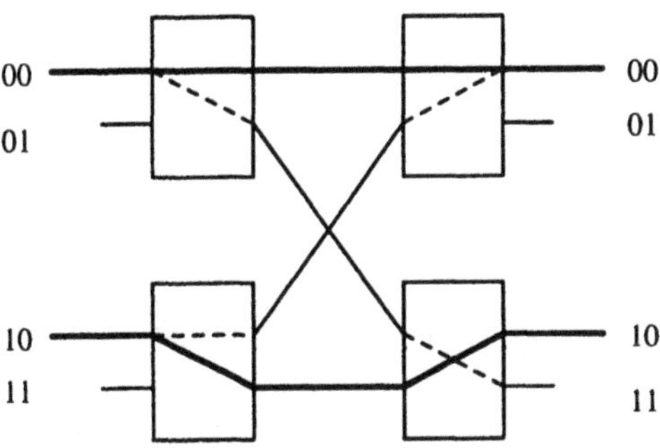

Figure 2: A 2×2 DBN.

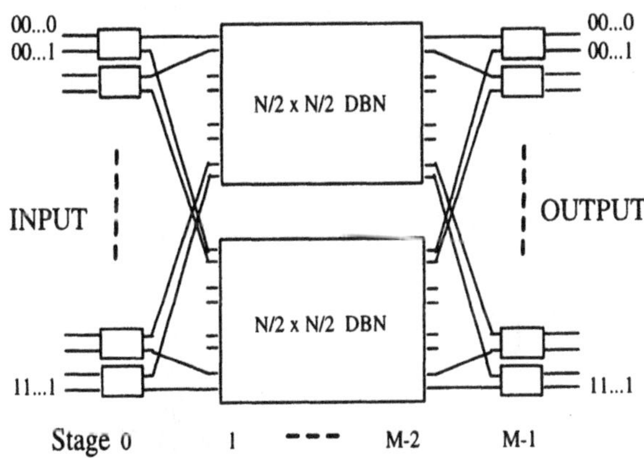

Figure 3: A recursively constructed $N \times N$ DBN.

Since signals go to only one of the two inputs of the SEs at the first and last stages, crosstalk-free is guaranteed at these two stages. Signals travel either in the lower half or the upper half in the other stages to avoid crosstalk.

A blocking MIN such as the Banyan may also be dilated in a similar way (see Figures 4(a) and (b)). A dilated Benes (or Banyan) network can realize the same set of permutations crosstalk-free as a regular Benes (or Banyan) can with crosstalk. In either case, the number of SEs (and links) in a dilated network, which is slightly larger than twice of the regular one, may be considered to represent the space cost for crosstalk avoidance. There are two variations of spatially dilated blocking networks: one is the Dilated

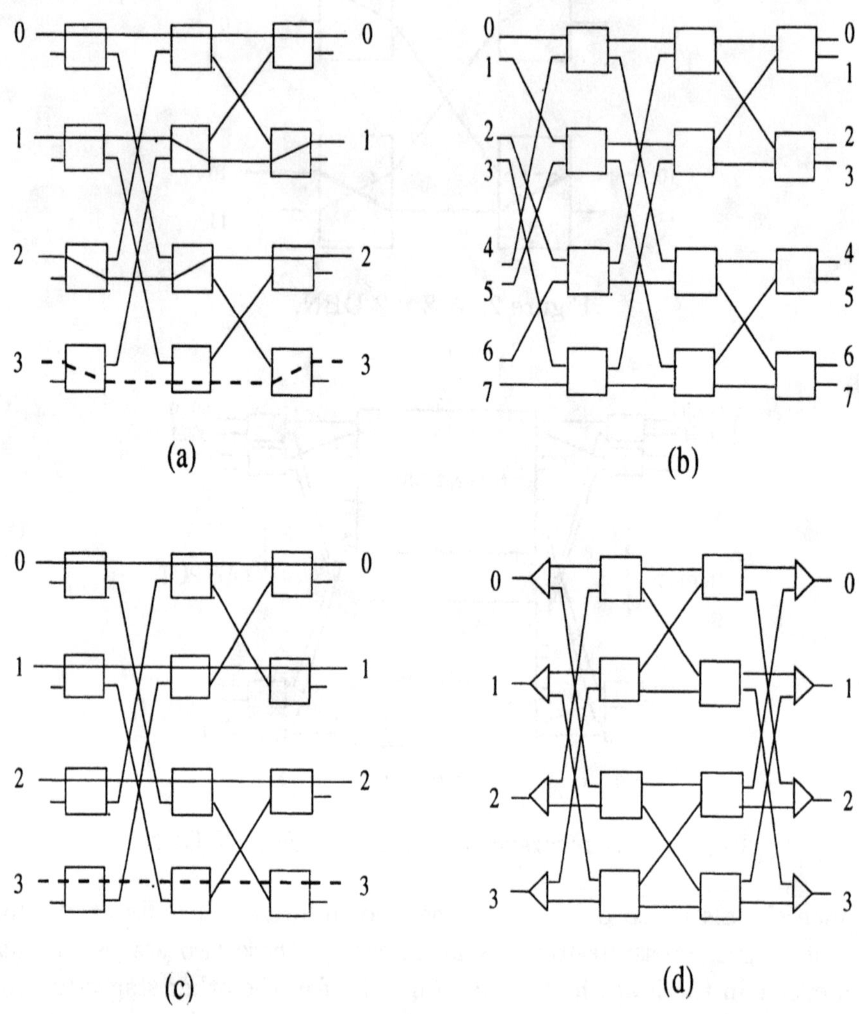

Figure 4: Four switching networks: (a) A 4×4 dilated Banyan. (b) An 8×8 Banyan. (c) A 4×4 Dilated Slipped Banyan (DSB) and (d) A two-plane 4×4 Banyan.

Slipped Banyan (DSB) [19] shown in Figure 4(c), and the other is a two-plane Banyan shown in Figure 4(d). In these space domain dilated networks, the crosstalk problem is avoided using more hardware.

It should be obvious that no permutation can be realized crosstalk-free in a regular MIN since it requires that at most one input and one output of any SE can be active at a time. However, using the time domain dilation approach proposed by Qiao, et al. in [14], a permutation may be partitioned into two (or more) sub-permutations such that each can be realized in a MIN crosstalk-free. In this way, a permutation can be realized in a regular MIN crosstalk-free in several rounds, one for each such sub-permutation, which will be referred to as a crosstalk-and-conflict-free (CF) mapping. The number of rounds (or CF-mappings) needed for any permutation, which is at least two, may be considered to represent the time cost for crosstalk avoidance.

3.2 Space-Time Trade-offs and Implications

As to be discussed in Section 4.2, it was shown in [14] that the number of permutations that can be partitioned into two CF-mappings in a regular Banyan exceeds the number of permutations that can be realized crosstalk-free (in one round) in a dilated Banyan. This seems to indicate that if the cost in time (in terms of the number of rounds) and in space (in terms of the number of SEs and links) were interchangeable, then the time domain dilation is more effective in realizing permutations than the space domain dilation. In addition, if hardware complexity is not a concern, then a two-plane Banyan may be used instead of the dilated Banyan since both have approximately the same space cost, but the former can be more powerful.

Time domain dilation is useful not only for avoiding crosstalk in realizing permutations in blocking MINs, but also for establishing an arbitrary set of connections that would normally cause conflicts in blocking or nonblocking MINs (dilated or not) by partitioning the set into several CF-mappings. Let T_d and T_u be the average time cost of a set of arbitrary connections in a dilated Banyan and an undilated (i.e. regular) Banyan, respectively. Analysis, which is to be presented later in Section 5.1, has shown that $T_u < 2T_d$. Since the space cost of the former, denoted by S_d, is more than twice of that of the latter, denoted by S_u, we have $S_u \cdot T_u < S_d \cdot T_d$, which implies that the time domain dilation may be more cost-effective provided that the cost in time and in space were interchangeable. An intuitive explanation is that the time domain approach integrates the solutions to the crosstalk and path conflict problems while space domain dilation only deals with crosstalk avoidance and still relies on time multiplexing techniques to resolve path conflicts.

The above discussions have the following implications on the implementation of optical MINs. First, whenever the limit on the network size is reached, the time domain approach may be used as a viable way to trade the maximal bandwidth available to each individual input and output pair for increased connectivity. Secondly, it is useful when future technology allows the transmission rate to scale up faster than the network size, or in other words, when the cost of increasing the bandwidth of each connection becomes as "cheap" as (or even cheaper than) the cost of constructing a network of twice of its original size.

For example, assume that one has a 16×16 dilated Banyan or a 32×32 Banyan, which has essentially the same hardware complexity. To connect 32 sources with 32 destinations using the former in the space domain dilation approach, two sources (and destinations) need to be multiplexed onto a single input (and output, respectively). An alternative is to use the latter as in the time domain dilation approach, thereby letting each source and destination connect to its own input and output, however, only one of the two sources (destinations) sharing the same SE can transmit (receive) at a time. Assume that the channel bandwidth (as determined by the transceiver rate) is 2.5 Gb/s, then theoretically speaking, in both cases, the bandwidth available to each source and destination pair would be 1.25 Gb/s when traffic is evenly distributed among all inputs and outputs. Based on the above cost-effective analysis, however, using the time domain dilation approach may achieve a better bandwidth utilization, or in other words, results in a higher effective bandwidth between a source-destination pair.

Continue the above example and assume that there is a need for network evolution in which the bandwidth for a source-destination pair can be increased to 2.5 Gb/s. Instead of trying to build a 32×32 dilated Banyan (or equivalently a 64×64 Banyan), which may not be feasible, the time domain dilation approach can be used, which deploys 5Gb/s transceivers but still uses the 32×32 Banyan. Clearly, it is possible that increasing the network size is feasible, and hence, the space domain approach is more desirable. However, if hardware complexity size is not a concern, a two-plane Banyan may be used instead as mentioned above.

4 Regular Communication Patterns

In many applications, especially in parallel computing systems, regular communication patters are required and frequently used. Hence, establishing connections in optical MINs quickly for regular communication patterns is critical. To help close the gap between the relatively slow electronic processing speed and the high bandwidth of an optical MIN, a connection paradigm called Reconfiguration with TDM (or RTDM) was proposed by Qiao and Melhem [15], of which the time domain dilation for crosstalk-avoidance is a special instance. Specifically, when the set of connections cannot be established in a MIN due to conflicts, it is desirable to partition it into a *minimum* number of CF-mappings. Once these CF-mappings and corresponding network configurations are determined, a sequence of control signals needed to set each SE appropriately can be stored in a cyclic shift-register. At run time, the MIN can simply go through a sequence of configurations under global synchronization without incurring much overhead that would have been introduced had complex electronic processing is involved.

In this section, we examine the ability of optical MINs, especially Banyans, to emulate (or embed) regular structures such as rings, meshes and trees, and to realize permutations.

4.1 Common Topology Embedding

Many popular topologies, such as rings, meshes, hypercubes, and binary trees have been proposed for parallel computing systems. A good amount of research have been carried out on these network topologies. If we can embed a topology into a MIN, many results achieved on the topology can be used on a MIN as well. Here, the key is to embed a topology efficiently in a MIN such that a direct link can be emulated in a MIN in as few time slots as possible. Several popular topologies have been embedded into an optical MIN efficiently.

The DSB network (shown in Figure 4(c)) was proposed to facilitate *stage control* of a dilated Banyan, in which all the SEs at the same stage are set to the same state (i.e. cross or straight), thus requiring only one control bit (e.g. [0] for straight and [1] for cross) per stage, and more importantly, one electronic driver circuit per stage. An $N \times N$ DSB can emulate a fully-connected network of the same size by applying the time domain approach (along with the space domain approach). Specifically, the set of $N(N - 1)$ connections in the fully-connected network can be partitioned into N CF-mappings as follows: the k-th CF-mapping, where $0 \leq k \leq N - 1$, contains connections from input i to output j as long as $i \oplus j = k$, where $\oplus$ is the bit-wise exclusive-OR operation. The control word (CW) to set the corresponding configuration in the DSB is $CW = [0]k \oplus [k]0$. By going through a sequence of N configurations corresponding to the N CF-mappings in a time division multiplexed fashion, the DSB provides the equivalent of the full-connectivity.

While full-connectivity may be reasonable for certain applications such as using the optical MIN as a hub in a local-area network environment, many programs in a parallel and distributed computing environment, however, exhibit communication locality and regularity and do not require the full connectivity. Accordingly, we can reduce the number of CF-mappings needed and thus increase the network bandwidth utilization in most cases.

Given that many algorithms have been proposed for popular structures such as rings, meshes, hypercubes and binary trees, optimal emulation of these structures in a Banyan by applying the time domain approach has been considered. For example, Qiao and Melhem [15] showed that while a fully-connected network requires $2N$ CF-mappings, a ring can be emulated with only two CF-mappings, a mesh with only four CF-mappings, a cube-connected-circle (CCC) with three CF-mappings, and a hypercube with only $\log N$ CF-mappings. In addition, a complete binary tree of $N - 1$ nodes can be emulated with four CF-mappings using an elaborate procedure involving nested recursions. Such an emulation is shown to be optimal when in-order labeling of the tree nodes is used, but may or may not be improved to involve only three CF-mappings when other labeling orders are used.

We will discuss issues related to realizing permutations in the next subsection and those related to establishing a set of arbitrary connections in Section 5.

4.2   Arbitrary Permutation Routing

As mentioned earlier, a dilated Banyan (and Benes) can realize the same set of permutation crosstalk-free as an Banyan (and Benes, respectively). In [14], a one-to-one but not onto mapping between the set of permutations realizable in a dilated Banyan (or Banyan) and the set of permutations that can be partitioned into two CF-mappings was developed, which showed that the latter contained more permutations than the former.

In this subsection, we approach this problem from a different angle and consider the permutation capability of undilated networks. An interesting question is: What is minimum number of passes required for realizing a permutation in such a network? In other words, we are interested in what types of partial permutations could be possibly realized crosstalk-free in an optical MIN. Recently, Yang, Wang and Pan [24] introduced a concept called *semi-permutation*. A semi-permutation is a partial permutation that ensures that there is only one active link passing through each input SE and output SE, and thus it has the potential to be realized crosstalk-free in an optical MIN. They have shown that any permutation can be decomposed into semi-permutations and the total number of semi-permutations for an even integer N is $2^N \cdot (\frac{N}{2})!$.

A simple algorithm for decomposing a permutation into semi-permutations is also described in [24]. The basic idea is to construct a bipartite graph for the given permutation between N inputs and N outputs of the network. Then, for each connected component of the graph, start from a vertex of this component in the input sets, traverse through an unvisited edge to the neighbor vertex in the output sets, back and forth until we return to the starting vertex. (During the traversing, a visited edge is marked "forward" if the traverse direction on this edge is from V_1 to V_2; and marked "backward" if the direction is opposite.) Finally, take all one-pair mappings corresponding to the edges marked with "forward", to form one semi-permutation; let the remaining one-pair mappings, corresponding to the edges marked with "backward", form another semi-permutation. It is easy to see that the complexity of the above decomposition algorithm is $O(N)$.

Thus, the problem of realizing a permutation in a crosstalk-free network can be transformed into the problem of realizing semi-permutations in the crosstalk-free network. However, it should be pointed out that introducing the semi-permutation concept in a network composed of 2×2 SEs can only guarantee crosstalk-free in the SEs in the input stage and the output stage of the network. In fact, realizing a semi-permutation in a single pass implies that there is only one active input on each SE in the input stage and only one active output on each SE in the output stage. To ensure the entire network crosstalk-free, we need to know if there exists a proper routing that can eliminate crosstalk in the SEs in the intermediate stages along different active paths. We look into this issue for two different types of networks: Banyan and Benes.

Due to the unique path nature of a Banyan network, a semi-permutation is routed through the network in a fixed switch setting. Consequently, some semi-permutations can be realized in a Banyan network in a single pass while others cannot. Yang, Wang, and Pan [24] showed that the number of semi-permutations that can be realized crosstalk-free in an $N \times N$ Banyan network in a single pass is $2^{\frac{3}{4}N} \cdot N^{\frac{N}{4}}$. By comparing the number of semi-permutations that can be realized in an $N \times N$ Banyan network and the number of all possible semi-permutations for an N-element set, we can see that there are a substantial amount of semi-permutations that cannot be realized in a Banyan network, especially when N gets larger.

A Benes network can be constructed by concatenating a Banyan network and a reverse Banyan network with the center stages overlapped. Electronic Benes networks are well known for being capable of realizing all possible permutations [22]. It is not surprising that Benes networks also have desirable properties to support permutations in optical networks. Yang, Wang, and Pan [24] have shown that any semi-permutation can be realized crosstalk-free in a Benes network in a single pass. They also gave an efficient algorithm for routing semi-permutations. The routing algorithm for a semi-permutation in an $N \times N$ Benes network is obtained by slightly modifying the decomposition algorithm described earlier in this subsection.

Now we know that any permutation can be decomposed into two semi-permutations and that semi-permutation can be realized in a Benes network in a single pass. Therefore, any permutation can be realized crosstalk-free in a Benes network in two passes. It should be pointed out that a permutation requires at least two passes in any $N \times N$ optical MIN due to the constraint of crosstalk-free in the input stage of SEs. In other words, two is the lower bound on the number of passes for any optical permutation networks under the constraint of crosstalk-free. It indicates that an undilated Benes network reaches this lower bound and realizes permutations optimally.

Gu and Peng [5, 6] also considered the permutation routing problem. Their goal is to minimize the number of wavelengths required for permutation routings by using (1) edge-disjoint paths and (2) node-disjoint paths on optical MINs. The edge-disjoint property ensures the nonblocking feature. The node-disjoint property, which includes the edge-disjoint property, poses a more strict constraint. The main purpose of ensuring node-disjoint is to avoid crosswalk. A good amount of research has been done on minimizing the number of wavelengths for permutation routings by edge-disjoint paths. However, relatively little has been done on minimizing the number of wavelengths for permutation routings by node-disjoint paths.

Gu and Peng's work [5] focused on butterfly networks, although it can be extended to other networks. A butterfly network of n stages has $N = 2^n$ inputs and outputs. Nodes at stage i ($0 \leq i \leq n-1$) are labeled by $< w, i >$, where w is an $(n-1)$-bit binary number that denotes the row of the node. Two nodes $< w, i >$ and $< w', i' >$ are connected if and only if $i' = i+1$ and either $w = w'$ or w and w' differ in the i'th bit from left. Figure 5 shows the butterfly network of three stages.

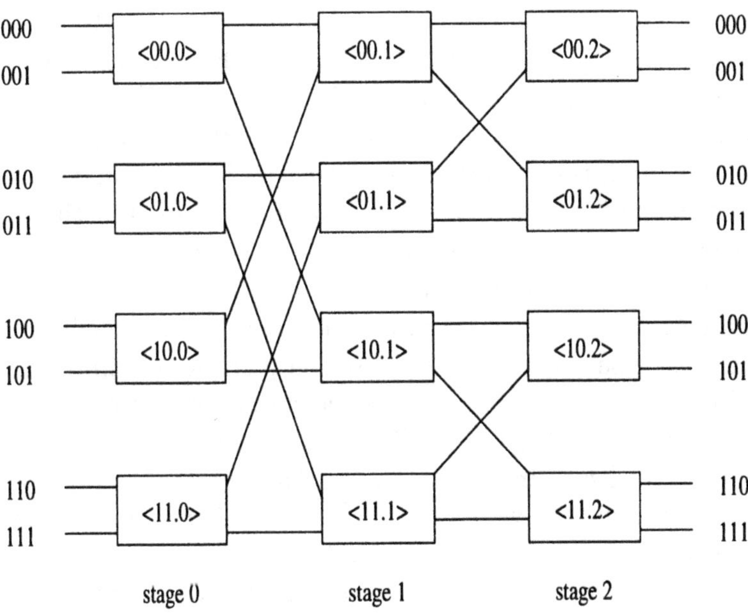

Figure 5: A three-stage Butterfly network.

It is shown that on optical butterfly networks, the problem of finding the minimum number of wavelengths for arbitrary partial permutations by node-disjoint paths is NP-complete. However, any partial permutation can be realized by $2^{\lceil n/2 \rceil}$ wavelengths and there exist permutations which require at least $2^{\lceil n/2 \rceil}$ wavelengths. Note that for arbitrary partial permutations by edge-disjoint paths, Agrawal [1] showed that any partial permutation can be realized by $2^{\lceil (n-1)/2 \rceil}$ wavelengths and there exist permutations which require at least $2^{\lceil (n-1)/2 \rceil}$ wavelengths.

On the relationship between edge-disjoint paths and node-disjoint paths, Gu and Peng [5] showed that for any partial permutation R in an optical butterfly network, if R can be realized in L (with $2^{k-1} \leq l < 2^k$ and $1 < k < n/2$) passed by edge-disjoint paths then R can be realized in less than $2^{k+1} + (n-2k)l$ wavelengths by node-disjoint paths.

4.3　Permutation on WDM MINs

Gu and Peng [6] also considered permutation routing on a WDM MINs. The optical bandwidth of the WDM all-optical network is the number of wavelengths supported by each link of the network. If the number of wavelengths for realizing a permutation request is at most the bandwidth of the network (W), then the permutation can be routed in one time slot. However, when the number of wavelengths needed is beyond the bandwidth, multiple time slots are required. A naive and straightforward approach is to partition a permutation R into R_1, R_2, ..., R_r, where r is the number of time slots derived by dividing the number of wavelengths by the bandwidth. Then realize each R_i by W wavelengths in each time slot. However, this approach is not efficient in minimizing the number of time slots.

An approach is used in [6] in which the message for every input-output pair is routed to an intermediate destination and the intermediate destination in the last time slot is the output of the pair. A deterministic routing algorithm is given that realizes any partial permutation routing on an optical MIN in at most $\lceil (\log N - 1)/(2\lfloor \log W \rfloor) \rceil$ time slots. Currently, the achievable bandwidth W in practical networks can be as large as 16. Hence, for any practical networks where $N \leq 2^{17}$, the number of time slots required on these networks is less than 2. It should be note that randomized algorithms exist that can realize with high probability any permutation in two rounds.

4.4　Specific Permutation Routing

The above subsection describes routing strategies for arbitrary permutations. Because it is a general strategy, it may not be the most efficient one. For certain class of permutations, better results exist. For example, Gu and Peng [5] has proved that there exist permutations which require at least $2^{\lceil \log N/2 \rceil}$ time slots to transfer them. This means that no efficient algorithm exists to solve a general permutation routing on an optical MIN. Gu and Peng [5] also gave a proof that any partial permutation can be realized in $2^{\lceil \log N/2 \rceil}$. Although it matches the lower bound for the number of time slots, it clearly uses too many time slots.

For certain specific class of permutations, better results exist. For example, Gu and Peng [5] presented a routing algorithm on bit-permute-complement (BPC) permutations. The class of BPC permutations is an important class in parallel processing. It includes perfect shuffle, unshuffle, bit-reversal, butterfly permutations in FFT algorithms, and segment shuffles [9]. Much work has been done on the routing for BPC permutations on other popular networks. The results presented in [5] show that for BPC permutations on the butterfly network, routing on optical MINs is as efficient as routing on their corresponding electronic MINs. The slowdown factor is only 2. Since several important MINs such as butterfly networks, Ω networks are all equivalent, their results hold on other equivalent networks as well. They also provide a polynomial time algorithm to find the minimum number of time slots for BPC permutations. Compared with the general routing strategy described before, this approach is much more efficient.

4.5 Relations with Electronic MINs

Assume that the set Ω consists of all permutations admissible to an regular electronic Ω network. Also let θ be the set of permutations realizable in two time slots on an optical Ω network. Qiao has proved that the size of θ is larger than the size of class Ω [14]. An interesting question left unanswered is how to characterize the set θ. Moreover, compared with Ω, how much larger is the set θ?

Recently, Shen, Yang, and Pan designed an $O(N \log N)$ algorithm to decide if a given permutation belongs to θ [18]. Since the set θ consists of permutations that can be realized in two time slots, the membership problem is equivalent to finding two CF groups for a given permutation. If we can find such CF groups, then the permutation belongs to θ. Otherwise, it does not. Their method basically follows the following strategy. First, they use a method called *optical window* to construct the conflict graph for a given permutation. This step involves checking each optical window for possible conflicts and adding conflict edges in the conflict graph for a permutation. The time needed is $O(N \log N)$. Then, a simple linear algorithm is used to check the two-colorability of the graph. If the graph is two-colorable, then the permutation belongs to θ. Normally, a graph coloring problem is an NP-problem. However, deciding if a graph is two-colorable can be done in linear time via depth-first traversal. Whenever a new node is reached, its neighbors are checked. If the number of different colors of its neighbors is larger than 1, we simply stop since it implies that the graph needs at least more than two colors to color it. Otherwsie, we color the node with a different color and continue the traversal. Since the construction of the conflict graph takes $O(N \log N)$ time, we can easily solve the membership problem for a given permutation in $(N \log N)$ time. Since the set θ is quite large and include many popular permutations, the above result implies that we can route these permutations efficiently on an optical MIN.

Using the above membership algorithm, Shen, Yang, and Pan [18] also proved that $\theta = \Omega+1$, where $\Omega+1$ denotes the set of admissible permutations to the extra-stage Ω network. An extra-stage Ω network is simply obtained by adding one more stage of shuffle in front of an Ω network. An extra-stage Ω network provides two paths between an input and an output and allows much more permutations to be realized. Gazit and Malek [4] presented an algorithm to compute the size of $\Omega + 1$. Their formula shows that the size of the $\Omega + 1$ set is much larger that the size of the corresponding Ω set. Therefore, the results proved in [18] implies that the set θ is also much larger the set Ω.

5　Establishing Random Connections

When emulating rings, meshes, hypercubes, and binary trees or realizing permutations, the communication patterns involved have certain regularity. In many applications, most communication patterns are irregular in nature. In this section, we discuss issues related to establishing a set of arbitrary connections in optical MINs, especially photonic Banyans.

5.1　Performance of Greedy Algorithms

We first describe an analytical model to evaluate the performance of a routing/scheduling algorithm which considers a given set of arbitrary connections in a random order, and which tries to establish as many connections in a round (i.e. CF-mapping) as possible. The model first uses a recursive procedure to calculate the probability that a new connection can be established in a round (i.e. CF-mapping) given some existing connections, and then a Markov process to determine the average number of rounds needed for a set of one-to-one random connections. For brevity, we will examine Banyan networks with individual switch control.

Consider an $N \times N$ Banyan in which $s < N/2$ connections have been established. Let Γ denote a new connection from a random input to a random output, $F(N, s)$ be the probability that Γ can go through the first stage, and $\Phi(N, s)$ be the (conditional) probability that Γ can subsequently go through one of the two subnetworks of size $N/2 \times N/2$. Then, the probability that

Γ can successfully be established is

$$\Pi(N, s) = F(N, s) \cdot \Phi(N, s) \tag{1}$$

As described in [16], $F(N, s)$ is simply $(N - 2s)/N$. In addition, $\Phi(N, s)$ can be derived from a general term calculated using the following recursive formula (by setting $n = N$ and $k = s$)

$$\Phi(n, k) = \sum_{j} F'(n, j|k) \cdot \Phi(n/2, j) \tag{2}$$

where $F(n, j|k)$ is similar to $F(N, s)$ and is the probability that Γ can go through the second stage of an $n \times n$ Banyan with k established connections, of which j connections are in the same subnetwork (A^n) as Γ (see Figure 6 for an illustration).

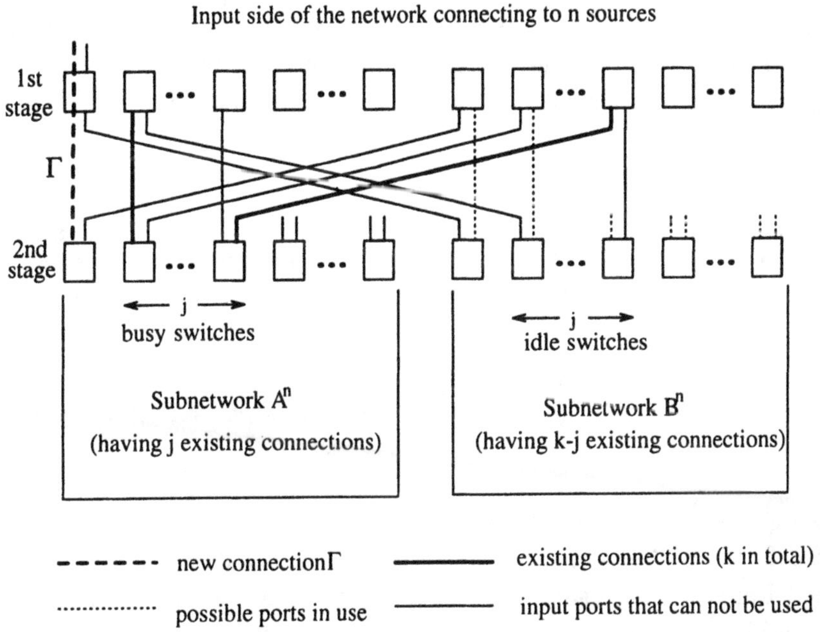

Figure 6: An illustration of the new connection Γ.

Based on $\Phi(N, s)$, a Markov chain can be constructed. Figure 7 shows a Markov chain of five states for an 8×8 Banyan, where state s means that the network currently contains s established connections. Starting off in state 0, a transition from state s to state $s + 1$ means that a new connection has just been established, while that from state s to itself means that the requested connection cannot be established due to either conflict or crosstalk. Accordingly, the transition probabilities, $\pi_{s,s}$ and $\pi_{s,s+1}$, are equal to $1 - \Pi(N,s)$ and $\Pi(N,s)$, respectively.

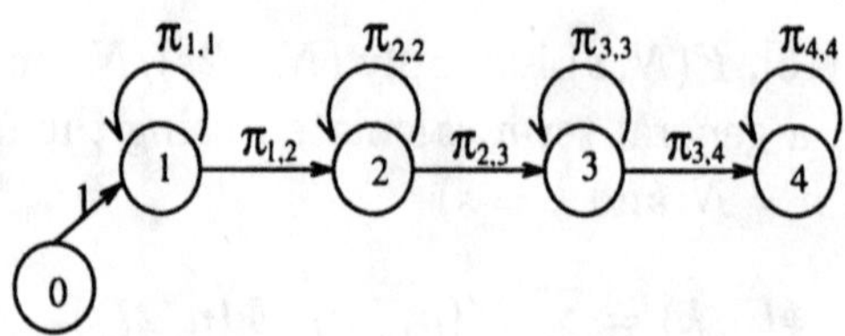

Figure 7: Markov process for an 8×8 Banyan.

To determine the average number of rounds needed to establish a set of r one-to-one random connections in an $N \times N$ Banyan, which we denote by $K(N,r)$, let $P(i|r)$ be the probability that i out of the r connections can be established in the first CF-round (where $1 \leq i \leq N/2$). This $P(i|r)$ is equal to the probability that the corresponding Markov process will be in state i after r transitions from the initial state 0, and can be obtained by raising the transition probability matrix of the corresponding Markov chain to the power of r [16].

Assume that i connections are established in the first round. Since the remaining $(r - i)$ connections are also random and would need, on average, an additional $K(N, r - i)$ rounds, the following recursive equation can be used to compute $K(N,r)$, with $K(N,0) = 0$:

$$K(N,r) = \sum_{i=1}^{\min(N/2,r)} P(i|r) \cdot [1 + K(N, r - i)] \tag{3}$$

Note that the analytical model described above is also applicable to dilated Banyans. Specifically, let $\Pi_d(N,r)$ and $K_d(N,r)$ be defined for the dilated Banyan in the same way that $\Pi(N,r)$ and $K(N,r)$ are defined for the (undilated) Banyan previously. Then one can show that $\Pi_d(N,s) = \Pi(2N,s)$ and $K_d(N,r) = K(2N,r)$.

Since $K(N,r) > K(2N,r)$, it is clear that the number of rounds needed using the time domain approach is larger than that using the space domain approach. However, as shown in Figure 8, both simulation and analysis show that $K(N,r) < 2 \cdot K_d(N,r)$, which means that the time domain approach can achieve a better space-time tradeoffs than the space-domain approach (see the discussion in Section 3.2).

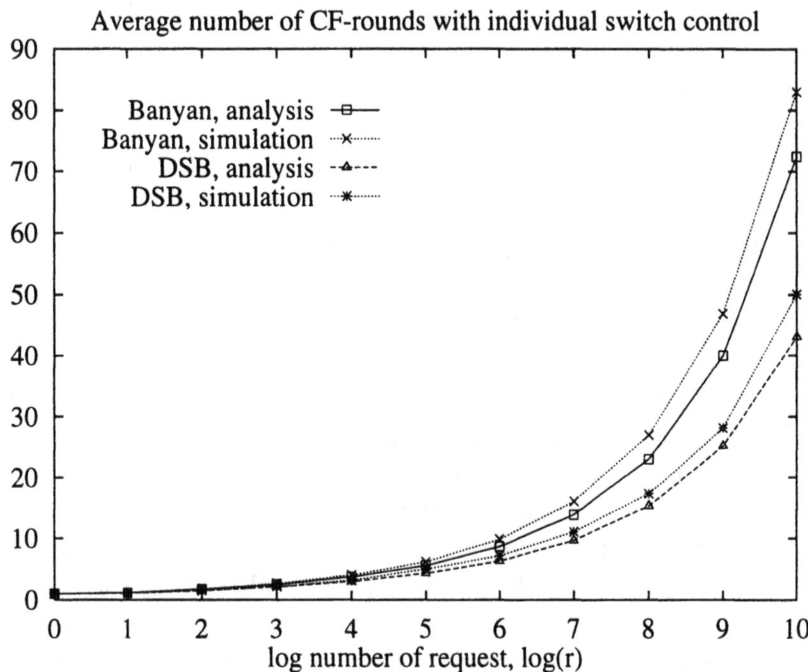

Figure 8: Average number of rounds in a 32×32 Banyan and DSB under switch control.

5.2 Heuristic Scheduling

The greedy algorithm tries to schedule as many messages as possible in single round without considering the overall picture of connection requests. It is always desirable to schedule an arbitrary set of connections in as few rounds as possible. By having a global view point on a set of connection requests, we may do a better scheduling job than the greedy aglgorithm. Such a scheduling problem may be transformed into a graph-coloring problem. Graph-coloring problem for general graphs is a NP-complete problem and cannot be solved efficiently, and thus using graph-coloring solution to derive optimal scheduling would be of little or no use in high-speed networks. As pointed out in [15], no optimal scheduling algorithm with a polynomial time complexity is available under individual switch control.

Several heuristic routing algorithms were reported by Denehy and Pan [3]. In their experiments, each member of a routing is a source and destination pair or path. This routing mechanism ensures that each destination will receive a message, but it allows for sources sending messages to several destinations, thereby encompassing permutation, broadcast, and multicast capabilities. By permuting the sources, it can be seen that N^N such routings exist. From this fact, it can easily be seen that an optimal solution requires exponential time.

The optical windowing methods described in [18] are used for permutation routing and are expanded to determine the admissibility of a multicasting routing. Basically, the optical window method is used to construct conflict graphs for a given set of messages. Their algorithm uses a graph coloring algorithm as a subroutine. Since the graph coloring problem is an NP-hard problem, they proposed two heuristic graph coloring algorithms in order to schedule the messages efficiently. Their heuristic algorithms find a valid coloring for an undirected graph with N vertices in in $O(N^2 \log N)$ time. Each vertex has an associated vector which indicates its currently available colors. The two algorithms differ in the selection of colors. The first selects colors by scanning the vector from the lowest bit. The second algorithm selects the color randomly. Both algorithms sort the vertices by decreasing degree to reduce the number of colors used. The idea is to ensure that messages with more conflicts with others are scheduled first. Once they are finished transmission, the remaining messages would have more freedom to transfer across a network.

A variable number of random routing simulations were performed on 4×4, 8×8, 16×16, 32×32, and 64×64 Omega MINs. From several experiments, it was determined that 100,000 test cases provided a stabilized average with negligible deviation from the previous results. The averages provided by the 100,000 test cases are plotted in Figure 9. From these results, it is seen that the use of the Low Bit Graph Coloring Algorithm consistently produces a better average than the Random Bit Graph Coloring Algorithm for the number of passes needed to realize a routing in an optical MIN.

5.3   Optimal Scheduling

Although optimal scheduling for a set of arbitrary connections in a Banyan network under individual-switch control is an NP-complete problem, it is possible to solve the problem in polynomial time for the same network under stage control. In this subsection, we describe algorithms for establishing a set of arbitrary connections in a Banyan network under stage control. As mentioned earlier, stage control is especially attractive to optical MINs because it requires less control and fewer electronic driver circuits. Under

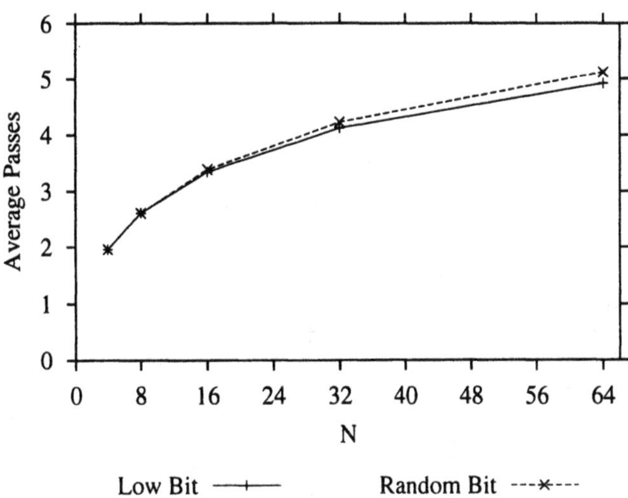

Figure 9: Average Number of Passes for Routings in an $N \times N$ Omega MINs.

stage control, it is possible to derive optimal scheduling algorithms having a polynomial time complexity. We start our descriptions with a heuristic algorithm called *Odd-Even* first. Then, we describe an optimal algorithm based on it. Both algorithms run in polynomial time and thus can be used in practice to reduce the number of rounds. Since it is difficult to derive analytical models to predict their performance, several simulations are carried out to evaluate these two algorithms.

In a Banyan, the control word for each connection is obtained by exclusive-*ORing* its input binary representation with its output binary representation. For example, the control word required for a connection from input $S = 7[111]$ to output $D = 5[101]$ is $W(S, D) = [111] \oplus [101] = [010]$ implying that stage 1 to stage 3 need to be set to straight($w_2 = [0]$), cross($w_1 = [1]$) and straight($w_0 = [0]$), respectively. Under this setting, input $7[111]$ is first connected to output port $(1 \oplus 0)11$ (i.e. $7[111]$) at stage 1, which in turn is connected to output port $1(1 \oplus 1)1$ (i.e. $5[101]$) at stage 2, and finally to output $10(1 \oplus 0)$ (i.e. $5[101]$) at stage 3.

From Figure 10, we observe that at every stage, the input and output ports are numbered in the same way. More importantly, the binary representations of the two input (or output) ports of a SE at stage k differ only in bit $n-k$. Accordingly, at stage k, input port $P_{in} = p_{n-1} \cdots p_{(n-k+1)} p_{n-k} p_{(n-k-1)} \cdots p_0$ is connected to output port $P_{out} = p_{n-1} \cdots p_{(n-k+1)} (p_{n-k} \oplus w_{n-k}) p_{(n-k-1)} \cdots p_0$. Accordingly, given two connections (S, D) and (S', D') having the same con-

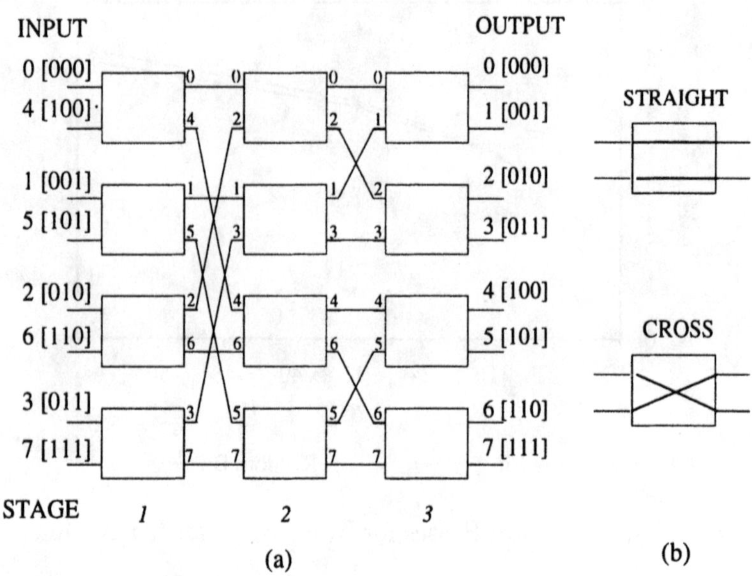

Figure 10: An 8×8 Banyan network and two states of a SE.

trol word W, they will use a common SE if and only if the *hamming distance* between S and S' is 1. More specifically, they will use a common SE at stage k if and only if (the binary representations of) S and S' differ in bit $n - k$.

For example, the two connections sharing any control word $W[w_2 w_1 w_0]$ that are from $S = 0[000]$ and $S' = 4[100]$, respectively, would use the top SE at stage 1. The two connections from $0[000]$ and $2[010]$, respectively, would use the top SE (or the third from the top) at stage 2 if they share control word $[0w_1 w_0]$ (or $[1w_1 w_0]$). Similarly, the two from $0[000]$ and $1[001]$, respectively, would use the top SE (the second, the third or the bottom SE) at stage 3 if they share control word $[00w_0]$ ($[01w_0]$, $[10w_0]$ or $[11w_0]$).

A heuristic algorithm called Odd-Even works as follows [17]. Partition the connections according to their required control words and their input parities. Specifically, we let $O[i]$ and $E[i]$, where $0 \leq i \leq N-1$, represent the group of the connections requiring control word i, but whose input parities are odd and even, respectively. The number of connections in $O[i]$ and $E[i]$, ranging from 0 to $N/2$, will be denoted by $|O[i]|$ and $|E[i]|$ respectively.

The heuristic algorithm establishes the connections requiring the same control word in one round, as long as they have the same input parity. Since the binary representations of the two inputs having the same parity differ by at least two (2) bits, the hamming distance between them is at least 2 (could be 4, 6 and so on). If $|O[i]| > 0$ (or $|E[i]| > 0$), then scheduling the connections in $O[i]$ (or $E[i]$) in one round guarantees that they will be SE-disjoint.

Because in a Banyan whose SEs are already set under stage control, there are exactly $N/2$ connections having even (or odd) input parity. Based on the description of the algorithm, all of the $N/2$ connections can be established in one round. Hence, the maximum number of SE-disjoint connections that can be established in one round under stage-control, $N/2$, can be reached using the Odd-Even algorithm. The time complexity of the algorithm is clearly polynomial of the number of connections to be established.

The Odd-Even algorithm just described may be too conservative since having the same parity is sufficient but not necessary for connections requiring the same control word to be SE-disjoint. For example, it is possible that connections having different input parities, such as $(0, 2)$ and $(7, 5)$, are SE-disjoint (in addition to requiring the same control word), and thus can be established in one round.

The optimal algorithm is an improved version of the Odd-Even algorithm through merging some connections in one pass. The optimal algorithm tries to reduce the number of rounds resulted from using the Odd-Even algorithm by merging $O[i]$ and $E[i]$ into one round. For each control word i ($0 \leq i \leq N-1$), if both $O[i]$ and $E[i]$ are not empty, an attempt is made to establish the connections in the two groups in one round. Specifically, $O[i]$ and $E[i]$ are merged if (and only if) every connection in $O[i]$ is found to be SE-disjoint with every connection in $E[i]$. The attempt to merge $O[i]$ and $E[i]$ is aborted as soon as a connection in $O[i]$ is found to share a SE with another connection in $E[i]$.

Note that merging $O[i]$ and $E[i]$ is the only possible way to reduce the number of rounds because given two different control words i and j, one cannot merge $O[i]$ with $O[j]$, nor $O[i]$ with $E[j]$, nor $E[i]$ with $E[j]$. Hence, it is obvious that the algorithm will result in a minimum number of rounds after the merging takes place and is optimal. Its time complexity is also polynomial of the number of connections to be established.

The average schedule lengths using each of the two algorithms in a Banyan are shown in Figures 11 and 12. For comparison purposes, the schedule length using the Greedy algorithm, which is a heuristic algorithm similar to the random algorithm analyzed in the previous subsection, is also shown. These results are for two network sizes, namely $N = 32$ and $N = 64$, respectively, although the results for other sizes are similar. In Figure 11, the number of connections to be established, R, is assumed to be less than a small percentage (about 10%) of the total number of possible connections, which is N^2. As can be seen, the Odd-Even algorithm performs the worst and the Greedy algorithm performs nearly as well as the Optimal algorithm. In addition, the number of rounds needed in a DSB is smaller than that needed in a Banyan.

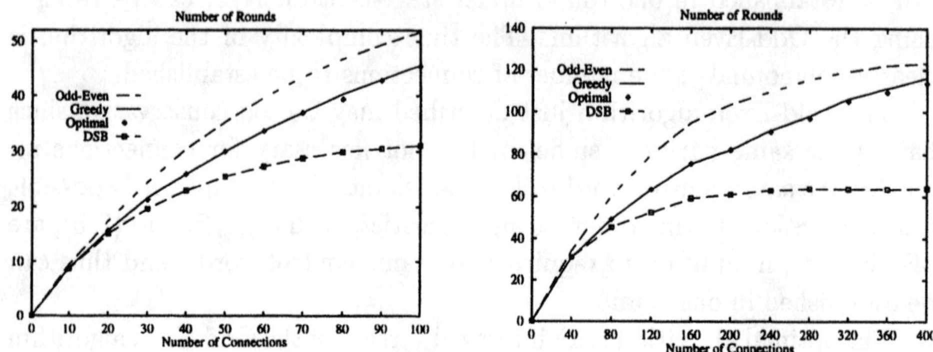

Figure 11: Simulation results for $N = 32$ (left) and $N = 64$ (right) with a small R.

In Figure 12, R is assumed to be large (ranging from 100 to 1000 when $N = 32$, and from 400 to 4000 when $N = 64$). As can be seen, the Odd-Even algorithm performs nearly as well as the Optimal algorithm since merging becomes rare and both algorithms result in a schedule length which approaches the maximum, which is 64 for $N = 32$ and 128 for $N = 64$.

However, the Greedy algorithm performs poorly. Specifically, we note that the schedule length using the Greedy algorithm will be about 1.8 times longer when $N = 32$, and 2.4 times longer when $N = 64$. This is because the Greedy algorithm allows several connections having one parity to be established in one round with several other connections having the opposite parity. This results in fewer than $N/2$ connections in one round and accordingly, more than $2N$ rounds for N^2 connections. Since the Optimal algorithm has a polynomial time complexity and is only a little more complicated than either the Greedy or the Odd-Even algorithm, it should be used when the load condition (i.e. the number of connections to be established) is unknown or varies greatly.

From Figure 12, we also observe that the schedule length in a DSB reaches its maximum of 32 at about the same time when the schedule length

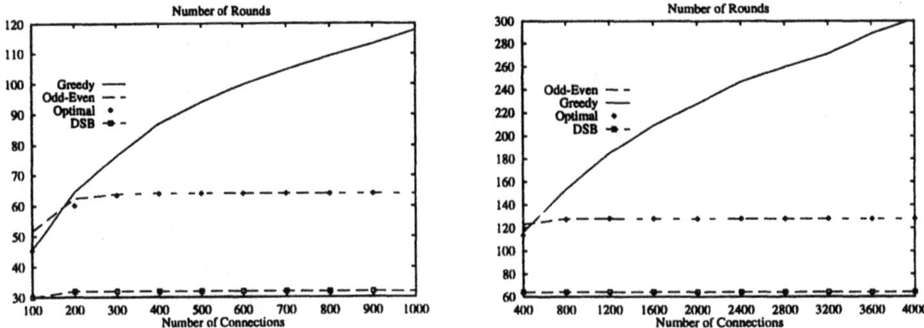

Figure 12: Simulation results for $N = 32$ (left) and $N = 64$ (right).

in a Banyan of the same size reaches its maximum of 64. Prior to that point, the latter is less than twice as long as the former for a wide range of R values as can be seen from Figure 11. This is similar to the case in which the connections to be established are random and may contain duplicates [16]. If we consider the number of SEs (and links) used in a network as a cost in space, and the number of rounds needed as a cost in time (or wavelength), then using a Banyan results in better space-time tradeoffs than using a DSB of the same size because the DSB has at least twice as many SEs and links as the Banyan. A practical implication is that when space cost is not a concern, one may use a two-plane Banyan instead of a DSB, both of which have approximately the same space cost, to reduce the cost in time (or wavelength) [16].

6   Crosstalk Fault Diagnosis

The fault-tolerance issue in a MIN is very important for the overall system reliability in communication or parallel computing environments. Electronic MINs can be made fault-tolerant by introducing redundancy in terms of extra stages, links or SEs with associated increase in cost and hardware complexity of the network [22]. These techniques can be equally applied to optical MINs. For optical MINs that are dilated in the space-domain, the fault-tolerance capability is automatically introduced since when a path going through, say, the upper half of the network, is damaged, an alternate path going through the lower half can be used. Obviously, the performance (e.g. throughput) of the entire network may be degraded since the alternate path will share some SEs with several other paths in the lower half of the network, and not all of them can be used at the same time for crosstalk-free communications. In such cases, the time domain dilation approach is again useful when combined with the space-domain dilation in the sense that these paths sharing some SEs can be used alternately.

However, for a MIN to be fault tolerant, the control algorithm first needs to detect and locate the faulty SEs in a MIN. Hence, fault diagnosis plays a crucial role in fault tolerance in a MIN. Numerous methods for diagnosing electronic MINs for multiprocessor communications and telecommunication systems have been studied under the common assumption that only the inputs and outputs of a MIN are accessible [22]. Most of the diagnostic methods studied so far assumed so-called *stuck-at-x* faulty SEs.

Unfortunately, these methods are no longer applicable to the diagnosis of optical MINs in which SEs are faulty in the sense that they generate excessive crosstalk but they can still change their states. Unlike a stuck-at-x electronic SE whose signal at any output port is either 0 or 1, a photonic SE will generate crosstalk whose amount can be anything in between a range, say -10 to -40 dBs (or 1% to 0.01% of the input power). Accordingly, diagnosing photonic SEs is different as it involves *quantification* of the amount of crosstalk of each SE, not just determining whether a 0 or 1 are at an output port of the SE. More specifically, four tests are needed for each photonic SE in which its state is set to either straight or cross, and either its upper input or its lower input is used for injecting a test signal. In each test, the crosstalk ratio, which is the power of the crosstalk signal over that of the output signal (see Figure 1(a) for an example) is determined to see if it exceeds a threshold value. As a result, a photonic SE may have up to four types of crosstalk-faults depending on whether it passes or fails each of these four tests.

What makes diagnosing photonic MINs more challenging is that not only may one crosstalk-faulty SE mask another crosstalk-faulty SE, as a stuck-at-x faulty SE may do to another stuck-at-x faulty SE in electronic MINs [13], but two non-crosstalk-faulty SEs may generate crosstalk as if one or both of them are crosstalk-faulty. This occurs when the crosstalk signals from the two SEs merge at the output, and their combined power is treated (incorrectly) as the power of the crosstalk signal from one SE. Similarly, the accuracy of the measurement will be affected if second-order crosstalk (which is generated when a first-order crosstalk signal goes through a photonic SE) merges with a first order crosstalk signal. Hereafter, we will refer to crosstalk-fault as simply *fault* in the context of diagnosing optical MINs.

Research in diagnosing optical MINs is, in contrast to its electronic counterpart, still in its infancy. Most of the approaches used in practice for diagnosing crosstalk in optical MINs are *ad hoc*, and little is known about systematic ways to perform the testing. In a pioneering work by Choy, et al [2], a method requiring a reasonable number of tests to detect and locate a (crosstalk) faulty SE in DBNs was proposed. However, the method may either fail to report an actual fault or issue a false alarm if multiple SEs can be faulty at the same time.

6.1   Diagnostic Method for DBNs

Qiao [13] recently proposed an efficient diagnostic method for an $N \times N$ DBN under a simplifying assumption that the power loss due to radiation, scattering, absorption as well as other undesirable effects within a SE and waveguide links connecting two stages in the DBN is a known constant. The method uses an integrated two-level process [13]. At level one is a rigorous and thorough procedure called *Test-All-Switches* (TAS), which takes two phases to complete. In phase one, $4N$ tests are performed while using a naive method, $4M \cdot N$ tests may be needed since there are $M \cdot N$ SEs in the DBN (where $M = 2 \log N$ is the number of stages in the DBN as shown in Figure 3). An example test for a 4×4 DBN is shown in Figure 13. As can be seen, the DBN is configured such that the power of the test signal as well as that of all the crosstalk signals generated go to distinct outputs and thus can be measured accurately. This can be accomplished by using a recursively defined algorithm called Complementary Subnets (or CS). Using CS, the SEs in a subnet used by the crosstalk signal generated at a preceding stage are set to a state complementary to the state of the SEs in the other subnet used by the test signal.

Based on these measures, step-by-step calculations with a time complexity of $O(N \log^2 N)$ are performed in phase two of the TAS procedure to obtain the crosstalk ratios of each and every SE in the DBN. These crosstalk ratios can then be used to identify any single faulty SE or any combination of multiple faulty SEs, each of which may have up to four types of faults, in the DBN. They can also be used to calculate the signal-to-crosstalk ratio (SXR) of the entire DBN.

Since the amount of time required for each invocation of the TAS procedure, which involves network configuration, test signal injections, power

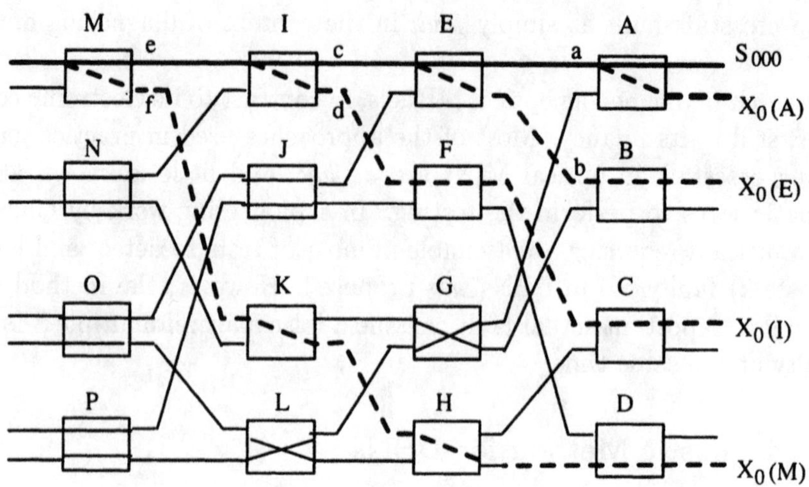

Figure 13: An example test using an appropriate configuration.

measurements and calculations, can be substantial, an unnecessary invocation of the TAS procedure introduces too much overhead, and thus should be avoided. This motivates us to devise an *"on-the-fly"* procedure called *Test-One-Path* (TOP) at level two. The TOP procedure is applicable when most of the SEs in the DBN are assumed to have not changed their crosstalk characteristics after a recent invocation of TAS. Only the SEs along a path are suspected of becoming faulty. The TOP procedure requires only one test to collect necessary measurements, and involves very little computation ($O(\log N)$ instead of $O(N \log^2 N)$) to determine if the suspicious SEs are actually faulty. The recursively defined CS algorithm is also used to configure the DBN appropriately for each TOP test.

Although TOP has a much lower overhead than TAS, it can not replace TAS because of the possible changes in the crosstalk characteristics of the SEs due to the changes in the control voltage, temperature, polarizations, and the aging of the materials. This is also why the TAS procedure may need to be carried out once in a while as a maintenance routine, and under automated (i.e. computer program) control. An important feature of this diagnostic process is that it is amendable to automatic test generation. Both TAS and TOP utilize a well defined algorithm to configure the network at the beginning of a test so that necessary measurements can be taken accurately. The input at which a test signal is injected, the proper network configuration for this test, and the outputs at which the signals are measured can all be determined by a computer program. In addition, the computation to be performed based on these measurements can also be carried out by a program. It is interesting to examine if the TAS procedure can be speeded up by reducing the number of tests needed (in phase one of TAS), by overlapping the tests with computation (i.e. the two phases), and/or by parallelizing the computation. Another interesting future work is to apply some of the ideas of the proposed diagnostic process to photonic switching networks with architectures other than DBNs.

6.2 Diagnostic Method for Dilated Blocking Networks

Hwang and Lee [8] proposed an algorithm for diagnosing crosstalk faults for a class of dilated blocking networks. The characteristic of dilated blocking networks, such as Dilated Omega Network, is that only one possible path can be established between any input-output pair. The impact of this characteristic crosstalk faults diagnosis is that the input signal and crosstalk signal do not merge into the same output switch. This property relaxes the condition for us to design the diagnosing crosstalk algorithm in comparison with dilated non-blocking networks studied by Qiao [13]. They proposed a two-level diagnosis algorithm to satisfy the need of this architecture. The first-order crosstalk is detected and located in the first level. On the other hand, the number of unused photonic switches grows in $O(N^2)$ as the size of network increases in $N \times N$ during the diagnosing crosstalk algorithm. This fact motivates them to speed up the test by paralleling two first-order crosstalk faults at the first-level algorithm. In order to measure second- or higher-order crosstalk, the second-level algorithm is called. Their diagnosing algorithm is applied to other dilated photonic isomorphic topologies such as Dilated Baseline Network and Dilated Banyan Network. After certain modification, their algorithm can also be used to diagnose crosstalk faults for other rearrangeable multistage networks.

7   Conclusions

As optical technology advances, more and more networks adopt optical technology for faster and higher bandwidth communications. In this chapter, we survey the major challenges encountered and approaches adopted in the research of optical MINs. As optical MINs are more widely adopted in communication networks and parallel/distributed computing systems [10], [21], many new research issues will emerge and demand for better design and analytical methods. For example, a subject of future research is to apply the Markov model in this chapter to other environments such as packet switched Banyan networks without internal buffers. Another topic is to predict the performance of optical MINs when used in an ATM environment, where traffic is highly bursty in nature. New traffic models such as self-similar stochastic models may be necessary in order to obtain realistic performance measures. In the diagnosis procedures described in this chapter, the effects of the nonuniform power losses in the SEs and waveguide links, which can only be approximated through experiments, have been largely ignored. Diagnostic methods to detect lossy (or faulty) waveguide links in an integrated photonic switching networks are needed, and deserve further investigation. How to design efficient parallel algorithms on systems with optical MINs is another interesting research topic. Implementing efficient broadcast operations in optical MINs is also a new challenge. Another direction is to provide optical networking solutions for next-generation Internet networks. Several solutions have been introduced so far providing optical transport of IP networks, such as IPoWDM (for IP over wavelength-division multiplexing). Several architectures [11] have been proposed to provide optical networking solutions, based on wavelength-division multiplexing and compatible with the IP world.

References

[1] D. P. Agrawal, "Graph theoretical analysis and design of multistage interconnection networks," *IEEE Transactions on Computers*, Vol. 32, No. 7, pp. 637-648, Sept. 1983.

[2] W. Choy, R. Thompson, and G. Berglan, "Detection and location of crosstalk faults in a class of optical switching networks," *IEEE/OSA J. Lightwave Technology*, to appear.

[3] T. E. Denehy and Y. Pan, "Multicasting on Optical Multistage Interconnection Networks," *Proceedings of the 11th International Conference on Parallel and Distributed Computing and Systems*, Boston, MA, Nov. 3-6, 1999, pp.101-106.

[4] I. Gazit and M. Malek, "On the number of permutations performable by extra-stage multistage interconnection networks," *IEEE Transactions on Computers*, Vol. 38, No. 2, 1989, pp. 297-302.

[5] Q.-P. Gu and S. Peng, "Wavelengths requirement for permutation routing in all-optical multistage interconnection networks," Proceedings of the 2000 International Parallel and Distributed Processing Symposium, May 1-5, 2000, Cancun, Mexico, pp. 761-768.

[6] Q.-P. Gu and S. Peng, "Efficient protocols for permutation routing on all-optical multistage interconnection networks," 2000 International Conference on Parallel Processing, Toronto, Canada, pp. 513-520, August 2000.

[7] H. Hinton, "A non-blocking optical interconnection network using directional couplers," *Proc. IEEE Global Telecom. Conference*, pp. 885-889, Nov. 1984.

[8] I. Hwang and S. Lee, "Diagnosing crosstalk faults in a class of dilated blocking optical multistage interconnection networks," *Information Sciences*, Vol. 124, No. 1-4, pp. 59-91, May 2000.

[9] J. Lanfant, "Parallel permutations of data: a benes network control algorithm for frequently used permutations," *IEEE Transactions on Computers*, Vol. 27, No. 7, pp. 637-647, 1978.

[10] K. Li, Yi Pan, and S.Q. Zheng, *Parallel Computing Using Optical Interconnections*, Kluwer Academic Publishers, Boston, USA, 1998.

[11] M. Listanti and R. Sabella, "Optical Networking Solutions for Next-Generation Internet Networks," Special issue in *IEEE Communications Magazine*, Vol . 38, No. 9, Sept. 2000.

[12] K. Padmanabhan and A. N. Netravali, "Dilated networks for photonic switching," *IEEE Trans. Communications*, vol. 35, no. 12, pp. 1357-1365, Dec. 1987.

[13] C. Qiao, "A two-level process for diagnosing crosstalk in photonic dilated Benes network," *Journal of Parallel and Distributed Computing*, vol. 41, no. 1, pp. 53-66, 1997.

[14] C. Qiao, R. Melhem, D. Chiarulli and S. Levitan, "A time domain approach for avoiding crosstalk in optical blocking multistage interconnection networks," *J. Lightwave Technology*, vol. 12. no. 10, pp. 1854-1862, Oct. 1994.

[15] C. Qiao and R. Melhem, "Reconfiguration with time-division multiplexed MINs for multiprocessor communications", *IEEE Trans. on Parallel and Distributed Systems*, vol. 5. no. 4, pp. 337-352, 1994.

[16] C. Qiao, "A Universal Analytic Model for Photonic Banyan Networks," *IEEE Trans. on Communications*, Vol. 46, No. 10, pp. 1381-1389, 1998 (a preliminary version appeared in Proc. of IEEE Infocom, pp. 822-829, 1996).

[17] C. Qiao and L. Zhou, "Scheduling switching element disjoint connections in stage-controlled photonic banyan networks", *IEEE Trans. on Communications*, Vol. 47, No. 1, pp. 139-148, 1999 (a preliminary version appeared in Proc. of ICCCN, pp. 232-237, 1996).

[18] X. Shen, F. Yang, and Y. Pan, "Equivalent Permutation Capabilities between Time Division Optical Omega Network and Non-optical Extra Stage Omega Network," *Proceedings of 1999 IEEE International Performance, Computing, and Communications Conference*, Phoenix/Scottsdale, Arizona, USA, pp. 356-361, February, 1999.

[19] R.A. Thompson, "The dilated slipped banyan switching network architecture for use in an all-optical local-area network," *J. Lightwave Technology*, vol. 9. no. 12, pp. 1780-1787, Dec. 1991.

[20] R.A. Thomspon and P.P. Giordano, "An experimental photonic time-slot interchanger using optical fibers as reentrant delay-line memories," *J. Lightwave Technology*, vol. LT-5, no. 1, pp. 154-162, Jan. 1987.

[21] C. Tocci and H.J. Caulfield, *Optical Interconnection – Foundations and Applications*, Artech House Publishers, 1994.

[22] A. Varma and C.S. Raghavendra, *Interconnection Networks for Multiprocessors and Multicomputers: Theory and Practice*, IEEE Computer Society Press, 1994.

[23] Y. Yang and G.M. Masson, "Broadcast ring sandwich networks," *IEEE Trans. Computers*, vol. C-44, no. 10, pp. 1169-1180, 1995.

[24] Y. Yang, J. Wang and Y. Pan, "Permutation capability of optical multi-stage interconnection networks," *Proceedings of the First Merged IEEE International Parallel Processing Symposium & Symposium on Parallel and Distributed Processing (IPPS/SPDP '98)*, Orlando, FL, March

1998, pp. 125-133 (Also appeared in *Journal of Parallel and Distributed Computing*, Vol. 60, No. 1, pp. 72-91, Jan. 2000).

OPTICAL NETWORKS - RECENT ADVANCES
L. Ruan and D.-Z. Du (Eds.) pp. 187 - 203
©2001 Kluwer Academic Publishers

Connection Management for Wavelength-Routed Optical WDM Networks

Byrav Ramamurthy
Department of Computer Science and Engineering
University of Nebraska-Lincoln, Lincoln NE 68588-0115
E-mail: `byrav@cse.unl.edu`

Lu Shen
Department of Computer Science and Engineering
University of Nebraska-Lincoln, Lincoln NE 68588-0115
E-mail: `lshen@cse.unl.edu`

Elie Sawma
Department of Computer Science and Engineering
University of Nebraska-Lincoln, Lincoln NE 68588-0115
E-mail: `esawma@cse.unl.edu`

Contents

1 Introduction

The Internet is growing at a tremendous rate today. New services, such as telephony and multimedia, are being added to the pure data-delivery framework of yesterday. Such high demands on capacity (bandwidth) could lead to a "bandwidth-crunch" at the core wide-area network resulting in degradation of service-quality. Fortunately, technological innovations have emerged which can provide some relief to the end-user struggling with the Internet's well-known delay and bandwidth limitations. At the physical layer, a complete overhaul of existing networks is envisaged from electronic media (such as twisted-pair and cable) to optical fibers [1]. Optical fibers employing the promising technique of wavelength division multiplexing (WDM) can support around 1000 times the capacity of their electronic counterparts. WDM allows the simultaneous transmission of several channels on the same fiber each on a different wavelength (frequency) [2, 3, 4]. WDM is already being deployed in commercial point-to-point fiber links including undersea installations. WDM-based optical networks are currently being tested in the U.S. (e.g., MONET, NTONC projects) [5] and Europe (RACE, ACTS projects) [6].

1.1 Wavelength-Routed Optical WDM Networks

Although current optical networks provide high bandwidth, they suffer from the electronic bottleneck, which happens in the network node for O-E-O conversion. To eliminate the electronic bottleneck, the next generation of optical networks will be capable of selectively routing and switching individual wavelengths, creating what is called a wavelength-routed optical WDM network. So optical technology can be used at a truly network level, instead of remaining a link level technology. In a wavelength-routed optical WDM network, each wavelength can be routed through the optical network at the optical switching nodes, removing the need for opto-electronic conversion and electronic routing. Data can remain in optical domain without requiring costly high-speed electronic equipment and eliminate the bottleneck due to O-E-O conversion at intermediate router nodes.

Critical pieces in enabling the deployment of wavelength-routed optical WDM networks are the intelligent optical components, such as optical crossconnects (OXC), wavelength add-drop multiplexers (WADM), and a feasible and effective control plane for WDM networks. The control plane of wavelength-routed optical networks is responsible for control and management of the optical networks, including configuration management, fault management and performance management [22]. In this chapter, we mainly focus on connection management, which is one of the critical functionalities of configuration management. Provisioning of connections, in a wavelength-routed WDM network, requires algorithms for route selection, and signaling mechanisms for requesting and establishing connectivity. An effective and feasible connection management method for WDM networks is based on a suitable network control scheme and RWA (routing and wavelength assignment) scheme. In the following sections, we will discuss connection management in more detail and compare different methods for implementing connection management through the control plane in optical networks.

This chapter is organized as follows. Section 2 compares different connection management schemes and discusses their advantages and disadvantages. Section 3 describes the results from some experiments to observe the effect of different connection management schemes and RWA algorithms on network performance. Finally, in section 4, we present our conclusions and discuss research topics related to connection management.

2   Connection Management

Connection management of optical networks can be implemented using different strategies. In this section, we compare the different connection methods from different points of view, using the critical metrics for connection management, such as blocking probability, connection setup time, network resource utilization, network resource cost etc.

Connection management schemes can be classified using several criteria. From the system architecture point of view, we can distinguish between centralized or distributed methods, that require different amounts of network resource knowledge. Based on the holding times of the lightpaths, there are static and dynamic connection management methods. Depending on the network requirements, network design engineers can choose different signaling methods: in-band or out-of-band.

2.1   Centralized and Distributed Approaches

Two kinds of connection management methods, Centralized and Distributed, can be employed in optical networks [25, 27]. In the centralized one, a central control center/central network management system for the whole network holds the global information of the network, such as the network topology, the link states, the wavelength usage on each link and the status of each network element. When a source node needs to transfer data, the request for connection is sent to the control center and then a route is calculated, according to the routing and wavelength assignment algorithm in the center, based on the global information of currently available network resources. Then, the control center will reserve the resources for the connection by notifying each node along the route. After the control center receives acknowledgment from each node, it will send a message to notify the source node to send data along with the reserved path to the destination. When a connection is finished, the control center will signal each node involved to release the selected wavelength.

In the distributed approach, each node along the route will be involved in making decisions on selecting the wavelength. The connection request goes through each node along the route and reserves the wavelengths based on the local information (or partial information of the network) at the node. After all the nodes on the route agree to the request, the source will start sending data along the reserved route. When a connection is torn down, the release request will be sent to the destination and release network resources being used at each node. A path is decided by coordination among the different nodes, not by a control center. And, the information stored at each node needs to be sent out to other nodes to reflect changes in the node status. It is called a distributed approach due to its distributed nature of operation.

Because the centralized approach does not need coordination from each node for computing a lightpath, it is simple to a implement and effective in small-sized networks. Another advantage of a centralized approach, over the distributed one, is its lower blocking probability, due to the fact that the central control center knows the updated information of the whole network.

But, the centralized method is vulnerable due to the presence of a single point of failure. If the control center crashes, the whole system cannot work properly. Another disadvantage with the centralized approach is its lack of scalability. When the network becomes large, the information needed to be stored at the control center becomes staggeringly large. Consequently, the computation time becomes large and the control center will become the bottleneck for the whole network. Moreover, since different sub-networks may be using equipment from different vendors, sub-networks may have their own control centers, which differ from one other. So if the network becomes large enough to contain many vendors' network elements, interoperability between sub-networks will be a problem.

Compared with the centralized approach, the distributed one is more suitable for a large-scale network. It leaves the decision of selecting the local wavelength to each node and distributes the computation task to each node on the route and thus it eliminates the bottleneck due to the control center and improves the reliability of the network. Different vendors' network elements can communicate with one other using well-known routing protocols and wavelength assignment protocols. So it also provides better scalability. But, the distributed approach increases network resource costs, because it requires changing the state of each node in the network and increases the complexity of the system. It also has a higher blocking probability than the centralized one, due to its distributed working fashion.

2.2 In-Band and Out-of-Band Signaling

A standard signaling system needs to be set up for conveying connection related information in optical networks. The availability of the signaling channel, which transports signaling messages, is of crucial importance in an optical network. The channel can be either in-band or out-of-band.

For in-band signaling, the signaling messages are carried on the same channel as the data, maybe using the overhead frames in the data channels. For example, the Synchronous Optical Network (SONET) overhead frames can be used to carry signaling messages. However, all-optical networks do not extract the content of signaling messages at the intermediate switching nodes and cannot implement O-E conversion to process the signaling messages. So it is impractical to implement in-band signaling for all-optical networks, although it could enhance the network resource utilization by making use of the overhead bytes. Out-of-band signaling employs a separate network for signaling message transportation. It can obtain high speed signaling for large volume of information at the expense of consuming more networking resources.

2.3   Path Multiplexing (PM) and Link Multiplexing (LM)

In dynamic wavelength routing networks, a link refers to a link/wavelength between two adjacent optical switching nodes. A lightpath refers to the optical path (route), and one/several chosen wavelengths from a source station to a destination station through intermediate optical switching nodes (OXC). We can regard a lightpath as a virtual pipe-line for transferring data across a WDM network.

When the same wavelength is used on each link along the lightpath, it is called PM (path multiplexing). When different wavelengths (or possibly the same one) are used in each link along the lightpath, it is called LM (link multiplexing) [18].

The additional condition required in PM is referred to as the wavelength continuity constraint. The bandwidth usage of PM is lower than that of LM. The restriction imposed by the wavelength continuity constraint can be avoided by the deployment of wavelength converters at each wavelength-routing node. [23, 26]

2.4   Different RWA algorithms for PM

A good wavelength assignment algorithm in a PM network can result in improved performance. Here, we list some proposed wavelength-assignment algorithms:

- Random wavelength assignment: Selecting wavelengths randomly from a set of available wavelengths.

- First-fit wavelength assignment [11]: Selecting the first available wavelength according to a predefined order of wavelengths.

- Most-used wavelength assignment [12]: Selecting the wavelength, which is used most in the available wavelength pool.

- Least-used wavelength assignment [12]: Selecting the wavelength, which is used least in the available wavelength pool.

Karasan and Ayanoglu [14] used network simulations to investigate the performance of the least-used, randomly selected, most-used, and first-fit wavelength assignment algorithms in networks with a single fiber on each link. Shortest path routing was used as the routing method. The blocking probability and networking utilization were used as the evaluation standard. They showed through experiments that the least-used heuristic provides the worst performance. The performance of random wavelength assignment was found to be slightly better than the least-used one, since it effectively balances the traffic load on each wavelength. They showed that the most-used wavelength algorithm, especially in networks with large number of wavelengths, could improve the network performance. But, the algorithm needs a global knowledge of the network status. First-fit wavelength assignment can achieve a good performance almost close to the most-used wavelength assignment algorithm. The benefit of the first-fit algorithm is that it only needs the state of the links along the lightpath, instead of global knowledge of network status as required by the most-used wavelength algorithm.

2.5 Static and Dynamic Assignment

In static WDM networks, lightpaths are assigned in advance and remain unchanged for a long period, perhaps several days. So the connection from one node to another node is fixed on some lightpaths during that period. If the bandwidth requirement of the traffic and the statistical properties of the traffic can be known in advance and remains unchanged for a long term, static WDM networks can work well. Although static WDM networks are not as complex as the dynamic one and reduce the processing time for computing lighpaths, the network bandwidth utilization is low under static connection configuration.

Dynamic configuration requires that lightpaths are established on demand. Each time, when a node needs to send data, it will request a lightpath from the network. The lightpath, route and wavelengths, is calculated for each request.

In a realistic network, lightpaths may be designed under a combination of both static and dynamic configurations, in order to obtain a more effective performance.

2.6 Source Initiated Reservation (SIR) and Destination Initiated Reservation (DIR)

In a Source Initiated Reservation (SIR) approach, wavelengths are reserved along the way when the connection request is sent from the source node to the destination node. In the DIR method, wavelengths are not reserved until the ACK of the connection request is sent back from destination node to source node.

For SIR, the network bandwidth, in terms of reserved wavelengths, is wasted during the reservation period. Using DIR can thus improve bandwidth efficiency. DIR brings obvious benefits for PM, since the request message gathers all the wavelengths usage on the way to the destination and then lets the destination select a wavelength based on the whole wavelength information on each node along the route. LM can also benefit from this approach. If one of the lightpath assignments fails, the destination can select another lightpath based on the gathered information. It reduces the processing time. However, DIR enhances the blocking probability, since it increases the chance of having inconsistent knowledge of wavelength usage at each destination node.

2.7   Dropping and Holding Schemes

In the dropping scheme [18], when the reservation fails at the intermediate node, NACK will be sent back to the source node and the request will be re-sent after a timeout. In the holding scheme, the intermediate node will not send NACK, instead it will buffer the request message and wait for wavelengths to become available at an intermediate node for a specified time interval. The holding scheme reduces processing time and propagation due to lower overhead for resending reservation requests.

2.8   Parallel and Sequential Reservation

Sequential reservation selects only one wavelength at each intermediate node. In a parallel method, the source node reserves all the available wavelengths and each intermediate node reserves the available sub-set of the reserved wavelengths. At the destination, a wavelength will be assigned and ACK will be sent back. The ACK is responsible for reserving that wavelength and releasing all the other wavelengths reserved previously. Parallel reservation makes sense only in PM. Because, in LM, any available wavelength could be selected at each link.

3   Experiments

In the above section, we introduced some connection management schemes and discussed their advantages/disadvantages under different conditions. The performance evaluation of each scheme is important for designing WDM networks. Some factors, such as the number of wavelengths and fibers, link propagation delay, network topology, processing time for the connection management etc., affect network performance as measured by different metrics, such as the blocking probability, throughput, utilization of the networks and so on. In this section, we first introduce a platform for optical network simulation and then discuss the impact of different network characteristics on the network performance.

3.1   SIMON (SIMulator for Optical Networks)

It remains difficult and expensive to evaluate the network performance in a real optical network. So it is a good choice to study optical networks using simulation tools, especially for the distributed control scheme, which is hard to evaluate. In this section, we give an introduction on SIMON - the acronym for a Simulator for Optical Networks [17]. It is a simulation package that provides wide range of functionalities from physical level of optical network simulation to high level routing algorithm modeling. SIMON is developed in C++ and runs on Unix machines. SIMON is a proven optical network simulation system. In the following section, we will discuss some our experiments on connection management using SIMON.

SIMON was originally developed by Byrav Ramamurthy for his research at UC Davis in 1997. The first version (SIMON 1.0) contains functionality for optical network simulation on physical device level except for wavelength conversion capabilities. The second version (SIMON 2.0) was enhanced from SIMON 1.0 by Jong Tae Lee at the University of Nebraska-Lincoln for his research on wavelength conversion and converter placement in 1998. SIMON 2.0 contains wavelength conversion functionalities including full, share-per-node, share-per-link, and hybrid wavelength conversion.

In our experiments [16], we studied the effects of various network parameters on the blocking probability, using the SICOM (Simulator for Connection Management [16]) based on SIMON.

Parameter	Value
Network Topology	Pacific Bell
Number of Calls	100000
Number of Wavelengths	8
Wavelength Assignment	FIRST AVAILABLE ON LINK
Routing Algorithm	Static Shortest Hop Path
Source-destination	Uniform distribution
Inter-arrival time distribution	Exponential distribution
Service time distribution	Exponential distribution

Table 1: Constant parameters used in the experiments.

3.2 Blocking Probability

Blocking Probability is defined as the ratio of the number of blocked connection requests to the total number of connection requests. It is a critical factor to evaluate the efficiency of WDM networks.

We presented the results from a few experiments to study the effects of various parameters on the blocking probability. Some parameters were kept constant in all of the experiments. These parameters are listed in Table 1. Other parameters were varied according to the experiments. In Table 2, we showed the values for the changing parameters. For each experiment, we varied the network-wide load between 20 and 50 Erlangs as shown in Table 2. The load is computed as the ratio of the arrival rate to the service rate. In all of the experiments, the mean service time is held constant at 1 second. The service rate, denoted by μ is computed as the inverse of the mean service time. Also, the arrival rate, denoted by λ, is the inverse of the mean inter-arrival time. The processing time refers to the time each node spends in processing the control data such as checking its Management Information Base (MIB) and its routing tables. The switching time refers to the time it takes for each node to configure its optical devices. The propagation delay refers to the time taken for the propagation of control data in the optical fiber.

We run the experiments for the Pacific Bell mesh network topology shown in Figure 1. The network consists of 15 nodes with an inter-node distance of 100 km. The mesh network is a set of interconnected rings as is typical of telecommunication networks. In all of our experiments, we as-

Parameter	Value
Network-wide Load (in Erlangs)	20, 25, 30, 35, 40, 45, 50
Transmission Time (in ms)	0, 0.5
Processing Time (in ms)	0, 0.01, 0.1, 1, 10, 20, 30, 40, 50
Switching Time (in ms)	0, 0.0005, 0.5, 1, 5, 10, 20, 30, 40, 50
Propagation Delay Parameter (ρ)	0, 1

Table 2: Variable parameters used in the experiments.

sume a uniform distribution of source-destination pair for each call and an exponential distribution of both inter-arrival time between calls and service time for each call. Also, we assume the nodes have full wavelength conversion. This means that every node is capable of converting any wavelength to any other wavelength on any of its links. The number of wavelengths is set to 8 in all of the experiments. For wavelength assignment, we use the FIRST AVAILABLE ON LINK scheme. We also use the shortest hop path method to route the calls. In all of the experiments, we measured the blocking probability of the network against the offered load.

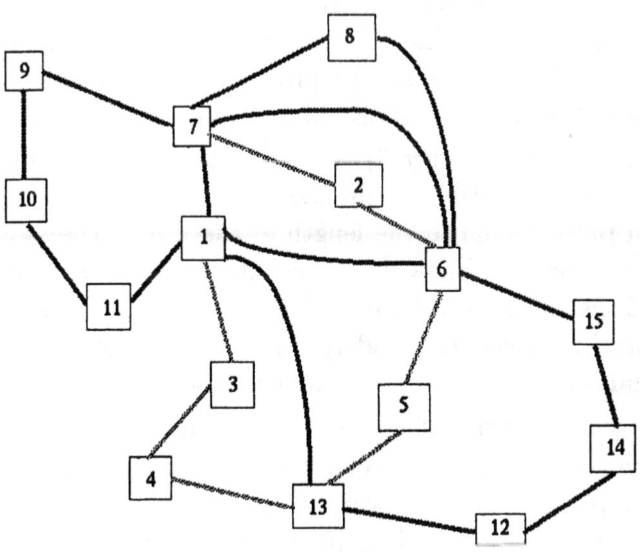

Figure 1: Pacific Bell mesh network of 15 nodes.

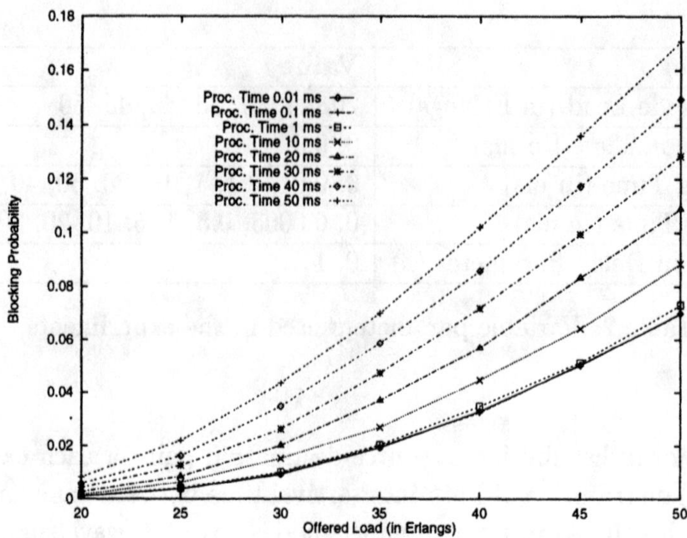

Figure 2: Blocking probability vs. load for the mesh network (Pacific Bell) for various processing times (Experiment 1).

In Experiment 1 (see Figure 2), we assume the following: transmission time of control data on every link is 0.5 ms, switching time at every node is 0.5 ms and propagation delay is proportional to the length of the links. We plot the blocking probability of the mesh network (Pacific Bell) vs. the offered load. For different plots, the processing time assumes the values: 0.01 ms, 0.1 ms, 1 ms, 10 ms, 20 ms, 30 ms, 40 ms and 50 ms.

In Experiment 2 (see Figure 3), we hold the following parameters constant: transmission time is 0.5 ms, processing time is 0.1 ms and propagation delay is again proportional to the length of the links. The switching time is varied; it assumes the following values: 0.5 μs, 0.5 ms, 1 ms, 5 ms, 10 ms, 20 ms, 30 ms, 40 ms and 50 ms. Note that the switching times of the proposed optical switches vary widely from the scale of nanoseconds to a few milliseconds depending on the technology used [2].

We notice that the processing times have a small effect on the blocking probability when their values are relatively small compared to the absolute value of the mean service time. For instance, for the processing times values of 1, 0.1 and 0.01 ms compared to 1000 ms for service time, we obtained almost the same blocking probability although the processing times varied by a multiplicative factor of 10. For the processing time values of 10, 20, 30, 40 and 50 ms, the blocking probability increased considerably although the

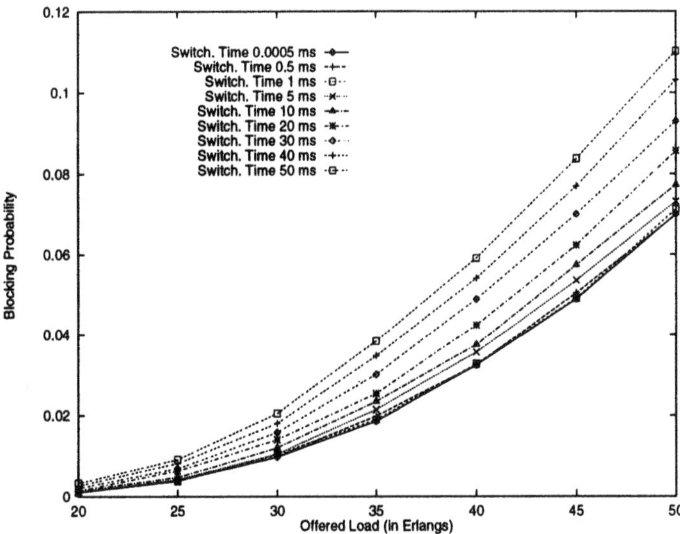

Figure 3: Blocking probability vs. load for the mesh network (Pacific Bell) for various switching times (Experiment 2).

processing times were increased by a factor of 10 each time. We noticed the same happening for switching times: the blocking probability is not affected much as long as the switching time remained relatively small compared to the mean service time. At a switching time of 50 ms, the blocking probability was significantly higher.

3.3 Throughput

Besides blocking probability, the throughput, the amount of data transferred in a time unit, is another typical factor used to evaluate the network performance. In [18], the effects on throughput under different network settings are evaluated.

The authors found that in all the cases in their experiments, the dropping scheme with the SIR has a slightly higher throughput than the holding scheme with SIR. And, the parallel reservation with SIR performs better than sequential reservation with SIR in a network with small propagation delay, but is worse in a network with large propagation delay.

In general, the DIR protocols have two useful features compared with SIR one. One is that it can reduce the bandwidth wasted and the effect is proportional to the propagation delay. Secondly, it could improve the network performance by making a better wavelength selection based on the wavelength usage information gathered from each node along the lightpath.

4   Conclusion

In this chapter, we studied different connection management methods in wavelength routed optical WDM networks. We compared different implementation on connection management, such as the distributed and centralized methods, PM and LM and in-band and out-of-band signaling. Different RWA algorithms were discussed and compared with one another. Tradeoffs between each method was given in detail. We also introduced the results from some experiments to observe the effect of different parameters on network performance, measured by blocking probability and network utilization.

In order for wavelength-routed optical WDM networks to become commercially viable, network management and control functions, including connection management, need to be addressed as an integral part of the networking architecture. The carrier sub-working group of the OIF (Optical Internetworking Forum) architecture working group is working on designing a framework and requirements for the User-Network Interface (UNI) and other interfaces to the Optical Layer Control Plane. IETF is also working on combining traffic engineering control with optical crossconnects [21]. Ori Gerstel in [19] proposed a hybrid method, combining both distributed and centralized signaling methods. Nasir Ghani, in [20], discussed a framework using Lambda-labeling switching technology in wavelength-routed optical WDM networks.

In this chapter, we did not attempt to cover all the issues in control and management of optical networks. Besides connection management, network survivability (fault management) and automatic topology discovery (configuration management) are key points for deployment of optical WDM networks. We refer interested readers to [22] and [24] for more information on control and management of optical networks.

References

[1] C. D. Chaffee, *The Rewiring of America: The Fiber Optics Revolution.* Orlando, FL: Academic Press, Inc., 1987.

[2] M. S. Borella, J. P. Jue, D. Banerjee, B. Ramamurthy, and B. Mukherjee, "Optical components for WDM lightwave networks," *Proceedings of the IEEE*, vol. 85, pp. 1274–1307, Aug. 1997.

[3] B. Mukherjee, *Optical Communication Networks*.   New York, NY: McGraw-Hill, 1997.

[4] R. Ramaswami and K. N. Sivarajan, *Optical Networks: A Practical Perspective*. San Francisco, CA: Morgan Kaufmann Publishers, Inc., 1998.

[5] J. P. Ryan, "WDM: North American deployment trends," *IEEE Communications Magazine*, vol. 36, pp. 40–44, Feb. 1998.

[6] E. Lowe, "Current European WDM deployment trends," *IEEE Communications Magazine*, vol. 36, pp. 46–50, Feb. 1998.

[7] *IEEE Communications Magazine*, Special section on Deployment of WDM Fiber-Based Optical Networks, vol. 36. Feb. 1998.

[8] P. E. Green, Jr, *Fiber Optic Networks*. Englewood Cliffs, NJ: Prentice Hall, 1993.

[9] C. A. Brackett, "Dense wavelength division multiplexing networks: Principles and applications," *IEEE Journal on Selected Areas in Communications*, vol. 8, pp. 948–964, Aug. 1990.

[10] J. P. Laude, *Wavelength Division Multiplexing*. Prentice Hall, 1993.

[11] I. Chlamtac, A. Ganz and G. Karmi, "Lightpath Communications : An Approach to High Bandwidth Optical WAN's," *IEEE Trans. on Commun.,,* vol. 40, pp. 1171-1182, Jul. 1992.

[12] I. Chlamtac, A. Ganz and G. Karmi, "Purely Optical Networks for Terabit Communication," *IEEE INFOCOM*, 1989.

[13] S. Subramaniam and R. Barry, "Wavelength Assignment in Fixed Routing WDM Networks," *IEEE Int'l Conf. Commun.*, pp. 406-410, Jun. 1997.

[14] E. Karasan and E. Ayanoglu, "Effects of Wavelength Routing and Selection Algorithms on Wavelength Conversion Gain in WDM Optical Net-works," *IEEE/ACM Trans. on Networking*, vol. 6, no. 2, pp. 186-196, Apr. 1998.

[15] J. Yates, J. Lacey and M. Rumsewicz, "Wavelength Converters in Dynamically Reconfigurable WDM Networks," *IEEE Communications Surveys*, 2nd quarter issue, 1999.

[16] E. Sawma, "Connection Management in Wavelength-routed Optical Networks", (MS Project Report), Department of Computer Science and Engineering, University of Nebraska-Lincoln, 2000.

[17] B. Ramamurthy, D. Datta, H. Feng, J.P. Heritage, and B. Mukherjee, "SIMON: A Simulator for Optical Networks", in *All Optical Networking, 1999: Architecture, Control and Management Issues* (J.M. Senior, C. Qiao, and S.Dixit, eds.), vol 3843, pp. 130-135, Proceeding of SPIE, Sep. 1999.

[18] Y. Mei and C. Qiao. "Efficient Distributed Control Protocols for WDM All-Optical Networks," In Proceedings of ICCCN, pages 150–153, 1997.

[19] O. Gerstel, "Optical Layer Signaling: How Much is Really Needed?", in *IEEE Communications Magazine*, Oct. 2000.

[20] N. Ghani, "Lambda-Labeling: A Framework for IP-Over-WDM Using MPLS," in *Optical Networks Magazine*, Apr. 2000.

[21] O. Awduche, Y. Rekhter, J. Drake and R. Coltun, "Muti-Protocol Lambda Switching: Combining MPLS Traffic Engineering Control With Optical Crossconnects," IETF Draft draft-awduche-mpls-te-optical-02.txt, Jan. 2001.

[22] M. Maeda, "Management and Control of Transparent Optical Networks", in *IEEE Journal on Selected Areas in Communications*, vol 16, no. 7, Sep. 1998.

[23] B. Ramamurthy and B. Mukherjee, "Wavelength conversion in WDM networking,", in *IEEE Journal Selected Areas in Communications*, vol 16 No. 7, pp. 1061–1073, Sep. 1998.

[24] N. Ghani, "Integration Strategies for IP over WDM," Optical Networks Workshop, Richardson, Texas, 2000.

[25] R. Ramaswami and A. Segall, "Distributed network control for optical networks," *IEEE/ACM Transactions on Networking*, vol. 5, no. 6, pp. 936–943, Dec. 1997.

[26] B. Ramamurthy, *Design of Optical WDM Networks: LAN, MAN and WAN Architectures*. Boston, MA: Kluwer Academic Publishers, Dec. 2000.

[27] H. Zang, L. Sahasrabuddhe, J. P. Jue, S. Ramamurthy, and B. Mukherjee, "Connection management for wavelength-routed wdm networks," in *Proceedings, IEEE GLOBECOM '99*, (Rio de Janeiro, Brazil), Dec. 1999.

OPTICAL NETWORKS - RECENT ADVANCES
L. Ruan and D.-Z. Du (Eds.) pp. 205 - 270
©2001 Kluwer Academic Publishers

Multicast Routing in WDM Optical Networks

N. Sreenath
Department of Computer Science and Engineering
Indian Institute of Technology, Madras, India 600036
E-mail: `sreen@hpc.iitm.ernet.in`

C. Siva Ram Murthy
Department of Computer Science and Engineering
Indian Institute of Technology, Madras, India 600036
E-mail: `murthy@iitm.ernet.in`

G. Mohan
Department of Electrical and Computer Engineering
National University of Singapore, Singapore 119260
E-mail: `elegm@nus.edu.sg`

Contents

1 Introduction

Traditional communication models have been one-to-one or *unicast*, and one-to-all or *broadcast*. Among these two extremes lies *multicast* which refers to the targeting of a data stream to a selected set of receivers. This model is used to characterize the communication patterns in a wide spectrum of applications such as replicated databases, command and control systems, distributed games, audio/video conferencing, and distributed interactive simulation. The following is a list of some applications that make use of multicast communication.

- *Multimedia*: A number of users "tune in" to a video or audio transmission from a multimedia source station.

- *Teleconferencing*: A group of workstations form a multicast group so that a transmission from any member is received by all other group members.

- *Databases*: All copies of replicated file or database are updated at the same time.

- *Distributed computation*: Intermediate results are sent to all participants in a distributed computation.

- *Real-time workgroups*: Files, graphics, and messages are exchanged among active group members in real time.

A good communication network which provides a multicast transmission system should have the properties like *high probability of delivery of information, low delay between source and destination* and *information hiding from the intermediate routers*. These properties can be achieved by using optical signals instead of electrical signals to transfer the information. This optical medium will also provide enormous bandwidth (tens of tera-Hertz). It is very difficult to exploit such a huge bandwidth as a single channel. Hence, the complete bandwidth is *channelized* using different wavelengths.

There are many possibilities to realize connectivity in multi-wavelength optical networks. The connectivity depends on the capabilities of network nodes. Based on the level of controllability in the network nodes, multiwavelength optical networks are classified as follows.

1. Broadcast-and-select networks

2. Linear lightwave networks

3. Wavelength routed networks

In the following sections the multicast routing in WDM networks is discussed. broadcast-and-select networks [1] and linear lightwave networks (LLNs) [2] are inherent choice for multicasting. However, in broadcast-and-select networks the limitation on number of wavelengths in a fiber and the receiver tuning latency make the multicast routing problem as a research issue. In Section 2, different techniques are discussed to overcome the above mentioned problems. In linear lightwave networks, the constraints are waveband inseparability and mutually independent source combining. Multicast routing problem in linear lightwave networks is addressed in Section 3. Nowadays, wavelength routed networks (WRNs) [3] are becoming popular. To support multicasting, the wavelength routed nodes should have the capability of optical splitting. If it is assumed that every node in the network has optical splitting capability then the problem of multicast routing boils down to the problem of multicast routing in conventional electronic networks. Due to their complex architecture split capable nodes are costlier than wavelength routing nodes without split capability. Hence, it is suggested that only a few nodes in the network are allowed to have splitting capability. The multicast routing problem in wavelength routed networks is addressed in Section 4. Recently, IP-over-WDM are receiving considerable amount of attention by the researchers. Routing in IP is based packet switching concept. However, wavelength routed WDM networks which are used for WAN environment are based on circuit switching concept. As said above, the splitting capability at nodes is useful to optically (power) split the incoming message. These capabilities of the nodes may not be considered by the existing IP multicast routing protocols. The issues related to multicasting in IP-over-WDM networks are discussed in Section 5. Finally, the summary of this chapter is given in Section 6.

2   Broadcast-and-Select Networks

In a broadcast-and-select network, the connectivity among the nodes is made in such a way, that the optical signal which is introduced into the network is broadcast to all nodes. It means that the signal propagates to all nodes in the network. This will avoid the possibility of reusing the optical spectrum. A broadcast-and-select network can be designed by connecting nodes using two way fibers to a passive star coupler. (An $n \times n$ star coupler combines all n inbound signals and broadcasts them on each outbound fiber.) When a node needs to transmit data, it tunes one of its tunable transmitters to a free wavelength and sends the data on its outgoing fiber. The passive star receives the data and broadcasts it to all nodes connected to it. Any node which wants to receive the data, will tune its receiver to the proper wavelength at the proper time. Thus, the broadcast function permits the complete optical connectivity among all pairs of nodes. Hence, broadcast star becomes a natural selection for creating multicast connections, by more than one node tuning its receiver to the same transmitter wavelength.

However, the constraint here is that two transmitters should not use the same wavelength at the same time. This is called *distinct channel assignment constraint*. In a large network, where the multicast connections are more in number, dedicating a wavelength channel is prohibitively expensive, and may exhaust the available optical spectrum. Hence, multiple access methods namely, *time division multiple access* (TDMA), *wavelength division multiple access* (WDMA), *sub-carrier multiple access* (SCMA), and *code division multiple access* (CDMA) are required to be employed through which a high logical connectivity can be achieved. In this, TDMA is a special case of WDMA, where TDMA operates on a single wavelength channel and WDMA operates on more number of wavelength channels. In SCMA, an additional level of modulation and demodulation is done electronically using microwave sub-carriers. In these three methods (TDMA, WDMA, and SCMA), the signals accessing a common medium are non-overlapping in time or in frequency domain and can be considered as orthogonal. In the same sense, CDMA is quasi-orthogonal, signals from different channels are approximately separable at the receivers. The idea in CDMA is to encode each bit in the data into a unique waveform. Every logical connection is allocated with a unique waveform.

2.1 Multicast Routing in Broadcast-and-Select Networks

In broadcast-and-select networks, the main constraint to provide connections for a multicast session is the availability of the receivers and the transmitters. If receivers are fixed at a wavelength, then each node requires a number of receivers equal to the number of wavelengths on a fiber. Hence, in [4, 5, 6, 7, 8] it is assumed that the tunable receivers and tunable transmitters are present in the network. But, in a large network, a multicast session with more number of destinations requires availability of a large number of receivers. A multicast request will be blocked even if one of the required receivers is not available. As a consequence of this, the throughput (*throughput* is defined as the ratio between number of requests blocked and number of requests generated) will reduce. In order to achieve reasonable throughput different methods are proposed in [4, 5, 6, 7, 8]. In [4], it is proposed to limit the number of destinations in a multicast session. Intuitively, if the number of destinations in a session are less, then it is highly probable that all the required receivers are available. As a consequence of this, the number of requests that is blocked will be less. Hence, the throughput will be high.

In [5], a multicast routing algorithm is presented. This algorithm mainly aims at finding the time at which all receivers become free. The multicast data is transmitted, only if all receivers are free. Here, the algorithm checks whether the transmission is possible immediately after the transmitter becomes free. If channel is not free, then the transmission is postponed till the channel becomes free. Once the channel becomes free, then it checks whether all receivers are available. It schedules the transmission when all receivers become free.

In the above methods [4, 5], optical bandwidth may not be used efficiently. The optical bandwidth can be time slotted, so that every node may transmit only in slots specified by a predetermined schedule. Using this method, better utilization of optical bandwidth can be achieved. Based on the above method, a multicast routing protocol is proposed in [6]. Here, the available bandwidth is divided into different time slots, and schedules for different traffic patterns (unicast, multicast, broadcast) are prepared. The multicast routing using this method is discussed in Section 2.2.

The above multicast routing algorithms [4, 5, 6] are assumed to have a single transmission of a message to all destinations belonging to a multicast group. If all receivers are not available at the same time, then some receivers need to wait for significant amount of time before receiving the message. This waiting time can be reduced by allowing the source to transmit the same message more than once. In [7], the receivers are grouped according to their available time so that a schedule can be prepared for transmitter and receivers. The advantages of partitioning the group of receivers and the scheduling of receivers is dealt in Section 2.3.

The above mentioned algorithms [4, 5, 6, 7] do not consider the tuning latency of the receivers. In general, tunable receivers take a non-negligible amount of time to switch between channels. In [8], a new method of multicast routing is proposed which is based on the concept of *virtual receiver*. The multicast routing based on virtual receiver concept is discussed in Section 2.4. A virtual receiver is a set of physical receivers that behave identically in terms of tuning. Using this virtual receiver concept, the original network with multicast traffic is transformed into a new network with unicast traffic. As a consequence of it, the scheduling algorithms that are devised for unicast traffic can work for scheduling of multicast traffic.

2.2 Time Slot Scheduling

In [6], a multicast routing algorithm using TDMA scheduling is described. Here, it is assumed that the network is operated in a slotted mode with a slot time equals to the packet transmission time plus the tuning latency. The slots which are allocated to transmit messages by a node are fixed. Hence, every node needs to buffer the message until its turn comes. The buffer at each node is organized into a set of queues. Each queue is associated with one of the $N - 1$ (where N is the number of nodes in the network) destinations. A *transmission schedule* indicates the slot at which a node can transmit. Based on whether traffic is unicast, broadcast, or multicast which is scheduled to a particular slot, the tuning of transceivers at corresponding nodes are planned. If it is an *unicast slot* (at which unicast message is planned to send), then only corresponding transmitter at the source node and receiver at the destination node are tuned to particular wavelength. Totally W s-d pairs can communicate at the same time, where W is the number of wavelengths available on a fiber. If a slot is a *broadcast slot* (at which broadcast message is planned to send), then only one node is allowed to transmit on a particular wavelength and all other nodes are tuned to receive. If a slot is a *multicast slot* (at which multicast message is planned to send), then one node is allowed to transmit and a set of nodes are tuned to receive. This paper [6] provides heuristics to schedule multicast messages on to different types of slots to achieve better channel utilization. Broadcast slots can also be made use to transmit multicast messages when multicast group includes more number of nodes. However, if the group size is a small fraction of the nodes in the network, then using broadcast slots for multicast messages will yield low bandwidth utilization. At the same time, if the group consists of a few destinations it is better to make use of unicast slots for transmitting the multicast messages. Hence, depending on the group size and the length of the session a decision involving the slot selection to transmit multicast message can be taken.

1. If multicast sessions are short and group involving is large, then both unicast slots and broadcast slot can also be scheduled to transmit multicast messages. The scheduling applied to this kind of situation is as follows. First, find the optimized schedules for individual traffic (for unicast traffic and broadcast traffic). Then, these two schedules are merged appropriately, in such a way that the resultant schedule satisfies the requirements of both traffic.

2. If sessions are for short duration and only few nodes are involved, then scheduling unicast slots for multicast transmission may give better bandwidth utilization. This type of scheduling for multicast connection can be viewed as scheduling many unicast connections.

3. If sessions are long then both unicast and multicast slots can be scheduled to transmit the multicast messages. To use multicast slots for multicast traffic, all the destination nodes need to be synchronized, so that all the nodes are tuned to the same wavelength. Here, only the nodes which belong to the multicast group are tuned.

To evaluate the usefulness of above algorithms, two performance metrics are defined, one is *wavelength throughput* (T) and the other one is *channel efficiency* (E). Wavelength throughput is defined as the average number of packets transmitted per unit amount of time (slot). This metric indicates the average number of wavelength channels used for communication. Second metric, channel efficiency is defined as the average number of times a packet is transmitted before it is received by all members of a multicast group. The value of this metric should be low for an efficient algorithm. Hence, an algorithm which is efficient with respect to the wavelength throughput may be inferior with respect to the channel efficiency. The third metric *multicast throughput* (D) can be defined [8] to reflect both wavelength throughput and channel efficiency. Multicast throughput is defined as $D = T/E$.

If a separate packet is transmitted to an individual destination then G number of repeated transmissions are required for the completion of a multicast transmission of a packet to all the receivers. Here G is defined as the number of members in a multicast group. This results in a low D value. Another way of sending packets is to schedule all the receivers of a multicast group in such a way that they simultaneously receive the multicast packet. This method results in an efficient multicast transmission. However, the transmissions to different multicast groups with common destinations can not be scheduled simultaneously. As a consequence of this, the wavelength throughput becomes low, and in turn multicast throughput becomes low. To achieve better performance a method is proposed in [7], which partitions the network into small groups and transmits a packet in multiple phases. The problems in this approach are a) tuning coordination among the receivers when there is non-negligible tuning latency, and b) the optimal splitting which is an NP complete problem. To overcome the above problems and to achieve better performance another technique is proposed in [8].

2.3   Partitioning the Receiver Group

In [7], it is shown that transmitting a multicast message multiple times to the subsets of a multicast group is more efficient than transmitting it a single time to the entire multicast group. Here, the multicast group is partitioned as many subgroups. One subgroup receives the multicast message earlier than the other subgroup. The members of the subgroup which have not received the message, require a separate transmission of same message by the source. The subgroup which receives the first message earlier than the other subgroup needs to wait for the next message. The objective is to minimize the average receiver waiting time, which is defined as the difference between the time that the receiver actually begins receiving a message and the time that the receiver becomes free to receive a message. By minimizing the average receiver waiting time, it is possible to achieve better receiver utilization. The low average receiver waiting time can be achieved by properly dividing a multicast group into subgroups. Hence, the actual problem boils down to the problem of partitioning the multicast group into subgroups. In this paper [7], some heuristics are provided to partition a multicast group. It is assumed that the time at which each receiver becomes free is known before partitioning the group into subgroups. And as the scheduling is done on-line, the proposed heuristics should be fast and simple. The heuristics are given below.

1. **Earlier available receiver** (EAR): This heuristic greedily schedules the first transmission as and when a receiver becomes available. If any of the receivers becomes available in the middle of the first transmission, then the next transmission starts immediately after the first transmission. Otherwise, the next transmission is scheduled as and when another receiver becomes available. The running time of this heuristic is $O(G)$, where G is the number of nodes in the multicast group. Consider the following example. Assume that there are 5 nodes in a multicast group numbered n_1 through n_5 and the schedule period is three units of time. Receivers at nodes n_1 through n_5 are ready at $t_1, t_2, t_8, t_9, t_{10}$ respectively. If EAR heuristic is applied to schedule the transmissions then n_1 transmits at t_1, n_2 transmits at t_4, n_3 transmits at t_8, n_4 and n_5 transmit at t_{11}. This is because, n_2 becomes ready when a schedule for n_1 is on. Hence, after the first transmission, a second transmission for n_2 is scheduled. In a similar way, a transmission for nodes n_4 and n_5 are scheduled after the transmission of n_3.

2. **Latest available receiver** (LAR): This heuristic first fixes the schedule of the latest receiver. Then it considers the next receivers which are ready just before the last receiver. If the schedule of this receiver conflicts with the previous scheduled receiver, then the present receiver

is also scheduled along with the previous receiver. If LAR heuristic is applied to the previous example, then first a schedule for node n_5 is fixed at t_{10}. If the transmission for node n_4 is scheduled at t_9, then this schedule conflicts with the previous schedule for n_5. Hence, n_4 is also scheduled along with n_5. Similarly, n_3 is also scheduled along with n_5. Similarly, n_2 and n_1 are scheduled at t_2. The running time for this heuristic is $O(G)$, where G is number of nodes in the multicast group.

3. **Best available receiver** (BAR): By considering every receiver this heuristic prepares a schedule for all receivers. When it considers receiver i, it is scheduled at the time when it becomes free. The receivers which are ready before receiver i are scheduled using LAR heuristic and the receivers which are becoming ready after i are scheduled using EAR heuristic. For this, the average waiting time is calculated. This process is repeated for all values of i. The schedule which gives lower average time is chosen. In the above example, for $i = 2$, n_2 and n_1 are scheduled at t_2, n_3 is scheduled at t_8, and n_4 and n_5 are scheduled at t_{11}. This heuristic's running time is $O(G^2)$, where G is the number of nodes in the multicast session.

In the simulation study, these three heuristics are compared using the metrics *average waiting time* and *average number of transmissions*. Among all three heuristics BAR performed well. EAR has more average waiting time than the other heuristics, and LAR offers a value in between BAR and EAR. Number of transmissions required is another metric to evaluate the performance of the above three heuristics. LAR requires less number of transmissions than other heuristics. BAR provides next to best, and it is same as the number of transmissions required for optimal scheduling, where exhaustive search is carried out to get a schedule with minimum waiting time. EAR requires more number of transmissions among all three heuristics.

2.4 Virtual Receiver

In [8], a technique for the multicast packet transmission is introduced based on the *virtual receiver* concept. A set of physical receivers that behave identically in terms of tuning is called as a virtual receiver. This concept is useful in a situation where tuning of receivers takes a non-negligible amount of time and different multicast groups may have some receivers in common.

The partitioning of physical receivers into virtual receivers transforms the the network with multicast traffic into a network with unicast traffic. As a consequence of it, the scheduling algorithms used to schedule the unicast traffic can be used to schedule multicast traffic. This is possible, because a set of destinations are considered as a single virtual receiver. For a given multicast traffic demand matrix, the aim here is to select a partition of receivers into virtual receivers so as to maximize multicast throughput. It is assumed that every node is equipped with a fixed transmitter and with a tunable receiver. As the division of receivers into a set of virtual receivers is an NP complete problem, heuristics are provided for virtual receiver set problem (VRSP). These heuristics make use of two operations *join* and *split*. Join operation specifies union of two sets into a single set, and split operation specifies the splitting of a set into two sub-sets. The first heuristic is *Greedy-join*, where the sets of receivers are greedily grouped to a virtual receiver set and tries to minimize the average number of repeated transmissions. The second heuristic is *Random-join*, where the receivers randomly join to form a set of virtual receivers. Similarly, the third heuristic is *Greedy-split*, which greedily splits the set of receivers into virtual receivers and fourth heuristic is *Random-split* which randomly splits the set of receivers into a set of virtual receivers. The technique based on the virtual receiver concept is guaranteed to give better performance as the decision of grouping physical resources into virtual receivers considers the total multicast traffic demand.

3 Linear Lightwave Networks

The optical spectrum on a single fiber is wide enough to carry many high speed connections on distinct wavelengths and each of them is called as a *wavelength channel* as shown in Figure 1(a). The assigned wavelengths must be spaced apart to avoid overlapping from the neighboring wavelength. Previously, due to the optical node technology, it was suggested that the channel spacing could be 100GHz. The bandwidth used for each wavelength channel is on the order of 1GHz. Hence, the total bandwidth used in the optical spectrum would be very less compared to the overhead of spacing in between two consecutive wavelength channels. But, the wider spacing (100GHz) of wavelength channels permits each wavelength to be routed individually. This is referred as *wavelength-selective routing* or *wavelength-routing*. The bandwidth allocated for each wavelength channel and the spacing between two consecutive channels is shown in Figure 1(b).

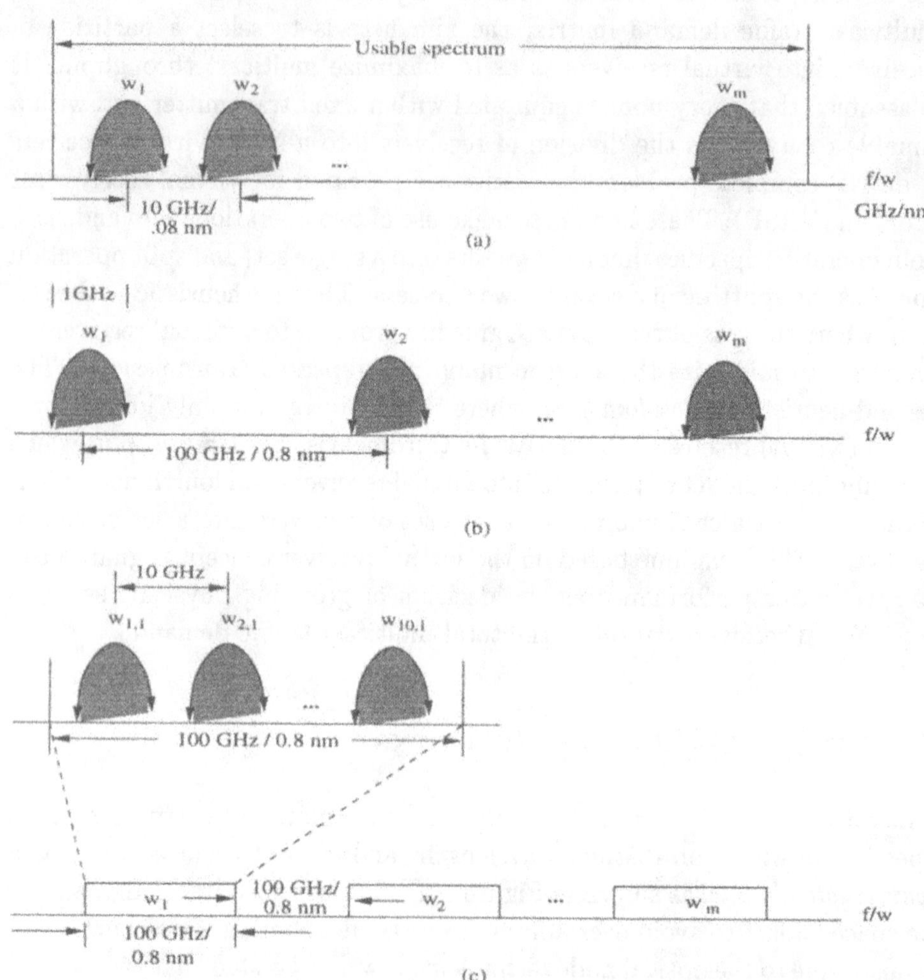

Figure 1: (a) Usable optical spectrum. (b) Wavelength channel spacing for wavelength routing. (c) Wavelength channel spacing for waveband routing.

To make use of optical spectrum more efficiently, a two-tier spectrum partition was proposed. First, the spectrum is partitioned into a coarser *waveband*. Each waveband is further divided into several wavelength channels. Each waveband can be routed independently at each network node. This is referred as *waveband selective routing* or *waveband-routing*. Each waveband can have the bandwidth of 100GHz and placed at the intervals of 100GHz. Figure 1(c) illustrates the spacing of wavebands. The network with the nodes having capability of switching individual wavebands is called as *waveband-routed networks*. Waveband-routing imposes a routing constraint that all wavelength channels sharing a common waveband "stick together" on the same optical path.

The optical nodes in LLN make use of *waveband-selective linear divider and combiners*. The discussion on waveband-selective switch is given below.

3.1 Waveband-Selective Switch

Figure 2 shows the realization of waveband-selective switch, to which there are n input and n output fibers. Each fiber assumed to carry m wavebands. There are three stages in this implementation. First stage corresponds to demultiplexing and third stage corresponds to multiplexing. Second stage corresponds to switching operation where it contains m switching elements, each one is operating on signals in one waveband. Although there are several wavelength channels in one waveband, the switch considers a waveband as the smallest recognizable entity. The switch independently controls each waveband. The switched signals are combined onto the output fiber. The switch can be a *linear divider and combiner* (LDC) which is useful in directing the input signal to more than one output fiber. This operation is called as *splitting* and is useful in multicast communications.

Linear Divider and Combiner

A 4×4 LDC is shown in Figure 3. Here, a 2×2 controllable coupler operates as power divider and combiner. Optical power which enters the coupler through the input fibers, is divided linearly. This divided power is combined at the combiner and leaves from the output fiber. By setting the coupler states, it is possible to vary the dividing ratio (δ_{ij}) and combining ratio (σ_{ij}). In Figure 3, δ_{ij} denotes the fraction of power from input port j directed by its divider to the combiner at output port i. σ_{ij}, denotes the fraction of power received from the divider serving input port j and combined onto output port i. The maximum power transfer between i and

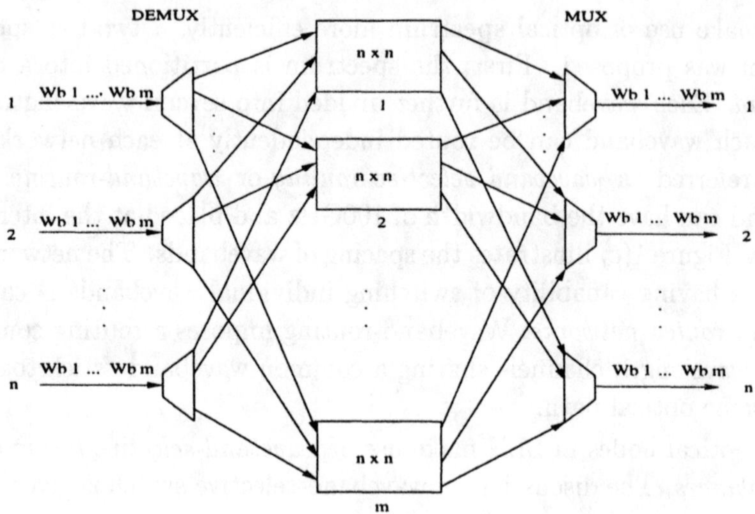

Figure 2: Architecture of waveband switch.

j is given by the factor a_{ij} using the following equation.

$$a_{ij} = \delta_{ij}\sigma_{ij} \tag{1}$$

In the ideal case of lossless couplers, the ratios are subjected to the following constraints.

$$\sum_i \sigma_{ij} = 1 \tag{2}$$

$$\sum_i \delta_{ij} = 1 \tag{3}$$

The LDCs are useful in controlling optical routing, combining and splitting. The splitting function of LDC provides multicasting of optical power from a single inbound fiber to several outbound fibers.

Properties of LLN

LLN has the following properties as it uses linear devices like LDCs for switching purposes [2].

1. Non-linear operations like frequency conversion, regeneration, and buffering are not permitted to perform on the optical signal between a source and destination.

2. Signals can be combined (linearly) on each line. These are termed as *interfering signals*.

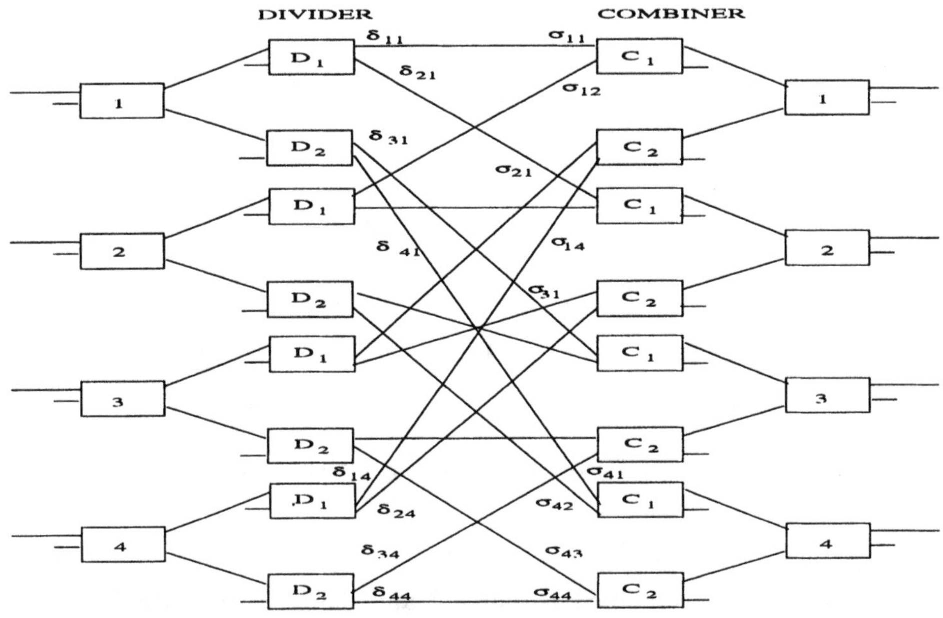

Figure 3: Linear divider and combiner.

3. As many wavelength channels exist, a multiple access scheme called *wavelength division multiple access* (WDMA) can be implemented. In such a case, the interfering signals are distinguished at the destination receivers.

3.2 Constraints on Routing Problem

Due to the above properties of LLN, the following constraints are imposed on LLN routing problem.

1. *Channel continuity*: Same wavelength is to be allocated between a source and a destination.

2. *Distinct channel assignment*: Two different connections sharing a link must be assigned distinct wavelengths.

3. *Inseparability*: The interfering signals can not be separated within LLN. This is because, LDCs operate on aggregate power (waveband) of an input fiber rather than on the individual wavelengths. Figure 4 describes the inseparability constraint. Node 1 is connected to node 4 on wavelength channel w_1 via routers A and B. Similarly, node 2 is

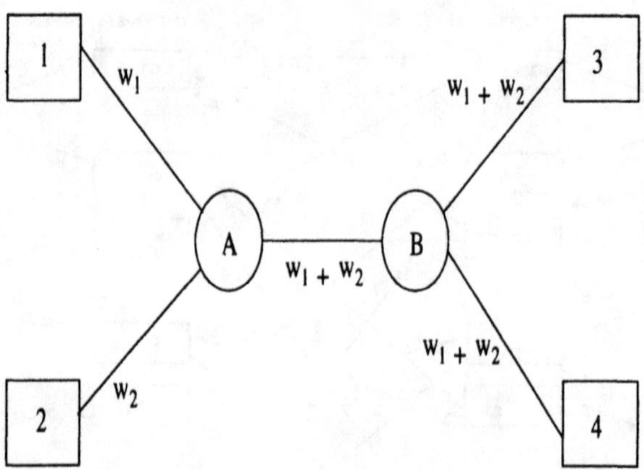

Figure 4: Illustration of wavelength inseparability.

connected to node 3 on wavelength channel w_2 via routers A and B. w_1 and w_2 are combined at router A and they together switched onto link $< A - B >$. Router B can not distinguish between w_1 and w_2 and it switches both wavelengths onto links $< B - 3 >$ and $< B - 4 >$.

4. *Mutually independent sources combining* (MISC) condition: Signals from mutually independent sources can only be combined onto the same fiber. It means, a signal is not allowed to split to take multiple paths in the network and then recombine on another link. Figure 5 illustrates the MISC constraint. Node 1 is connected to node 5 on wavelength channel w_1, node 2 is connected to node 4 on wavelength channel w_2 and node 3 is connected to node 6 on wavelength channel w_3. At router A, both signals (w_1 and w_2) are combined and placed onto link $< A - B >$, and this signal is divided at router B. As a consequence of it, both links $< B - C >$ and $< B - D >$ carry the same signal. At router D wavelength w_3 is combined with w_1 and w_2. Node 4 receives w_1, w_2, and w_3 and extracts signal on wavelength channel w_2. But, router E forwards the combined signal onto link $< E - F >$ and router F in turn switches the same onto link $< F - G >$. At router F, the signal on link $< C - F >$ and the signal on link $< E - F >$ are combined which is a violation of MISC condition. This results in a source interfering with itself, thereby garbling its information.

5. *Color clash*: A color clash will occur when wavelength allocation for a new connection results in combining on the same link with two or

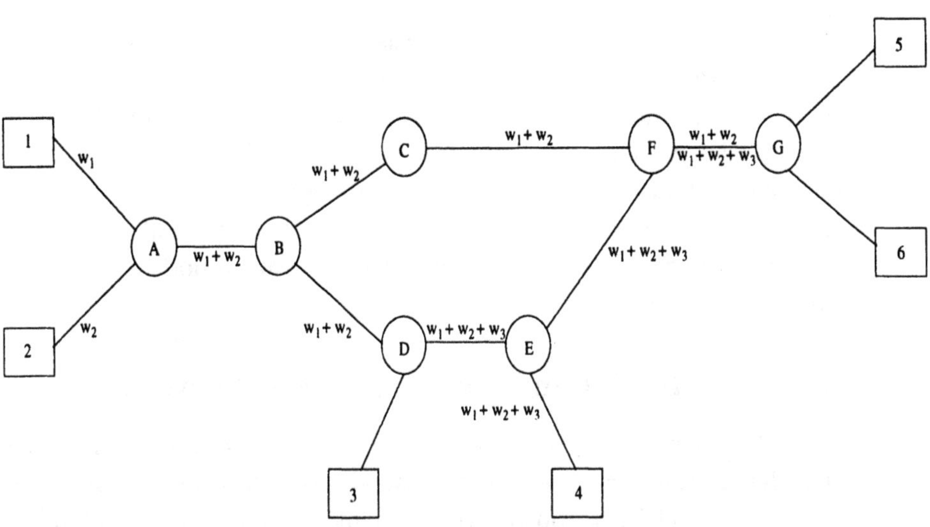

Figure 5: Illustration of mutually independent sources combining (MISC) condition.

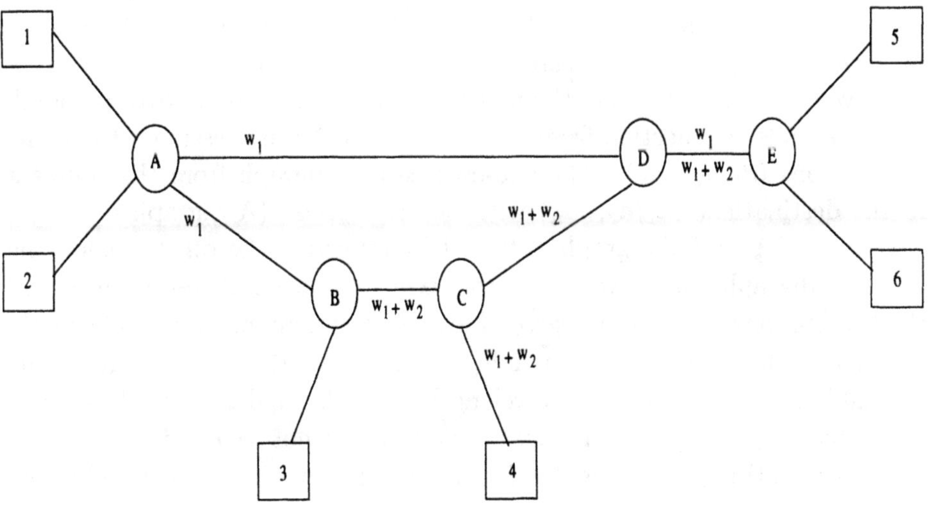

Figure 6: Illustration of color clash.

more connections for which the same wavelength is already allocated. Figure 6 illustrates a situation where color clash will occur. Here, node 1 is connected to node 5, node 2 is connected to node 4 and node 3 is connected to node 6. As the routes between node 1 and node 5 and node 2 and node 4 are link disjoint, these two are allocated same wavelength w_1 and connection between node 3 and node 6 are allocated wavelength w_2. This combined signal is added with w_1 at router D. This will violate the distinct channel assignment constraint resulting in a color clash. Hence, connection between node 3 and node 6 cannot make use of this route.

3.3 Multicast Routing in Linear Lightwave Networks

This section deals with multicast routing problem where a multicast source and a set of destinations are given. The constraints discussed in the previous section make the multicast routing problem more complicated and more calls will be blocked. In particular, with respect to the multicast routing, the signal may propagate in a wider range. As a result, it becomes very difficult to verify MISC violation or color clash. To restrict the propagation of signals to the unintended links, and to avoid color clash, the network is divided into several trees. These trees are link disjoint, thus avoiding possible MISC violation problem. A node may be present in more than one edge disjoint tree. Among these preselected trees, one tree is selected in such a way that, the selected tree has all destinations and source of the given multicast session. From the selected tree, a *sub-outtree* is found, which provides connection from the source of multicast session s to all the destinations $(d_1, d_2, ..d_n)$. The sub-outtree is a digraph from the source s to the destinations $(d_1, d_2, ..d_n)$ with no *semicycles*. (A digraph is said to have a *semicycle* if the graph obtained by removing the directions on the edges of digraph has a cycle.) In this tree, source s has in-degree 0, and the destinations have an in-degree of exactly 1 and an out-degree of exactly 0 and all other nodes have an indegree of 1. All destinations should be reachable from the source s. Wavelengths are allocated to the sub-outtree, while ensuring that the color clash condition is not violated. Thus, the multicast routing problem is divided into the following three subproblems.

1. Preselecting the edge disjoint trees in the given network, *pre-select-trees*.

2. Choosing one of the preselected trees, *tree selection*.

3. Checking for color clash and channel allocation, *channel allocation*.

3.3.1   Pre-Select-Trees

In the first subproblem, the given network is divided into edge disjoint trees with at least one spanning tree. In every tree at most one edge is allowed between any two nodes. This decomposition is done only once at the time of network design. Different approaches used for this decomposition are given below.

1. MAX trees: The given network is decomposed into the maximum number of edge disjoint spanning trees and the remaining edges are decomposed into the smallest number of edge disjoint trees. This can be done by making use of the method *partitioning of matroids* [9]. In [10], it is stated that using this method to partition the given network results in spanning trees with large diameter. As a consequence of it, each connection takes a longer path. This results in consuming more resources of the network and also results in many unintended connections. These unintended connections may interfere with some other new connections. Hence, trees with larger diameter are undesirable.

2. DENSE Trees: An alternative approach is to decompose the network into more edge disjoint densest spanning trees (trees with smaller diameter) and decompose the remaining edges into the smallest number of edge disjoint trees. This approach results in a fewer number of trees than that are obtained using MAX trees approach.

3. MAX degree DENSE trees: To achieve more number of DENSE trees a heuristic *modified breadth first search algorithm with parameter d* (MBFS-d) is used. This heuristic attempts to find spanning trees having a degree less than or equal to $d + 1$, where d is the degree of the node, i.e., number of edges incident on it.

The edge disjoint trees that are obtained using the above methods are referred to as pre-select-trees. *Tree LLNs* are built from these edge disjoint trees. This can be achieved by replacing every edge in each pre-select-tree by two fibers in both directions. For every connection request, an attempt is made to accommodate it in one of these tree LLNs. This ensures that no MISC violation can occur.

3.3.2   Tree Selection

An attempt is made to allocate a tree LLN to the given multicast session from the set of pre-select-trees. While choosing a tree, availability of transceivers at the source and at the destinations are taken care. Two methods are proposed in [10] to allocate sessions to the trees.

1. *Smallest component tree* (SCT): In this method all pre-select-trees are sorted in the ascending order of the number of nodes that are present in each tree. In the same order, the trees are verified whether all required nodes for the multicast session are present or not. First tree which satisfies the above condition is considered as probable tree for the given multicast session. The basic idea in this method is to use the smaller non-spanning components as much as possible, and to take away the load from the more utilized spanning trees.

2. *Minimum interference tree* (MIT): In this, the tree selection is based on the least amount of interference that it may create on its path from the source to the destinations. In these two methods, MIT is expected to perform better than SCT. This is because, the multicast tree constructed using MIT will affect the later sessions less than the tree constructed using SCT.

The above algorithms will not take care of dynamic addition of a node into the multicast tree, which is necessary for expanding the existing multicast tree. In MIT and SCT, a pre-select-tree which contains all nodes of a session is considered and appropriate wavelengths are allocated. If there is no such pre-select-tree, then that multicast session will be blocked. In [11], two methods are proposed to add a node to a tree. These methods consider the pre-select-trees in the sorted order of the number of nodes that are present in each tree. These methods do not consider whether the pre-select-tree contains all members of the session or not. In the chosen pre-select-tree if all members of the multicast session are not present, then this algorithm will try to include each node (say new-node) into the tree. If it is not possible to include any node then next pre-select-tree is chosen for processing. If any node of the session is not included even after processing all the trees then that particular session will be blocked. These methods can also be used to include a new node into the multicast session. The proposed methods are deferred in selecting the direction of links between new-node and the existing tree.

1. Method-1: In this method it is assumed that, at the time of link connection (disconnection), the links in both directions between the source and the destination must be joined (lapsed). Shortest paths from all the nodes of the tree to a new node (*new-node* are computed. The path which has minimum length in terms of number of hops is chosen for connecting new-node to the tree.

2. Method-2: In this method the transmission direction is considered with respect to source node s. If a source s of multicast session is not included then the shortest paths from all the nodes present in the tree to the node s are computed and the path with minimum hop length is chosen. The connection from node s to tree is formed. To add a destination which is not present in the existing tree, shortest paths from the nodes present in the tree to new-node are computed. The path with minimum hop length is chosen for the connection. While choosing the path the direction of the path is verified from the source to new-node.

It is naturally expected that, the resources required to establish the multicast session using the methods proposed in [11] will be less than that with the methods MIT and SCT. This is because, the paths are established only in required direction. As a consequence of it, less number of sessions will be blocked.

3.3.3 Channel Allocation

Wavelength channels are allocated to the chosen sub-outtree. While allocating the wavelengths, the color clash condition is taken care. To choose the wavelength, a heuristic *MIN* is applied. This heuristic chooses a wavelength which is least used among all the available wavelengths. If no wavelength is available on the present sub-outtree, then using the tree selection algorithm another sub-outtree is selected. If no sub-outtree is available then the request is blocked.

4 Wavelength Routed Networks

A WDM network employing wavelength-routing consists of optical wavelength routing nodes interconnected by point to point fiber links in an arbitrary topology [12]. The optical routing nodes in WRN do have the capability of switching a wavelength individually. A routing node is also known as a *wavelength cross connect* (WXC). End nodes with a number of optical transmitters and receivers are attached to these routing nodes. An optical WXC can be realized by using wavelength multiplexers, wavelength demultiplexers, and optical switches. A wavelength routing node may have the capability to tap a small amount of optical power from the wavelength channel which is forwarded by that node. The tapped optical power may be used by the local node. This type of node is called as a *Drop and Continue node* or simply a *DaC-node* [13].

A message reaching on an incoming link of WXC at some wavelength can be routed to any of the outgoing links using same wavelength or a different wavelength. In a network without wavelength converters, a message is sent from one node to another node using a wavelength continuous route (called as a *lightpath*) without requiring any opto-electrical conversion and buffering at intermediate nodes. a lightpath (LP) is an optical path established between two nodes, created by the allocation of same wavelength through the path [14]. An LP is a high-bandwidth pipe, carrying data up to several gigabits per second and an LP is uniquely identified by a physical path and a wavelength. The requirement that the same wavelength must be used on all the links along the selected path is known as *wavelength continuity constraint*. Two LPs can use the same fiber, if and only if they use different wavelengths. The lightpaths in a network constitute an *optical layer*.

A WDM network with WXC nodes having wavelength conversion capability is called as a wavelength-convertible network. A node with wavelength conversion capability is called as a *wavelength converting node* (WC-node). A wavelength converter is a device which can convert an optical signal on a wavelength to another wavelength. Thus, wavelength conversion provides the flexibility of using different wavelengths on different links of a lightpath. The conversion capability of a node is characterized by a conversion degree d, $1 < d \leq W$, where W is the number of wavelengths present in a fiber. A node having degree of conversion d can convert an incoming wavelength to d different wavelengths. A node is said to have *full conversion capability* $(d = W)$ if it can convert any wavelength to any other wavelength. A node is said to have *limited (or partial) conversion capability* $(d < W)$ if it can convert a wavelength to a subset of available wavelengths. Since convertible nodes are expensive, it is desirable to have few nodes as convertible nodes. A network with only a few WC-nodes is said to be one with *sparse wavelength conversion* [15].

To support multicasting, in a conventional network (electronic network), all nodes are assumed to have the capability of buffering an incoming message and transmitting it onto more than one output link. To support multicasting, nodes in a WDM network need to have light (optical) splitting capability. A node with splitting capability can forward an incoming message to more than one output link. If a network has splitting capability at all nodes then it is referred to as the network with *full splitting capability*. In a network with full splitting capability, a single tree can be generated to include all the destinations of a multicast session, like in a conventional electronic network [16, 17, 18, 19]. It means that, the source of a session may transmit data only once to communicate to all destinations. The tree-like structure which provides a single transmission to communicate to a set of destinations is called as a *multicast tree* or simply a *tree*. In a multicast tree all destinations connected together with source as the root. The source needs to transmit a message only once, to communicate to all the destinations belonging to the same tree. A split capable node is very expensive due to its complex architecture [13]. Hence, only a subset of the nodes in a network are assumed to be split capable nodes. A network with a few split capable nodes is called as the network with *sparse splitting capability* [20]. In a network with sparse splitting capability, it may not be possible to include all destinations of a session in one multicast tree. Hence, a set of trees are constructed to include all destinations of a multicast session. It means that the source needs to transmit the multicast data onto more than one channel may be on different fibers or on different wavelengths. The set of trees corresponding to a single multicast session is called as *multicast forest* [13, 20, 21]. The concept of optical splitting is explained below.

Figure 7 illustrates the benefit of using split capable nodes for a multicast session. Node 1 is the source of a multicast session, and nodes 2 through 5 are the destinations of the session. In the first case, shown in Figure 7(a), it is assumed that node 2 does not have splitting capability. It can forward the message received from node 1 to only one node (say node 3). Hence, it requires a separate path from source (node 1) to node 4. Here, the same link $< 1 - 2 >$ is shared by two paths and therefore, two wavelengths are used on the link $< 1 - 2 >$. Thus, source node, needs to transmit a multicast message on links $< 1 - 5 >$ using wavelength W_1, $< 1 - 2 >$ using W_1, and on $< 1 - 2 >$ using W_2. This can be viewed as three multicast trees are generated with source node (node 1) as the root of the trees. Links $< 1 - 2 >$, $< 2 - 3 >$ constitute the first multicast tree, links $< 1 - 2 >$, $< 2 - 4 >$ constitute second multicast tree, and link $< 1 - 5 >$ constitutes the third multicast tree. If node 2 has splitting capability, then it can multicast to node 3 and node 4 as shown in Figure 7(b). It should be noted that,

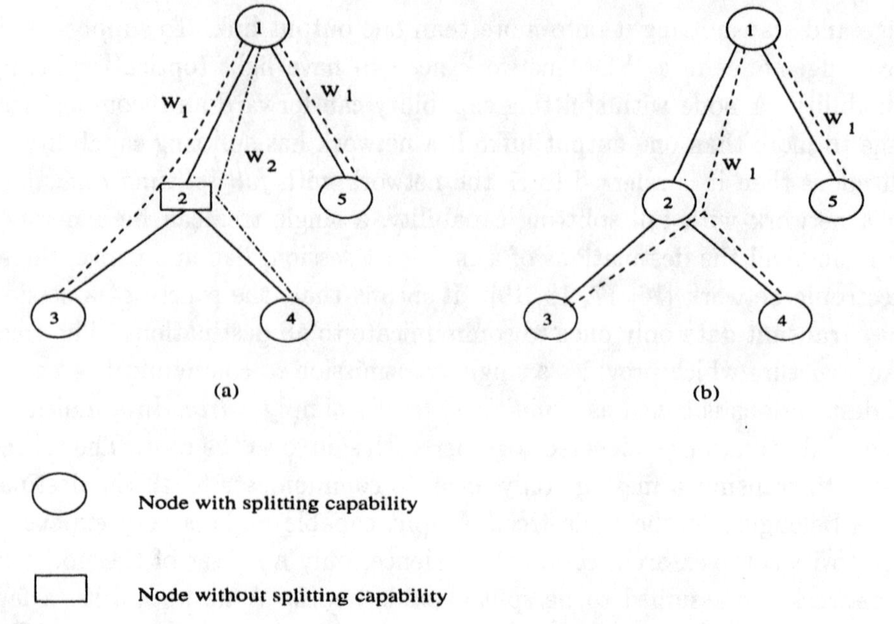

Figure 7: An example of multicast routing and wavelength assignment. Node 2 has (a) no splitting capability. (b) splitting capability.

only one wavelength is used on link $< 1 - 2 >$ in this case. Hence, source node needs to transmit a multicast message to only on links $< 1 - 2 >$ and $< 1 - 5 >$. Thus, only two trees need to be generated. Links $< 1 - 2 >$, $< 2 - 3 >$, and $< 2 - 4 >$ constitute a multicast tree and link $< 1 - 5 >$ constitutes the second multicast tree.

A network, with sparse wavelength conversion and sparse splitting capability, consists of nodes with different capabilities. A node may have splitting capability and/or wavelength conversion capability. In general, a node with split capability is called as *multicast capable node* or *MC-node*. A node with only splitting capability is called as a *split-node*, and a node with only wavelength conversion capability is termed as a *wavelength conversion node* or *WC-node*. A node having both splitting and wavelength conversion capabilities is called as a *virtual source* (VS). A VS node can transmit an incoming message to any number of output links on any wavelength. This phenomenon can be viewed as a source transmitting a message on any wavelength to as many immediate neighbors as needed. This is the reason behind calling such a node as a virtual source, when it has both the splitting and wavelength conversion capabilities. A split-node and a VS node are used for expanding the tree. A part of the tree which is expanded from a split-node or VS node is called as a *subtree*. The difference between a split-node and a VS node is that a split-node can not support more than one connection on the same outgoing link, whereas a VS does support by using different wavelengths. It means that the subtree spawned from a VS may use the same physical link which is used by the existing connection but on a different wavelength. A node without splitting capability is called as *multicast incapable node* or *MI-node*. An MI-node may have the capability of drop and continue (DaC) or it may have no capability. A node without any capability is an ordinary wavelength router, and it can either drop a message or can switch a message, but not both operations. Hence, for consistency, such a node can be called as *drop or continue* (DoC).

4.1 Node Architectures

A network may consist of a set of MC-nodes and a set of MI-nodes. The complexity of different node architectures depends on whether a node has splitting capability or not. Obviously, an MC-node will have more complex architecture than an MI-node as MC-node requires power splitters. In this section, the architectures of different nodes are studied.

4.1.1 Architecture of an MC-node

An MC-node consists of optical power splitters. To support multicast, the input signal needs to be transmitted onto various output connections. These connections may be on different links connected to the node or on different wavelength channels of the same fiber. If a node has L input links with a single fiber per link, and each fiber consists of W number of wavelengths, then to support multicast the input signal needs to be selected by $L \times W$ number of switches. Consider the architecture of an MC-node shown in Figure 8 which has three input links, three output links, and two wavelengths per fiber. This node can split the incoming signal onto 3×2 number of output connections. Here, the signals on each input link are split into three signals using 1×3 splitters. This is to select output links from the set of links connected to this node. To choose the wavelengths onto which the input signal needs to be transmitted, 1×2 splitters are incorporated. The output of these splitter stages are fed as input to a set of space-division (SD) switches. In this figure, the input signal is split into six (3×2) signals. For each signal one 3×1 SD switch is provided, so that, the SD switch selects one signal out of three input signals. Next stage consists of tunable filters

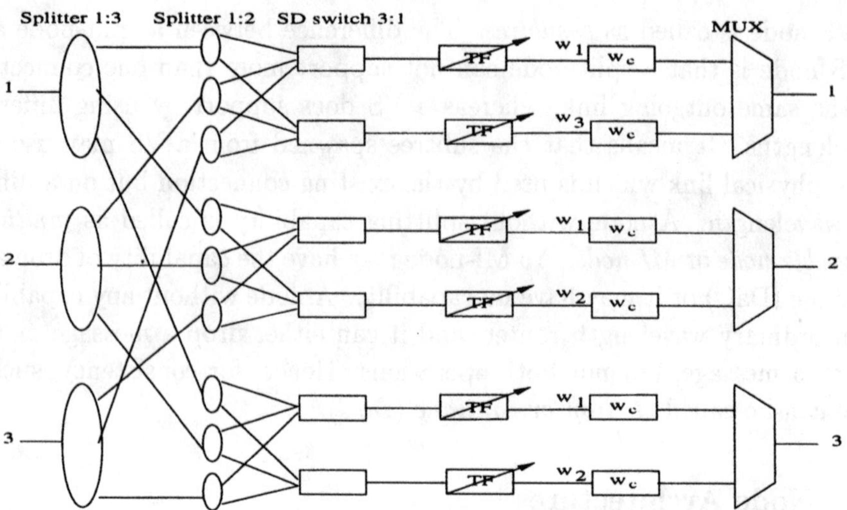

Figure 8: Architecture of an MC-node.

(TF) to extract signal on particular wavelength. After TF-stage there is
a wavelength converter connected to every tunable filter. This is to avoid
wavelength clash at the output link. However, this wavelength conversion
stage is an optional one. An MC-node need not perform wavelength conversion. Finally, for each output link one multiplexer is present, so that the
signals on various wavelengths are multiplexed and transmitted onto the
same fiber. In general, for an MC-node with L input links, M output links,
and W wavelengths per fiber, there are L number of $1 \times M$ splitters, $L \times M$
number of $1 \times W$ optical splitters, $M \times W$ number of $L \times 1$ space-division
switches, $M \times W$ number of tunable filters and wavelength converters, and
M multiplexers. Power amplification is required to compensate the power
loss due to splitting operation. Alternatively, one can use semi-conductor
optical amplifiers (SOAs) to turn "on" or "off" each of the output (split)
signals.

4.1.2   Architecture of an MI-node

A node without splitting capability is called as multicast incapable node
or an MI-node. MI-nodes may tap (or drop) a small fraction of signal
and switch the remaining signal to one of its neighbor nodes. The tapped
(dropped) signal may be used by the local station. This type of nodes which
can tap or drop the signal passing through it is called as a DaC-node. The

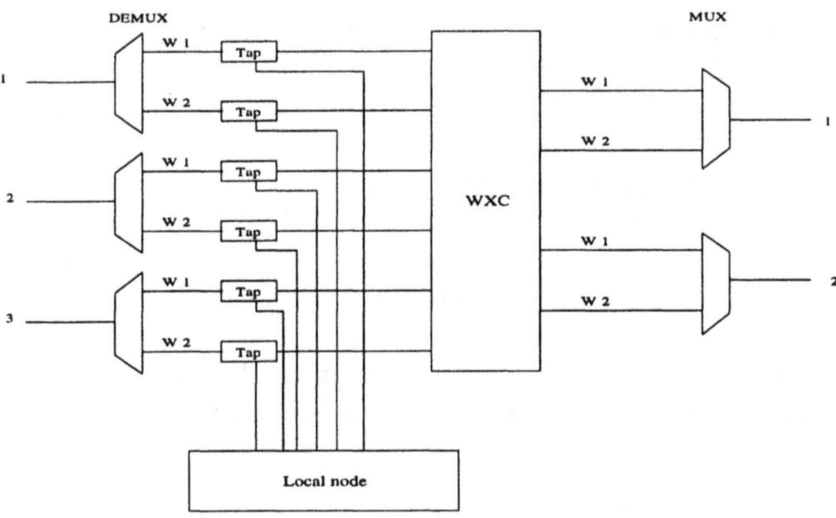

Figure 9: Architecture of a DaC-node.

architecture of a DaC-node is shown in Figure 9. Here, the input signal is demultiplexed and each of these demultiplexed signals is fed as input to the "tap" (drop) module which taps 5% of signal. Then, like in a conventional wavelength routed node, the signal is switched using SD switches.

4.2 Multicast Tree Generation

Multicasting is the process of transmitting data by a source to a set of destinations. Instead of transmitting packets from a sender to each receiver, the routes between source and receivers can share some links. In conventional networks, multicast route determination is traditionally formulated as a problem relating to tree construction [22]. The reasons to adapt tree structure for multicast communication are listed below.

1. The source need not send a packet to individual destinations.

2. The packets are transmitted in parallel to various destinations.

3. The tree structure minimizes data replication, since, the packet is replicated optically by routers only at branch points in the tree.

In a network with full splitting capability, a single tree can be generated to include all the destinations of a multicast session, like in a conventional electronic network [16, 17, 18, 19]. Hence, in a network with full splitting capability, the constraints to generate a multicast tree would be minimizing the number of transceivers and wavelengths in a fiber. In the next section, multicast tree generation algorithms for a network with full splitting capability are described.

In a network with sparse splitting capability, it may not be possible to include all the destinations of a session in one multicast tree. Hence, a set of trees are constructed to include all the destinations of a multicast session. It means that the source needs to transmit the multicast data onto more than one channel and may be on different fibers or on different wavelengths. The set of trees corresponding to a single session is called as multicast forest. In Section 4.2.2, multicast tree generation algorithms for a network with sparse splitting capability are described.

4.2.1 Multicast Tree Generation in a Network with Full Splitting Capability

In a full splitting capable network all nodes in the network have split capability. Hence, a single tree can be generated to route multicast traffic. The tree generation is similar to the tree generation in conventional networks. Apart from finding a path, in a conventional network, it is necessary to allocate bandwidth and buffers. But in optical networks, the constraints are number of transceivers and wavelengths. Hence, the tree generation algorithm should consider the availability of these resources while generating a multicast tree.

In [16], the problem of minimizing the number of transmitters and receivers that are required to generate a multicast tree is considered. It introduces the concept of *light-tree*. A light tree is a point-to-multipoint optical path established in the network created by allocating the same wavelength on every link of the tree. The concept of light tree can be implemented by incorporating optical multicasting (splitting) capability at all nodes of a network in order to reduce the average packet hop distance and the total number of transceivers in a network. Thus, a light tree provides single hop communication between a source node and a set of destination nodes. A solution is provided in [16], for routing unicast traffic and broadcast traffic using light trees. To carry unicast traffic, the virtual topology design problem is formulated based on light-tree concept. The proposed optimization problem has one of the below mentioned objective functions.

1. Minimize the network-wide average packet hop distance.

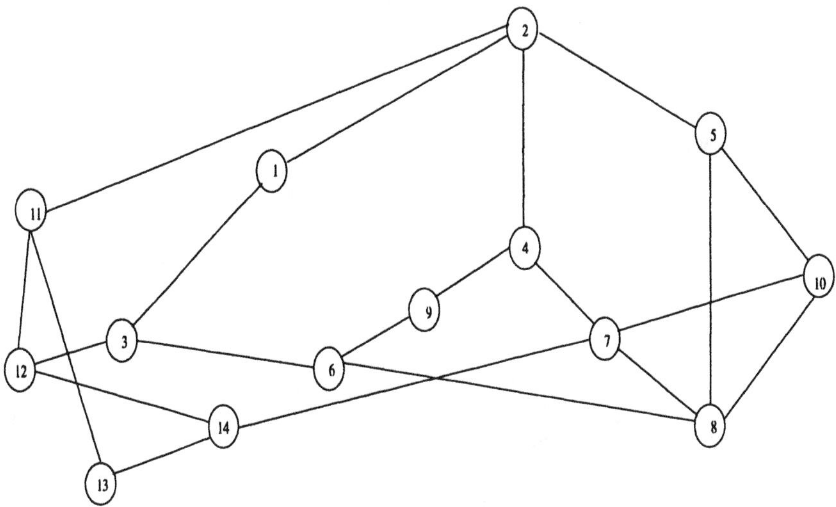

Figure 10: An example network (NSFNET).

2. Minimize the total number of transceivers in the network.

In [16], it is demonstrated that the average packet hop distance for the virtual topology based on light-tree is less than that of virtual topology based on lightpath concept. It is also demonstrated that the number of transmitters and receivers required for the virtual topology based on light-trees is less than that of the number of transmitters and receivers required for the virtual topology designed based on lightpath concept. For broadcast traffic, minimization of the number of transceivers is considered as the objective function. The light-tree concept can also be applied to the multicast traffic.

An example network is shown in Figure 10, where node 7 is the source and nodes 2, 6, and 9 are the destinations. Assume that only one wavelength is available. To provide communication among these nodes, using lightpath based solution, it is required to provide lightpaths between node 7 and node 4, node 4 and node 2, node 4 and node 9, and node 9 and node 6. To establish these 4 lightpaths, totally eight transceivers are required as shown in Figure 11(a). The packet traffic destined to node 6 should be received by node 4 and retransmitted it onto link $< 4 - 9 >$. Node 9 receives this traffic and retransmits onto link $< 9 - 6 >$. Each such retransmission is referred as an *electronic hop* or a *hop*. Hence, the traffic destined to node 6 needs 3 hops. If light-tree based solution is incorporated, a single light-tree is established to connect all these nodes as shown in Figure 11(b). At node 4, the data need not be converted into electronic form. The optical signal

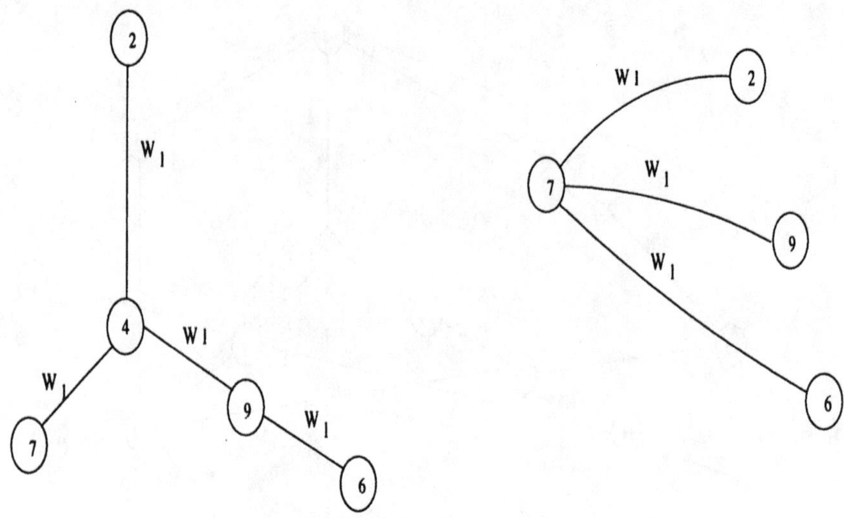

Figure 11: (a) Connections using lightpaths. (b) Connections using light tree.

is split and transmitted onto links $< 4 - 2 >$ and $< 4 - 9 >$. This signal
is tapped by node 9 and optically forwarded onto link $< 9 - 6 >$. Now,
one transmitter at node 7 and totally three receivers at node 2, node 9, and
at node 6 are required. This is because light-tree based solution does not
utilize electronic switch at node 4 or node 9. Packet traffic for every source
destination pair needs only one hop.

In [17], the problem of *multicast routing and wavelength assignment*
(MCRWA) is considered. Here, it is assumed that all nodes have the splitting capability. However, the splitting capability of a node is fixed to some
value, which is called as *splitting fanout*. If a network has splitting fanout of
f, then any node in the network can split the incoming message and transmit
to f output ports. To select multicast trees, two approaches are proposed,
namely, *static* and *dynamic*. In the static approach, a pre-determined set
of trees is computed to support the given multicast sessions. In this set of
pre-determined trees, there may be a single tree corresponding to a session
(*fixed routing*), or there may be more than one tree (*alternate routing*) corresponding to a session. In static approach, current state of the network
is not taken into consideration while computing the multicast trees. In the
dynamic approach, current utilization of the links is considered. When a
multicast session request arrives, a route is checked for availability. If it is
not available then the request is blocked. To generate a multicast tree *minimum cost shortest path* proposed in [23] is used. Two policies are considered

for connection establishment. In the first policy, connection is established only if the wavelength is available from the source to all destinations. This is called as *full destination blocking policy*. The performance of the this policy is evaluated using the metric *session blocking probability*, which is the probability that a session is blocked due to non-availability of resources. In the second policy, connection is established even it is not possible to provide wavelength to some of the destinations. This policy is referred to as *partial destination blocking policy*. The performance is evaluated using the metric *destination blocking probability*, which is defined as the probability that a destination is blocked. The simulation results show that the dynamic routing scheme results in lower blocking probability than the static scheme.

In [18], a new split node architecture is proposed. The split node (MC-node) described before, which is referred here as *splitter-and-delivery node* or as node with *SaD switch*, has a shortcoming. All signals entering a node with SaD switch faces power splitting regardless of their type (unicast or multicast). A unicast signal which is entering a node with SaD switch faces K-way splitting (where K is number of neighbors of the node), even though it is not destined to more than one neighbor node. This indiscriminate power loss results in unnecessary need for amplification due to which the complexity of node, and network control and management increases. This can be avoided by keeping only a few split components in the node. This type of node is referred to as *multicast-only splitter-and-delivery node* or as a node with *MOSaD*. All sessions which need to be split can make use of these shared split components. As a consequence of sharing split components, some multicast requests may be blocked due to unavailability of split components in a node. However, it is shown that the blocking probability of the network with MOSaD nodes is nearer to the network with SaD nodes.

The upper bounds on the number of wavelengths needed to support multicasting in an optical network have been derived in [19]. The problem of multicast routing is analyzed with two different situations. First situation considers dynamically arriving multicast session requests with a constant set of destinations. The second situation deals with dynamic addition and deletion of nodes from the destination set. For both scenarios, different network topologies are constructed and wavelengths required to construct the topologies are computed.

4.2.2   Multicast Tree Generation in a Network with Sparse Splitting Capability

In a sparse splitting WDM network, only a few nodes have the optical power splitting capability. In these networks, a node may have splitting capability and/or wavelength conversion capability. Hence, a node with splitting capability may be either split-node or a VS node. Any node, by default, is assumed to have the DaC capability. Hence, any node without splitting capability may be either a DaC node or DaC node with wavelength conversion capability. A DaC node with wavelength conversion capability is referred to as *WC-node*.

The objectives of multicast tree generation algorithms are to minimize one or more among the number of wavelength channels, number of wavelengths in a fiber, delay, and setup time. *Number of wavelengths* is defined as the maximum number of wavelengths required on any link to construct a forest for a given session. Each branch of the forest corresponds to a link in the network. To construct a forest one wavelength has to be reserved on each link, that is being used for constructing the forest. A wavelength on a link is referred to as a *wavelength channel* or simply a *channel*. The *number of wavelength channels* or simply *number of channels* refers to the total number of wavelength channels on a forest. *Delay* refers to the average number of hops from source to the destinations of multicast session, and *setup time* is defined as the maximum amount of time required to construct the multicast tree. Two basic approaches are proposed to generate a multicast tree/forest which are mentioned below.

1. Source-rooted approach: Multicast tree is constructed with source of a session as the root of the tree. The objective here is either to minimize total cost of the tree or to minimize individual cost of the path between the source and a destination. Here, cost of the path is defined as the number of wavelength channels (hops) used by the path. Depending on the objective there are two methods to construct a multicast tree.

   (a) Source-based tree generation: While constructing the multicast tree, these algorithms aim to minimize the cost of individual paths from source to each destination. Algorithms based on this method are described in the next section.

   (b) Steiner-based tree generation: While constructing the tree, these algorithms aim to minimize the total cost of the tree. This problem is referred to as minimum Steiner tree problem and is shown to an NP-complete problem [24, 25]. Hence, heuristics are proposed in [13, 20, 21] which try to construct minimum Steiner tree for the given multicast sessions. Algorithms based on this method are described in Section 4.4.

2. VS-rooted approach: In these algorithms, the tree is constructed with a special capable node (VS) as the root of the tree [26]. The main advantage in these algorithms is that the setup time is less when compared to the Source-rooted algorithms. These algorithms exploit the capabilities of the nodes in a sparse splitting network. Hence, in many situations it results in better wavelength channel utilization than using Source-rooted algorithms. An algorithm based on this method is described in Section 4.5.

4.3 Source-Based Tree Generation

In these algorithms, the destinations are added to the multicast tree in a shortest path to the source of multicast session. These algorithms provide computationally simpler solution to the multicast tree generation. The below mentioned algorithms consider the *hop count* as the measure of path length to find the shortest path between two nodes. It means that the shortest path between two nodes is defined as the one with minimum number of hops. These algorithms assume that the network has sparse splitting capability.

4.3.1 Re-Route-to-Source

This algorithm is proposed in [13]. Here, it is assumed that all MC-nodes have wavelength conversion capability. A spanning tree is generated using conventional minimum spanning tree algorithm, by assuming split capability at all nodes of the network. Multicast tree is generated by pruning the spanning tree to delete all branches that are not leading to any destinations. In a network with sparse splitting capability, only few nodes are MC-nodes. Hence, the multicast tree generated using the above process may consists of some MI-nodes which are used as *branching points*. Generally a branching point is assumed to be an MC-node, so that the data can be split and distributed to more than one node. As MI-nodes are also assumed as branching points, the tree need to be reorganized. The tree is verified whether every branching point in the spanning tree is an MC-node or not. This is performed in breadth-first order from the root of the tree, which is the source of the session. If any MI-node is acting as a branching point to more than one node, then all downstream branches except one branch from the node are disconnected. These affected children are connected to the source by using reverse shortest path to the source. If there is an MC-node, which is a member of the tree, in this shortest path then the affected children are connected to it.

An example is shown in Figure 12. Here, node 1 is the source of multicast session and nodes 2 through 6 are the destinations. The pruned spanning tree is given in Figure 12(a). Node 2 is assumed to be an MI-node but it is acting as branching point to nodes 3 through 5. Node 2 cannot distribute to all the nodes connected to it, as it is an MI-node. Hence, node 4 and node 5 are disconnected from node 2. These two nodes find their shortest paths to source via node 2. As node 6 is an MC-node, which is present in these shortest paths, node 4 and node 5 are connected to node 6, as shown in Figure 12(b).

4.3.2 Re-Route-to-Any

This algorithm is a variation of Re-Route-to-Source algorithm. The tree is constructed in a similar way as in Re-Route-to-Source. In this algorithm, the affected children are allowed to connect to any other node which is already present in the multicast forest. The closest node in the tree is chosen to connect the affected children.

An example is given in Figure 12. Here, node 1 is the source of a multicast session and nodes 2 through 6 are the destinations. The pruned spanning tree is given in Figure 12(a). Node 2 is assumed to be an MI-node, but it is acting as branching point to nodes 3 through 5. Node 2 cannot distribute to all nodes connected to it, as it is an MI-node. Hence, node 4 and node 5 are disconnected from node 2. Node 4 finds a shortest path to node 6 and establishes a connection to it. Node 5 finds node 4 as the nearest node in the tree, and establishes a connection to it as shown in Figure 12(c).

4.3.3 Member-First

In the previous algorithms, first a multicast tree is generated by assuming splitting capability at all the nodes. Later, the tree is adjusted to reflect the splitting capability of the nodes. In *Member-First* algorithm, the availability of MC-nodes is checked while generating the tree. Here, the multicast forest is constructed one tree at a time. To construct a tree each link is added one at

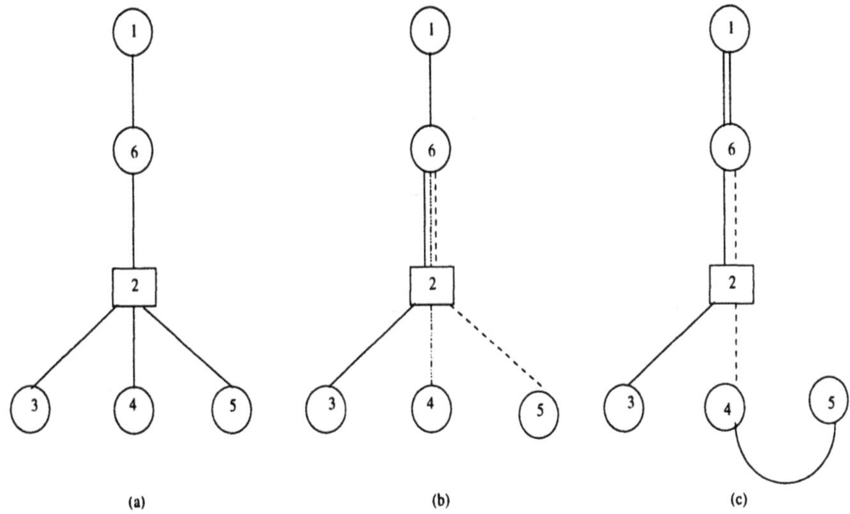

Figure 12: (a) Spanning tree. (b) Re-Route-to-Source. (c) Re-Route-to-Any.

a time as in Dijkstra's algorithm for constructing a spanning tree. However, the links which are leading to the destinations are given higher priority than the links which are leading to the nodes that are not destinations. To implement the priority, the concept of *fringe links* is introduced. The links which are being considered for possible inclusion in the present tree are called as *fringe links*. These fringe links are organized as a *priority queue*. The priorities of two links are compared only if the lengths of the paths from the source to reach these links are equal. Higher priority is given to the link which is present in the path to a destination node than the link present in the path to a non-destination node. While adding a higher priority link it may be necessary to delete some links. If all members are included in the forest, then the branches which are not leading to any destination are pruned.

In Figure 13, node 1 is considered as source and node 8, node 10, node 11, and node 13 are the destinations. First, node 1 which is source of the session is included in the tree. When node 1 is included in the tree, the fringe link set contains $< 1-2 >$. As there is only one link in the fringe link set, node 2 is included in the tree. In a similar way, node 3 is also included in the tree. After including node 3, the fringe link set contains $< 3-4 >$ and $< 3-5 >$. Equal priorities are assigned to these two fringe links, as these two links are leading to destinations. Node 4 is considered to include in the tree. Fringe link set is updated with links $< 4-6 >$ and $< 4-7 >$

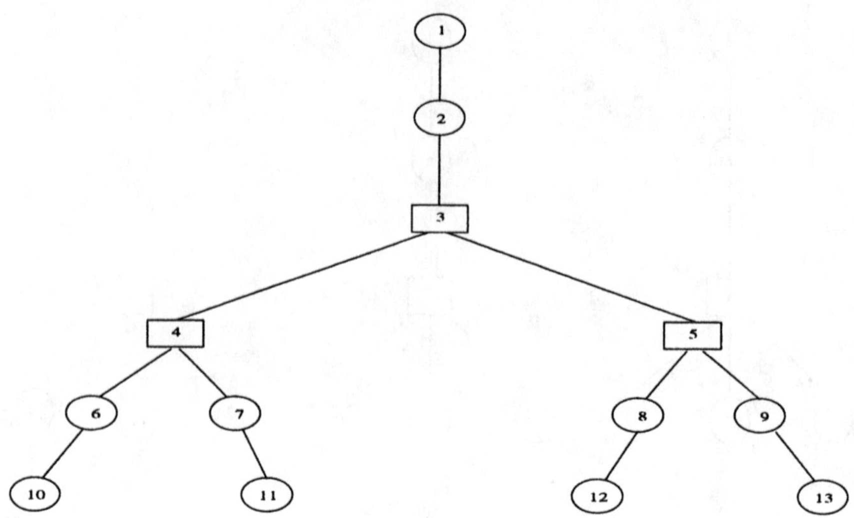

Figure 13: Illustration of Member-First algorithm.

after adding node 4 to the tree. Next link in the priority queue is $< 3 - 5 >$. Node 5 is added to the tree, and links $< 5 - 8 >$ and $< 5 - 9 >$ are required to be included to the fringe link set. Since node 8 is a destination node, higher priority is given to the link $< 5 - 8 >$ and added at the front end of the queue. Node 9 is added at the rear end of the queue. Now, node 8 is required to be included in the tree. It finds a shortest path to source as $< 8 - 5 - 3 - 2 - 1 >$. Since, node 2 is an MC-node, node 8 connects to node 2. In this path, node 3 and node 5 are MI-nodes. Hence, all the links that are connected to these nodes and fringe links corresponding to these nodes are deleted. Node 4 is disconnected from the tree and links $< 4 - 6 >$, $< 4 - 7 >$, and $< 5 - 9 >$ are deleted from the fringe link set (priority queue). As a consequence of this, priority queue becomes empty. Another tree is constructed to include destinations which are yet to be included in the forest.

4.4   Steiner-Based Tree Generation

In these algorithms, destinations are added to the existing multicast tree one at a time in such a way that the total cost of the tree is minimized. To add a node to the tree, it is required to find the shortest path to all nodes in the tree. The path that has minimum value among the computed shortest paths, is chosen to select the node to which the present node can

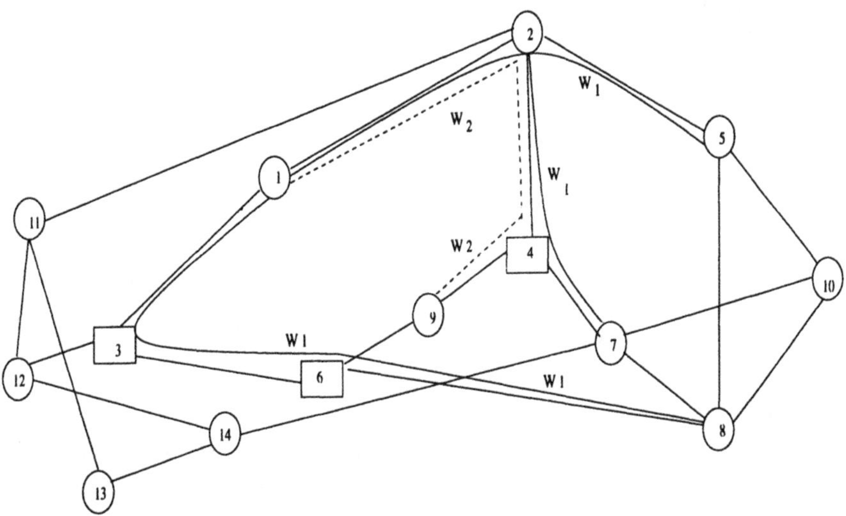

Figure 14: Illustration of Member-Only algorithm.

be connected. These algorithms are computationally expensive. Hence, heuristics are provided to choose a node to which the present node can be connected.

4.4.1 Member-Only

This algorithm is same as the algorithm proposed in [20] and is similar to Member-First algorithm. However, in this algorithm, a multicast tree is constructed by adding the destinations one at a time. In every iteration one destination is added to the tree. As the algorithm works only on destinations of the session, there is no necessity of pruning. While generating a multicast tree, the algorithm tries to include as many destinations as possible in one multicast tree. If a destination is at an equal distance (in terms of number of hops) to more than one node in a tree, then this algorithm connects the destination to one of the nodes which is chosen arbitrarily. If there is a node that can not be included, then the same procedure is recursively called to construct another multicast tree. At the time of constructing the tree, the splitting capability of the nodes is taken into account.

The aim here is to construct a multicast forest $F(s, D)$ for a given multicast source s with a set of destinations D, so as to reduce the number of wavelengths per fiber and number of wavelength channels per multicast forest. It is assumed that the network has sparse splitting capability. The algorithm tries to include as many destinations as possible in one multicast tree. In order to include all destinations in a multicast session, it may be necessary to create more than one tree rooted at the source of the session.

Let V denote the set of nodes which are useful in expanding the tree, and V' denote the set of nodes which are not useful in expanding the tree. In addition, let UV be the set of unvisited destinations that have not been included in the forest $F(s, D)$.

Initially V' is empty, V and T contain only s, and UV contains all destinations of the session. The destination nodes in UV are considered one by one in the nondecreasing order of their distances to source s. Consider a destination node u from UV. Find the shortest path from u to every member of V but not traversing through any node contained in set V'. Select a node $v \in V$ which is nearest to u. There may be more than one node in the tree at an equal distance from node u. In such a case, select a node v arbitrarily, and connect u to v along the shortest path and add it to tree T. If node v is an MI-node then delete it from the set V and add it to set V'. Add all MI-nodes on the path of $< u - v >$ to V' and MC-nodes to V. The destinations are added to the tree T one by one until no more destinations can be added to it. If a destination can not be added to the present tree then a new tree is constructed using the above procedure.

An example network is shown in Figure 14 to illustrate the working of Member-Only algorithm. In Figure 14, node 1 is the source of a multicast session, and nodes 2 through 9 are the destinations. Node 3, node 4 and node 6 do not have split capability, node 2 has both the split and conversion capability and all other nodes have split capability only. Initially the tree contains only node 1. It then tries to add as many destinations as possible to the multicast tree. Node 2 and node 3 find their shortest paths to node 1 and are directly connected to node 1. Node 4 and node 5 find their shortest paths to node 2, and node 6 to node 3. Node 3 is transfered from V to V' as it is an MI-node and its capability is already utilized. Next node 7 is considered. Set V contains node 1, node 2, and nodes 4 through 6. Node 7 finds that it is nearer to node 4 than any other node. Hence, node 7 is connected to node 4. As node 4 is an MI-node, it is deleted from set V and placed in set V'. Next node 8 is considered. It is at an equal distance to both node 5 and node 6. Since, the selection is done arbitrarily, node 6 may be chosen for establishing a connection and node 6 is deleted from set V and placed in set V'. Next, node 9 is considered. As node 4 and node 6 are not available for expanding the tree (as they are not present in set V) node 9 can not be included in the present tree. One more tree is constructed to

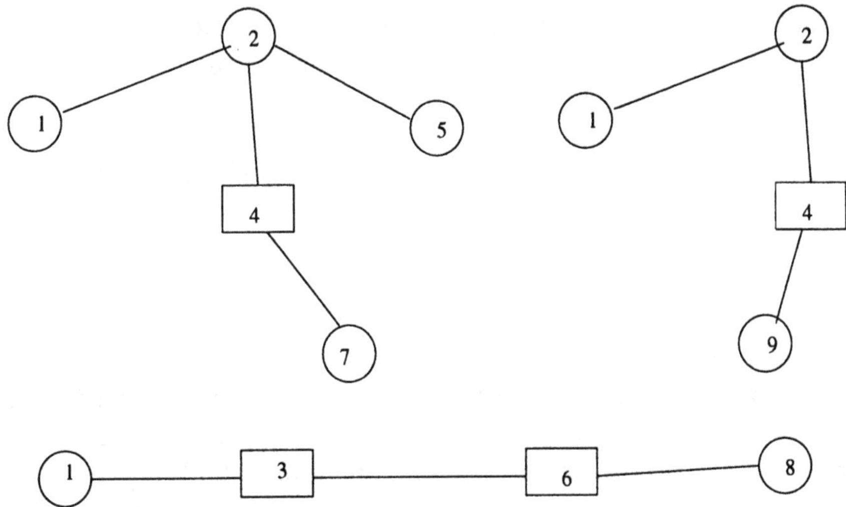

Figure 15: Forest generated by Member-Only.

include node 9 in the multicast session. Node 9 is connected to source, via node 4 and node 2. Totally three trees are formed for this multicast session as shown in Figure 15.

To study the performance of the above four algorithms (Re-Route-to-Source, Re-Route-to-Any, Member-First, and Member-Only), in [13], three metrics namely *number of wavelengths, number of wavelength channels,* and *delay* are used. Let S and C be the fraction of nodes that have split capability, and wavelength conversion capability, respectively. Let P be the mean Poisson distribution used to generate the number of multicast sessions, and G be the fraction of nodes that are destinations. The performance of above said metrics are studied by varying S, C, P, and G, one at a time and keeping all others at some constant value. In [13], the following conclusions are drawn from the simulation study.

1. Among all four algorithms, Member-Only requires least number of wavelengths and wavelength channels.

2. Re-Route-to-Source requires more wavelengths and wavelength channels than all other three algorithms.

3. Member-Only results in longest delay and Re-Route-to-Source results in shortest delay than all other algorithms.

4. Number of wavelengths per fiber increases with increase in P.

5. As S value increases, the values of all metrics (number of wavelengths, number of wavelength channels, and delay) decrease. This is due to more subtrees forming from a split-node.

6. As G value increases, the values of all metrics also increase.

7. Number of WC-nodes (the value of C) does not effect any of the performance metrics.

Shortcomings of the Above Algorithms

The performance in terms of the number of wavelengths per fiber and number of channels per forest (considered as the cost of the forest) for the Source-based algorithms (namely Re-Route-to-Source, Re-Route-to-Any, and Member-First) is poorer than Member-Only algorithm. On the other hand, the performance in terms of the delay from the source to the individual destinations is poorer for Member-Only algorithm than the Source-based algorithms [13]. However, the delay in optical networks is normally very low as a message is transmitted from source to destination optically without requiring any buffering or optical/electrical-electrical/optical (O/E-E/O) conversion at the intermediate nodes. Hence, it is preferred to minimize the cost of the forest than minimizing the delay on individual paths.

The Member-Only algorithm proposed in [13], includes as many destinations as possible in a multicast tree. If a destination is at an equal distance (in terms of number of hops) to more than one node in a tree, then this algorithm connects the destination to one of the nodes which is chosen arbitrarily. This algorithm is illustrated in Figure 14. Consider node 1 as source and nodes 2 through 9 as the destinations of a multicast session. To construct the multicast forest, three trees are required (Figure 15) with node 1 as root of the tree and it consumes two wavelengths and ten wavelength channels. These required resources can be reduced by employing heuristics, such as *Capability-Based-Priority* and *Spawn-from-VS*.

- In Figure 14, assume that nodes 2 through 7 are connected to the multicast tree. Next, node 8 is considered for connection. It has two options, either to connect to node 5 or to node 6. In such a situation, Member-Only algorithm chooses a node arbitrarily. Node 8 is connected to node 6. Next, node 9 is considered. As the DaC capability of node 4 and node 6 is exhausted, node 9 should be connected to the source using another tree. Alternatively, if node 8 is connected to node 5, then node 9 could have been connected to node 6. In this example, it results in saving of two channels. Hence, it is better to defer the usage of the nodes which are not having any special capabilities (considered as resources to be used only once).

- After the tree is constructed, if there is any node that cannot be included, the algorithm is recursively called to construct the second multicast tree from the source. Any node with both the capabilities (splitting and wavelength conversion) can transmit to it's children on the same link using different wavelengths. Hence, instead of constructing a tree from the source for the remaining nodes it may be advantageous to connect to the node having both the capabilities in the already existing tree. If such a node is not available then the destination node can be connected to the source. For example, in Figure 14 node 9 can be connected to node 2 (having both the capabilities) instead of connecting it to the source. In such a case, it results in saving of one channel.

4.4.2 Capability-Based-Connection

Capability-Based-Connection is proposed to construct a multicast tree, which has two heuristics namely *Spawn-from-VS*, and *Capability-Based-Priority* [21].

Spawn-from-VS Heuristic

A node having both splitting and wavelength conversion capabilities is called as a virtual source (VS). A VS node can act like a source to spawn a new (sub) tree. The usefulness of a VS is observed in a situation wherein two different wavelengths on the same outgoing link are used by the paths to different destinations. The benefit of VS is described below.

Consider a multicast session with node 1 as source and node 4 and node 5 as destination as shown in Figure 16(a). Here, the capability of the VS (node 2) is not considered. Node 4 is connected to node 1 via node 3 and node 2. Next, node 5 is considered for establishing a connection. Since the capabilities of node 2 (it can split on different fibers, because wavelength conversion capability is not considered) and node 3 (DaC capability) are exhausted, node 5 requires a separate connection from the source. As a result, this session requires a total number of six channels. Suppose that node 2 is a VS. In such a case, node 5 can be connected to node 2 via node 3 instead of connecting to the source as shown in Figure 16(b). Note that

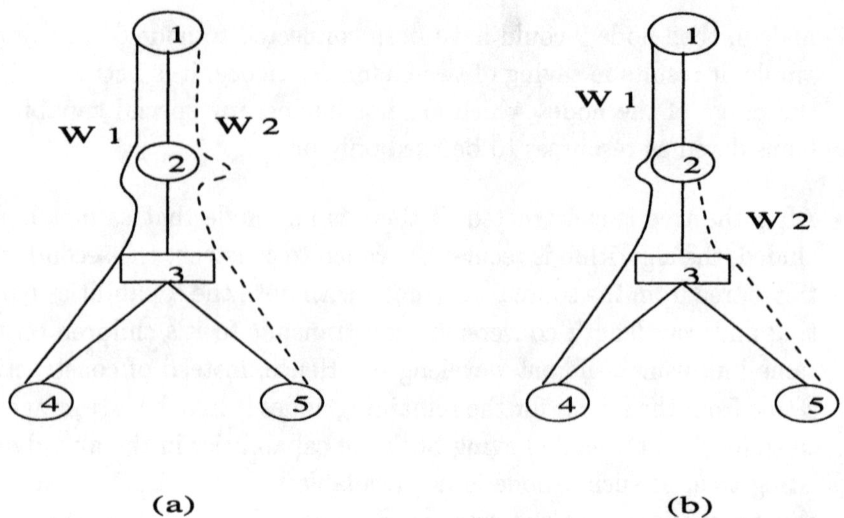

Figure 16: Benefit of virtual source in terms of savings in number of channels.

the multicast session requires only five channels.

Figure 17 depicts the benefit of VS in terms of number of wavelengths required. It shows a situation where there are two sessions. The source of the first session is node 1 and corresponding destinations are node 5 and node 6. The source for the second session is node 2 and the destination is node 7. In Figure 17(a), the capability of VS (node 3) is not considered. Node 5 is connected to the source (node 1) using w_1 via node 4 and node 3. Since the capability of node 4 is exhausted, node 6 requires a separate connection using wavelength w_2 from the the source. For the second session, node 7 is connected to its source, node 7, via node 1 and node 3. This connection requires a new wavelength w_3 as link $< 1 - 3 >$ carries 3 connections. In Figure 17(b), the capability of VS, node 3, is taken into consideration. Here, node 6 is connected to node 3 instead of node 1. Consequently, only two wavelengths are required on link $< 1 - 3 >$ as it carries only two connections.

Capability-Based-Priority Heuristic

The network is assumed to have different capable nodes namely split, wavelength conversion, drop-and-continue (DaC) and VS nodes. The nodes in the network are assigned some priority depending on the capabilities they have. This priority is used when a destination needs to be included in the tree and it is equally away from more than one node in the tree.

Priorities for different nodes are assigned as follows. VS nodes are assigned the highest priority followed by split-nodes, WC-nodes and DaC-

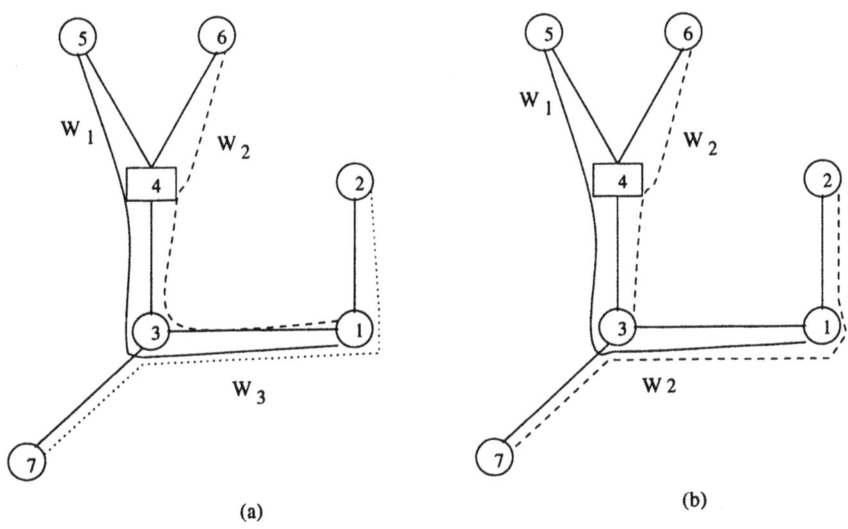

Figure 17: Benefit of VS in terms of savings of wavelengths.

nodes in the decreasing order of priority. A split-node has the capability of transmitting an incoming message on more than one outgoing link, whereas DaC-node and WC-nodes can transmit to only one outgoing link. It is better to defer the use of a DaC-node as it can be used only once. Hence, higher priority is assigned to the split-node than any DaC-node and WC-node. If there is a situation, where a node is at an equal distance to a split-node and to a DaC-node, then the split-node is chosen for connecting the node. A WC-node is given higher priority than a DaC-node, because of the flexibility to use any wavelength.

The benefit of giving higher priority to a split-node over a DaC-node is illustrated in Figure 18. Consider a multicast session with source as node 1 and destinations as node 2 through node 5. In Figure 18(a), the priorities of the nodes are not considered. Node 2 and node 3 are directly connected to the source, node 1. Node 5 is connected to node 2 using wavelength w_1. Since the DaC capability of node 2 is exhausted, node 4 requires a separate connection from the source using wavelength w_2. Node 6 is connected to node 3. This session requires a total number of six channels and two wavelengths per fiber. In Figure 18(b), the priorities of the nodes are considered. Here, node 5 is connected to node 3, where node 3 is a split-node. Node 4 can now be connected to node 2. Consequently, only five channels and one wavelength is required for the above session.

Tree Generation Algorithm

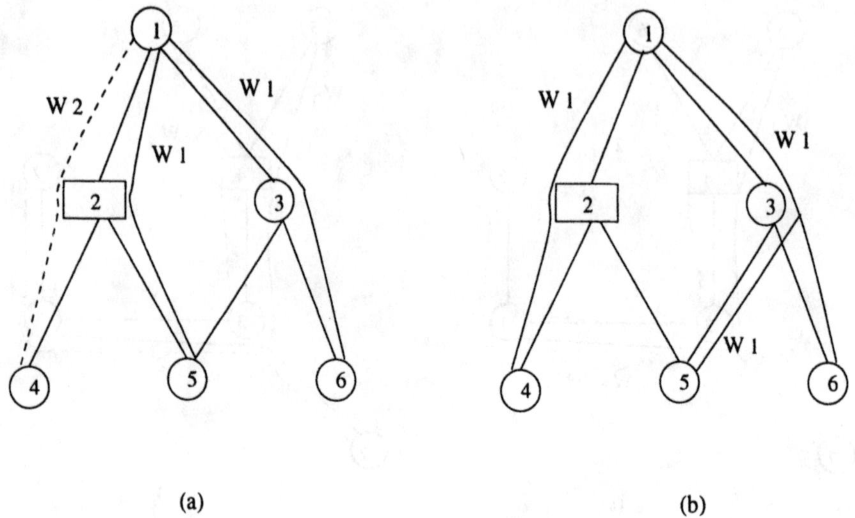

(a)   (b)

Figure 18: Benefit of giving higher priority to split-nodes.

The aim of the tree generation algorithm is to construct a multicast forest $F(s, D)$ for a given multicast source s with a set of destinations D, so as to reduce the number of wavelengths per fiber and number of wavelength channels per multicast forest. It is assumed that only a few nodes in a given physical network have splitting capability. Those nodes which do not have splitting capability can not have more than one immediate child in a multicast tree. The algorithm for constructing the tree makes use of the heuristics Spawn-from-VS and Capability-Based-Priority. It tries to include as many destinations as possible in one multicast tree. In order to include all destinations in a multicast session, it may be necessary to create more than one tree rooted at the common source.

Let V, V', and Z be the sets that contain a subset of only those nodes that are already included in the current multicast tree, where V denotes the set of nodes which are useful in expanding the tree, V' denotes the set of nodes which are not useful in expanding the tree and Z represents the set of VS nodes. In addition, let UV be the set of unvisited destinations that have not been included in the forest $F(s, D)$ and T be the set of nodes included in a multicast tree.

Initially V, V' and Z are empty, T contains only s and UV contains all destinations of the session. The destination nodes in UV are considered one by one in the nondecreasing order of their distance to source s. Consider a destination node u from UV. Find the shortest path from u to every member of V but not traversing through any node contained in set V'. Select a node $v \in V$ which is nearest to u. There may be more than one node in the tree at an equal distance from node u. In such a case, select the one with the highest priority (Capability-Based-Priority heuristic). Let d_{uv} be the distance from node u to node v. Let d_{us} be the distance from node u to source s.

If $d_{us} \geq d_{uv}$ then connect u to v along the shortest path and add it to tree T. If node v is a DaC-node or WC-node then delete it from the set V and add it to set V'. Add all DaC-nodes and WC-nodes on the path of $< u - v >$ to V', add all split-nodes to V, and add VS nodes to set Z.

If $d_{us} < d_{uv}$ or no such v exists then find a shortest path from u to every member of Z. (Z contains VS nodes present in the tree. This is to implement Spawn-from-VS heuristic.) Let w be the node which is nearest to u and let d_{uw} be the distance between u and w. If $d_{us} > d_{uw}$, then choose node w to establish a connection to node u. Otherwise, choose source s to provide connection to u. Add all DaC-nodes and WC-nodes on the path of $< u - w >$ (or on the path $< u - s >$) to V', add all split-nodes to V and add VS nodes to set Z.

The destinations are added to the tree T one by one until no more destinations can be added to it. If a destination can not be added to the present tree then a new tree is constructed using the above procedure.

Let N be the number of nodes in the network and N_D be the number of destination nodes for a multicast session. In order to add a destination node to the multicast tree, shortest paths are found from this destination node to every node in set V. A conventional shortest path finding algorithm such as Dijkstra's algorithm can be used for this purpose. It requires $O(N^2)$ time in the worst case. This process is repeated for each of the N_D nodes. Therefore, the worst case computational complexity of the algorithm is $O(N_D \times N^2)$.

An example is shown in Figure 19 to illustrate the working of Capability-Based-Connection algorithm. Here, node 1 is the source of a multicast session, and nodes 2 through 10 are destinations. Nodes 3, 4 and 6 do not have split capability, node 2 has both the split and conversion capability and all other nodes have split capability only. Initially the tree contains only node 1. It then tries to add as many destinations as possible to the multicast tree. Nodes 2 and 3 find their shortest paths to node 1 and are directly connected to node 1. Nodes 4 and 5 find their shortest paths to node 2, and node 6 to node 3. Next node 8 is considered. It is at an equal distance to both nodes 5 and 6. Since node 5 is a split-node and has higher priority than node 6, node 8 is connected to node 5. In a similar fashion,

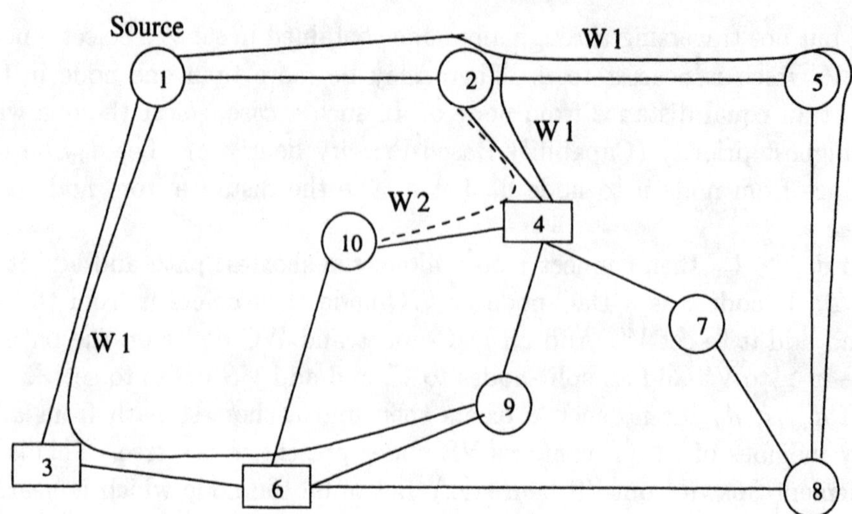

Figure 19: Illustration of Capability-Based-Connection multicast routing algorithm.

node 7 is connected to node 4 and node 9 is connected to node 6. Node 10 is considered last. It cannot be included in the multicast tree since both of its neighbors, node 4 and node 6 do not have splitting capability and their DaC capabilities have already been exhausted. Node 10 is connected to the nearest virtual source node 2. This is because, node 10 is nearer to node 2 than the source node. The link $< 2 - 4 >$ has already been utilized to connect node 4. Hence, node 2 provides a connection to node 10 on a different wavelength. The path from node 2 to node 10 forms a new subtree rooted at node 2.

In [21], the performance of Capability-Based-Connection algorithm is studied and compared with the Member-Only algorithm [13, 20].

1. There is a gain in the number of wavelengths and the number of wavelength channels compared to Member-Only algorithm.

2. The gain in number of wavelength channels reaches to the highest value when 40% of nodes in the network are split nodes.

3. In this algorithm wavelength conversion capability of the nodes is considered at the time of tree generation. As the number of WC-nodes increases, the possibility of split-node becoming a VS increases. This results in gain in number of wavelengths and number of wavelength channels.

Shortcomings of the Source-Rooted Approach

For a given multicast session, Source-rooted algorithms construct a set of trees with an objective of either minimizing the total cost of the tree (Steiner-based tree generation algorithms) or minimizing the individual cost of the path between the source and a destination (Source-based tree generation algorithms). Here, the cost of a path is defined as the number of wavelength channels (hops) used by the path. Source-rooted approach has the following limitations.

- In a wide-area network the destinations of a session are distributed over the globe. Hence, the delay incurred in constructing the tree will be very high.

- There should be a simple procedure to add and delete a node from the session. Deleting or adding a destination to the existing session may change the structure of the tree.

- If a link or a node fails, the tree may need to be re-constructed.

- For every multicast session a set of trees needs to be constructed. The computation overhead to construct a Source-rooted multicast tree for every multicast session is very high.

- VS nodes are used to split the optical signal only if they present in the path computed by the multicast tree construction algorithm. An example is illustrated in Figure 20. Consider node 1 as source and

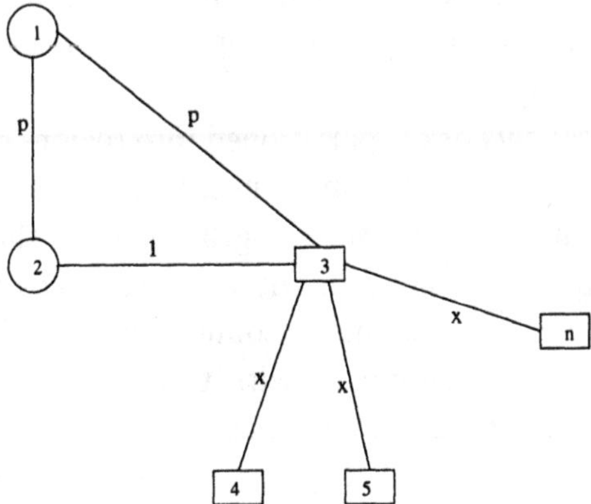

Figure 20: An example showing excess wavelength usage using Source-rooted approach.

nodes 4 through n as destinations of a multicast session. It is assumed that node 2 is the only VS node present in the network. It is also assumed that the distance between node 1 and node 2 is same as the distance between node 1 and node 3 which is equal to p number of hops. The distance between node 2 and node 3 is one hop. The distance between node 3 to every destination is x number of hops. The distance between any two destinations is assumed to be greater than $x + p$ number of hops. In Source-rooted approach, to construct multicast trees, the destinations, node 4 through node n, find their shortest paths to node 1 via node 3. As a result, to construct a multicast forest, Source-rooted approach requires $n \times (x + p)$ number of wavelength channels. If the tree is constructed via VS node (node 2) then the total number of wavelength channels required would be reduced to $p + n \times (1 + x)$.

4.5   Virtual Source Based Trees

In this section a new tree construction algorithm is discussed which is based on VS-rooted approach [26]. This algorithm overcomes the above limitations of Source-rooted approach. This approach is referred to as *virtual source based multicast approach* or simply *VS-based approach*. Some nodes in the network are chosen as VS nodes. These VS nodes are interconnected in such a way that a lightpath is established between every pair of VS nodes. The network is then partitioned into regions based on the vicinity of the VS nodes. These regions and the interconnectivity among the VS nodes is used while the multicast tree is constructed. Thus, the multicast routing works in two phases, namely, network partitioning phase, tree generation phase.

Network Partitioning Phase

The given physical network is partitioned into regions based on the vicinity of the VS nodes. A VS node can transmit an incoming message to any number of outgoing links on any wavelength. Assume that some nodes are chosen as VS nodes. Every node in the network needs to find a shortest path to the nearest VS and establish a connection to it. Thus, the network is partitioned into a set of trees each with the root as a VS node. Hence, the VS nodes are chosen in such a way that they satisfy the following conditions. VS nodes need to have high degree of connectivity and could partition the network into disjoint sets of nodes. The average distance (in terms of number of hops) from a VS node to the set of nodes connected to it is approximately equal to the average distance from any other VS node to its corresponding set of nodes.

A VS node acts as a multicast session distribution point to a set of nodes which are connected to it. Once the VS nodes are identified, then the paths between all VS nodes are computed. Every VS establishes connections to all other VS nodes. As a result of it, the network can be viewed as a set of interconnected VS nodes, and the remaining nodes in the network grouped into trees each with the root as a VS node.

Tree Generation Phase

Given a source and a set of destinations of a multicast session, the aim here is to generate a multicast tree. This phase makes use of the connectivity provided in the previous phase. As a result, the setup time for establishing the multicast session becomes low. The procedure to establish a multicast session is described as follows. The source of the multicast session first establishes a connection to a VS node. The VS node which is chosen by the source to establish a connection is referred to as *primary virtual source* (PVS). PVS is unique for a session. Any VS node which has not been allocated to a session can be chosen as PVS. Here, a VS with least distance (number of hops) from the source has been chosen as PVS for that session. Same wavelength must be available on all the links along the path from the source to the PVS. All other VS nodes which have one or more destinations of a session in the set of nodes connected to it are referred to as *secondary virtual source* (SVS) nodes. PVS and SVS nodes of a session establish connections to the destinations. Same wavelength must be available on the links along the path from a VS node to a destination connected to it. Thus, a multicast tree can be generated using pre-established trees from PVS and SVS nodes.

We now illustrate the working of multicast routing approach. A set of multicast sessions are given in Table 1, the physical network is shown in Figure 21. The VS nodes need to be selected in this network. The nodes with high degree of connectivity can be the candidate nodes for VS nodes. Here, node 4 has 6 neighbors (degree 6), nodes 9 and 13 have 5 neighbors (degree 5), node 6 has 4 neighbors (degree 4), nodes 1, 2, 3, 5, 8, 10, 12, 15, 16, and 17 have 3 neighbors (degree 3) and nodes 7, 11, and 14 have 2 neighbors (degree 2). To select a set of nodes as VS nodes it is required that those nodes should have high degree of connectivity and are evenly distributed over the network. Here, nodes 4, 6, 9, and 13 are chosen as VS nodes.

The network needs to be partitioned based on the vicinity of the selected VS nodes. Every node in the network finds the nearest VS node to it. Here, node 1, node 2, node 14, and node 15 are one hop away from VS node 4.

Table 1: Multicast sessions.

Session	Source	Destinations
1	8	1, 3, 5, 9, 11, 15
2	16	2, 4, 8, 10, 12
3	3	1, 2, 5, 7, 10, 11
4	1	2, 4, 10, 16, 17

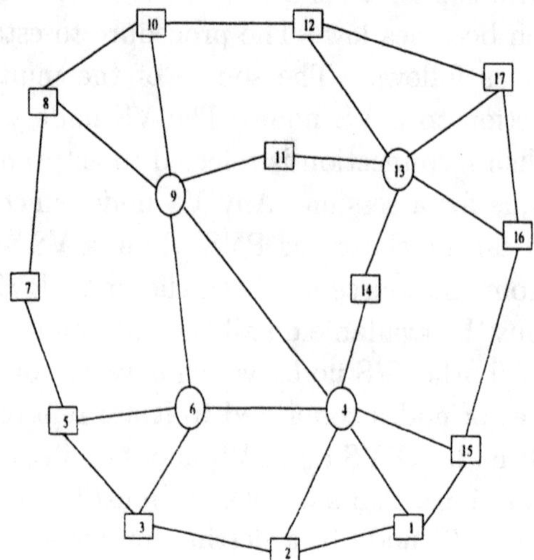

Figure 21: An example network.

Hence, these nodes can be connected to node 4. In a similar way, every node establishes a connection to its nearest VS node. The resulting trees are shown in Figure 22.

Multicast trees for the sessions given in Table 1 need to be generated. Consider the first session in which node 8 is source, and node 1, node 3, node 5, node 9, node 11, and node 15 are destinations. The source node needs to select a VS node as PVS. Node 9 is only one hop away from node 8 and node 9 has not been allocated to any other session. Hence, node 8 selects node 9 as PVS for the session. Shortest path between node 8 and node 9 is computed. One destination (node 11) is connected to PVS and remaining destinations are connected to VS nodes 4 and 6. These two VS nodes can be viewed as SVS nodes. The tree for the first session is shown in Figure 23(a).

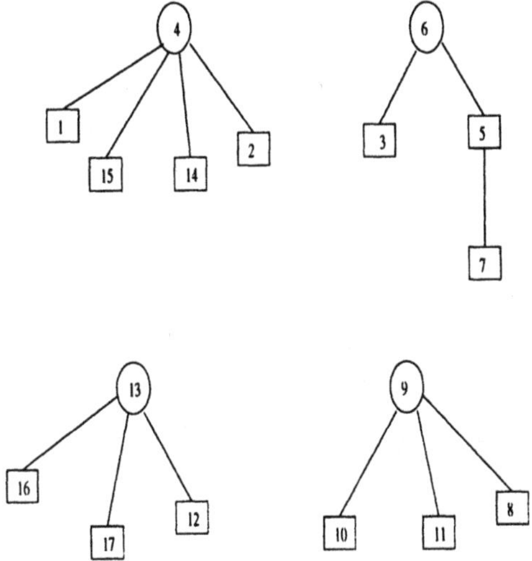

Figure 22: Trees for all VS nodes.

In a similar way the trees for the remaining sessions can be constructed as shown in Figure 23(b) and in Figure 24. Wavelength allocation to the trees for the above sessions is shown in Tables 2 and 3.

Advantages

- In this approach, source need not know about the location of destinations.

- The setup time for multicast tree is very less compared with Source-rooted multicast tree construction. The setup time in VS-based approach can be computed as follows. The connections are from source to PVS, PVS to SVS nodes, and SVS to the destinations. Hence, the time taken to construct multicast tree is sum of the time taken to connect from source to PVS, time taken to connect from PVS to SVS nodes, and the maximum time taken by any VS among all VS nodes to connect to their destinations. However, the connection is already established between PVS to SVS nodes. Hence, the setup time for constructing a multicast tree is time taken to connect from source to PVS plus maximum amount of time taken by any VS nodes among all VS nodes.

- There is a maximum of three light hop distance from source to any destination. Hence, fairness among destinations is achieved.

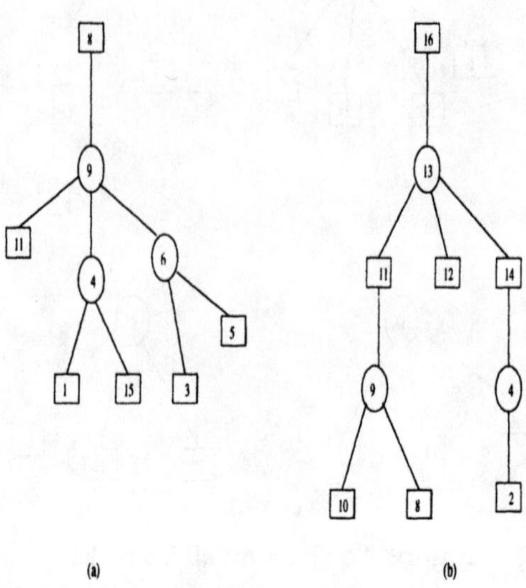

Figure 23: Multicast trees for (a) session 1. (b) session 2.

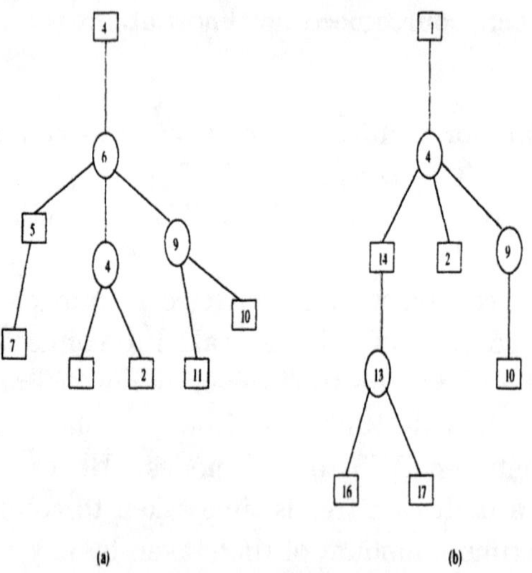

Figure 24: Multicast trees for (a) session 3. (b) session 4.

Table 2: Lightpaths for session 1 and session 2.

LP	Path	WL		LP	Path	WL
8-9	8-9	w1		16-13	16-13	w1
9-4	9-4	w1		13-9	13-11-9	w1
9-6	9-6	w1		13-4	13-14-4	w1
9-11	9-11	w2		13-12	13-12	w1
4-1	4-1	w2		9-10	9-10	w3
4-15	4-15	w1		9-8	9-8	w1
6-3	6-3	w1		4-2	4-2	w3
6-5	6-5	w2				

Table 3: Lightpaths for session 3 and session 4.

LP	Path	WL		LP	Path	WL
3-6	3-6	w1		1-4	1-4	w1
6-4	6-4	w1		4-9	4-9	w1
6-9	6-9	w1		4-13	4-14-13	w1
6-5	6-5	w1		4-2	4-2	w1
6-7	6-5-7	w1		13-16	13-16	w1
4-1	4-1	w1		13-17	13-17	w1
4-2	4-2	w2		9-10	9-10	w1
9-10	9-10	w2				
9-11	9-11	w1				

- The procedure of dynamic addition or deletion of members in the group is simple in VS-based approach. The node which wants to join in the multicast session can be connected to its nearest VS.

- There are multiple VS nodes for every session. Hence, fault tolerance can be achieved even though some VS nodes fail.

Limitations

- A VS can act as PVS for only one session at a time. As a result of it, the number of sessions active at any time is equal to the number of VS nodes in the network. However, a PVS can be made use for more than one session, (a) if more number of wavelengths are reserved for providing connection among the VS nodes and (b) multiplexing techniques can be used to utilize a wavelength channel more effectively.

- As number of VS nodes increases, the overhead due to the resources reserved for paths between VS nodes also increases.

The performance of VS-rooted approach can be compared to Source-rooted approach as follows. In VS-rooted approach, all VS nodes in a network are identified and connectivity among these VS nodes is provided. When a multicast session request comes, the tree is constructed in parallel from every VS node. This results in low setup time which is the maximum amount of time required to construct the multicast tree.

In Source-rooted approach, the setup time can be computed as follows. The construction of a multicast tree can be either from the source (*source initiated tree construction*) or from the destinations (*destination initiated tree construction*). In the source initiated tree construction method, the source node needs to find the availability of wavelengths on the shortest paths to destinations and configure the switches at all intermediate nodes. In the destination initiated tree construction method, each destination finds the availability of wavelengths on the shortest path to a node in the tree and establishes a connection to that node. In these two methods, the number of switches to be configured is equal to the number of wavelength links (channels) used to construct the tree. Hence, in Source-rooted approach, the setup time to construct a multicast tree can be computed as the number of wavelength links (channels) used in the tree.

For smaller group sizes, Source-rooted approach is slightly advantageous than VS-rooted approach in terms of the number of wavelengths required in a fiber. As the group size increases, in Source-rooted approach, the rate at which the number of wavelengths increases is higher than that in the VS-rooted approach. Hence, for larger group sizes VS-rooted approach is advantageous than Source-rooted approach. The effect of group size on the number of wavelengths is explained as follows. To achieve the connectivity among all VS nodes, some number of wavelengths are reserved. It is assumed that, these many number of wavelengths are available on all other physical links. As group size increases, these wavelengths are utilized for connecting the destinations to their VS nodes. Hence, the extra wavelengths required will be less. In VS-rooted approach the generation of multicast tree is distributed among all VS nodes. The tree, for which the VS as root, has connections to fewer number of destinations. As a result, a link in the network is shared by fewer number of multicast trees. In Source-rooted approach, as group size increases, more number of trees (which has source as root) share a physical link. As a consequence of it, the number of wavelengths required also increases.

In VS-rooted approach, when there are fewer VS nodes then the overhead due to connectivity among the nodes is low. As the number of VS nodes increases, the overhead due to connectivity among VS nodes increases. Hence, number of wavelength channels required will be more. In Source-rooted approach, as the number of VS nodes increases, the possibility of encountering a VS node in tree construction is more. Hence, the number of wavelength channels required will be low.

The effect of number of VS nodes is as follows. In Source-rooted approach, as the number of VS nodes increases, the possibility of encountering a VS node in tree construction is more. Hence, the number of VS nodes increases, the number of wavelengths required decreases in Source-rooted approach. In VS-rooted approach, more number of wavelengths are required when more number of VS nodes are present in the network. This is because of more number of wavelengths required to interconnect all VS nodes.

5 IP-over-WDM Networks

Recently, Internet-based services are expanding its applications based upon the *Internet Protocol* (IP) and are driving the need for more bandwidth. WDM can support many hundreds of gigabits per second on a single fiber. Hence, there is a growing interest in integrated IP-over-WDM networks. In this combination, IP layer is viewed as the common revenue-generating convergence sublayer and optical (WDM) layer is viewed as bandwidth-rich transport layer. All services provided by IP networks are need to be mapped to IP-over-WDM networks. One such service is multicasting. In this section, a brief review of IP multicast routing protocols is given, then multicast routing in IP-over-WDM networks is discussed.

5.1   IP Multicast Routing

IP multicast routing protocols are classified into two groups namely, dense mode and sparse mode, depending on the destination group size and the distribution of destinations that they support. Dense mode refers to an environment where group members are densely packed, and sparse mode refers where the group members are distributed across many regions. Alternatively, multicast routing protocols are classified as source-based and shared-tree based, depending on the root of the multicast tree. The examples for dense mode multicast routing protocols are *Distance vector multicast routing protocol (DVMRP), Protocol independent multicast for dense mode (PIM-DM)*, and *Multicast open shortest path first (MOSPF)*. The examples for sparse mode multicast routing protocols are *Protocol independent multicast for sparse mode (PIM-SM)* and *Core based trees (CBT)*.

In DVMRP, a separate tree is constructed for each source and a set of destinations. The path between the source and a destination is chosen in such a way that the path is reverse of shortest path from the destination to the source. First, the multicast data is broadcast by the source using reverse path forwarding. Then, using PRUNE messages, pruning of branches is done to delete the branches that are not leading to any destinations. Nodes can explicitly join in the existing tree by using GRAFT messages. The nodes which run this protocol are considered as *multicast capable* (MC) nodes. The nodes that are not supporting DVMRP are termed as *multicast incapable* (MI) nodes. MC nodes could able to store the multicast data and can forward the same on more than one outgoing link. MI nodes are by-passed using *IP tunneling* concept.

MOSPF is extension of OSPF to construct a multicast tree. Every node (router) in the network maintains the topology information. Multicast group membership is flooded to all nodes in the network. This protocol is based on link state protocol. Each node computes the shortest paths to all other nodes using link state information. Each node maintains the information of every multicast group in its link state database. This will allow a node to compute a pruned tree without explicit flooding and pruning.

PIM receives its name because it is independent on the mechanisms provided by any unicast routing protocol. PIM makes a clear distinction between a multicast routing that is designed for dense mode environment, and one that is designed for sparse mode environment. PIM for dense mode (PIM-DM) is similar to DVMRP. However, PIM-DM relies on the presence of some unicast routing protocol to adapt changes in the topology, whereas DVMRP makes use of its own unicast routing protocol. PIM for sparse mode (PIM-SM) employs the concept of *rendezvous point* (RP) where receivers meet sources. The source of a session selects a primary RP and set of alternative RPs. Only one RP acts as active RP for a multicast group. Every member of the multicast group connects to the primary RP using explicit JOIN messages. The source of the multicast session uses the RP to announce its presence and to find a path to members that have joined the group. The tree that is formed using above mechanism is a RP-shared tree. The RP-shared tree provides connectivity but does not optimize the delivery path. To optimize the delivery path a destination has option of switching to the source rooted shortest path tree as soon as it starts receiving data from the source.

Core based tree (CBT) constructs a single delivery tree that is shared by the members of a multicast session. A core based tree uses one or a set of nodes as the core of multicast tree. A node that wishes to join a session need to send JOIN message to the existing tree for that multicast session. All members that are already included will forward the JOIN message to the core of that tree. If the source of the session is not a member of the tree then it needs to transmit the multicast data to one of the members of the tree, so that the data is distributed to all members of the session.

5.2 WDM Multicast Routing

Multicasting in IP-over-WDM networks can be done by using either IP multicast routing or by WDM multicast routing. In IP multicast routing, each IP router on a multicast tree makes copies of a data packet and transmit a copy to each down stream router. This requires O/E-E/O conversion at all routers on the tree, and results in inefficient use of resources. O/E-E/O conversion can be avoided by using a virtual topology. The first drawback in such a scheme is the control overhead involved in establishing a multicast tree for a session with shorter duration. The solution is to provide *optical burst switched networks* (OBS) [27]. Here, no wavelength is dedicated to a multicast tree and the multicast data is transported using OBS. In OBS, the path is setup as and when some bursty traffic need to be transported. The second drawback in using virtual topology is, high bandwidth consumption for the sessions with large multicast groups. Because, it requires an unicast path is to be setup between source and every destination. The solution for the second drawback is to provide optical splitting capability at the routers (optical wavelength cross connects). However, optical splitting is a costlier operation. Hence, only few routers have the optical splitting capability. In such a scenario, some routers (nodes) in the network are multicast capable (MC), and some nodes are multicast incapable (MI). With respect to IP multicast, MC nodes are the nodes where multicast routing protocol is running, and IP datagrams are replicated when the paths to different routers diverge. MI nodes are the nodes where multicast routing protocol in not running. While constructing a multicast tree, MI routers are by-passed using IP tunneling. However, an MI router with respect to IP may have the optical splitting capability. Because of this, even though a node supports optical splitting it may not be considered while constructing a multicast tree. This results in poor bandwidth utilization and some times non-delivery of data to some destinations may also occur. Hence, the IP multicast routing protocol need to consider the capabilities of the nodes. But, it may require some modification in the existing IP multicast routing protocols. Two different approaches are proposed based on whether to provide modifications in the existing routing protocols or not, *Not-Modify-IP* and *Modify-IP* [28]. In the rest of this section multicast routing in OBS, multicast routing using Not-Modify-IP, and Modify-IP are discussed.

5.2.1   Multicast Routing in OBS Networks

In OBS, the main overheads are control packets and guard bands. Three different schemes are proposed to minimize the above said overheads, namely, *separate multicast (S-mcast)*, *multiple unicast (M-ucast)*, and *tree-shared multicast (TS-mcast)*.

In S-mcast, each multicast session constructs its own tree along which the assembled bursts carrying multicast data for the multicast group is delivered. In each session, IP packets arriving within a given assembly time are assembled into a burst. After the assembly time is over, the burst is transmitted along the multicast tree. To generate the tree, the algorithms which are using source-rooted approach may be used.

In M-ucast, a copy of the multicast data can be assembled together with unicast data for the same destination. Here, the control overhead is reduced since the guard bands can be shared by both unicast and multicast traffic in a longer burst. Also, the number of control packets may also decrease. This scheme is particularly useful when the multicast data is only of few packets.

In TS-mcast, the degree of membership overlap is exploited, and a set of shared trees are generated to transport the multicast data. Here, the set of multicast sessions originating from a router is divided into a number of subsets based on some strategy. Each subset is called as *multicast sharing class (MSC)*. IP packets belonging to the multicast sessions in a MSC are assembled together to form bursts. This forms longer bursts and reduces overheads in transporting bursts. The members in each MSC are decided based on the way the group members of the multicast session overlap. Three strategies are proposed in [29], *equal coverage, super coverage*, and *overlapping coverage*, to decide the members of MSC. Equal coverage strategy groups the sessions which have same members. Hence, the tree generated by one session can be made use by the other session. In super coverage strategy, all sessions are considered which have members as the subset of one of the sessions. The tree formed for the session which is superset of other sessions (with respect to membership of multicast group) is shared by all other sessions. Here, the sessions are added, to form one MSC, one by one until the gain of sharing keeps increasing. However, in OBS networks a burst may be blocked at the intermediate nodes due to contention for a limited number of wavelengths. A blocked burst may be buffered in a *fiber delay line* (FDL) or it may be dropped in buffer less OBS networks. To recover a lost burst, a protocol for reliable multicast is proposed in [30]. This protocol operates at WDM layer and requires some additional functions, like sub casting and maintaining burst state, that need to be performed at the WDM switch.

5.2.2 Not-Modify-IP

In this approach the following assumptions are made with respect to the capability of the switch. A multicast controller on each optical switch controls the WDM layer multicast and it knows the splitting capability of the local node. The membership information known by an IP controller is also available to the multicast controller. Consider a situation where the multicast diverges at a node, say S1. If the switch at node S1 is not multicast capable (not having optical splitting capability) then some downstream nodes can not get connected to the tree, say one such a node is S2. These affected nodes need to get unicast path from any node in the reverse path from the source.

If DVMRP is used as the IP multicast routing protocol, then three options are suggested to connect those nodes that are not able to be connected to the multicast tree due to MI nodes present in the WDM optical layer. The first option is initiated by the parent of the node, that could not be included in the tree. In this example S1 will initiate the process of getting a connection from its upstream router for S2. The second option is initiated by the relative of affected node S2. The meaning of *relative* here is, a node which is a neighbor of affected node S2 but it is not parent of node S2. The third option is initiated by the affected node itself. It tries to get a unicast path from a node in the path from the source of the session.

If MOSPF is used as the IP multicast routing protocol, then every node has full information about the multicast group membership. A node receives a multicast message only through the shortest path from the source. Hence, if a node could not get connection then a unicast path is set up from a node (which is an MC node) in the shortest path.

However, the above protocol to setup the connection is a source initiated one. It is possible that a volatile tree is generated in the source initiated process. In [31], a receiver initiated tree setup protocol is proposed. This protocol also considers the various capabilities of the nodes like splitting, wavelength conversion, Dac, and limited add/drop capabilities.

5.2.3 Modify-IP

IP multicast routing protocols can be modified in such a way that the splitting capability of the switches is considered while generating multicast tree. By adding the knowledge of splitting capability of local switch, each IP controller is able to construct the entries into the routing table which reflect the capability of that switch. If MOSPF is used as the IP multicast routing protocol then the source rooted algorithms can be made use to generate the multicast tree.

6 Summary

Multicasting is the ability to transmit information from a source to multiple destination nodes and is becoming an important requirement in high-speed networks. As WDM technology matures and multicast applications become increasingly popular, supporting multicast routing at WDM layer becomes an important and yet a challenging topic. In this chapter, the multicast routing in various WDM networks namely, broadcast-and-select networks, linear lightwave networks, and wavelength routed networks was discussed.

Broadcast-and-select networks are inherent choice for providing multicast routing. However, the number of wavelengths and the receiver tuning latency are considered as the main problems. In this chapter, some solutions for the above problems were discussed. In linear lightwave networks, the constraints for multicast routing are waveband inseparability and mutually independent source combining. The multicast routing and channel allocation strategies in linear lightwave networks were also discussed in this chapter.

In wavelength routed networks, to support multicast routing the nodes in the network need to have optical power splitting capability. However, a node with optical splitting capability is an expensive one. Hence, it is suggested to have splitting capability only at a few nodes. Depending on whether all nodes have splitting capability or a few nodes have splitting capability, the networks are classified as networks with full splitting capability or networks with sparse splitting capability. The multicast routing in these two types of networks was also discussed in this chapter. In the networks with full splitting capability, if every node has wavelength conversion capability then the multicast tree generation is similar to the multicast tree generation in conventional electronic networks. The goals here were to reduce the number of transceivers, and to design efficient routing and wavelength assignment for multicast traffic (which is also termed as MCRWA problem). In a network with sparse splitting capability, only a few nodes in the network have splitting capability. If wavelength conversion capability is also added to these nodes then these nodes can transmit on any number of output links and on any wavelength. These nodes are termed as VS nodes. These VS nodes need to be exploited to the maximum possible extent. The algorithm Capability-Based-connection mainly concentrates on exploiting the capability of VS nodes which are present in the already constructed tree. The *virtual source based trees* was proposed to construct the multicast tree using the knowledge of the presence of VS nodes in the network. Here, VS nodes are used for constructing the tree in parallel. This results in reduced setup time.

In [32], a set of desirable properties for multicast routing are listed. Among these properties, *high probability of delivery of information, low delay between source and destinations*, and *information hiding from the intermediate routers* can be achieved by using the optical signal to transfer the data (optical networks) from the source to the destinations. The *scalability* property (when more multicast requests arrive) can be achieved using WDM, as more number of wavelengths can be accommodated in a fiber. *Fault toler-*

ance and *multicast tree flexibility* can be dealt by using VS-rooted approach. Multicast tree flexibility means a node may join a session or may leave from the session dynamically. Another important property for a good multicast routing is that the source need not know the location and identities of multicast group. This can be achieved by using distributed implementation of tree generation with VS-rooted approach. The multicast routing algorithm should be independent of routing algorithms used in different domains of a WAN environment, which is termed as *routing algorithm independence*. This aspect requires attention for further research.

Presently, the research in developing algorithms for routing both unicast and multicast traffic is active. In [33], scheduling of multicast traffic and unicast traffic in a broadcast-and-select network is dealt. In [34], scheduling of unicast traffic and multicast traffic is dealt in a wavelength routed network with full splitting capability. This problem is yet to be addressed in a network with sparse splitting capability.

Recently, IP-over-WDM networks, have received considerable amount of attention. In a wavelength routed WDM network, one wavelength is assigned on every link of multicast tree. This is similar to circuit switching concept. On the other hand, IP is based on the packet switching concept. In an IP multicast routing protocol, IP multicast packets are broadcast from the source. Any *multicast capable* (MC) router buffers the incoming multicast message and forwards the same onto the required output links. A router in an IP network is called as MC router, if appropriate IP multicast protocol is running in it. An MC router can forward an IP multicast packet to another MC router using IP-in-IP tunneling to by-pass *multicast incapable* (MI) routers. In [28], two approaches are proposed for IP multicast routing in WDM networks. The algorithms based on the first approach, require the existing IP multicast routing protocol to be modified. The protocol is modified in such a way that, the multicast tree/forest generated by the algorithms mentioned in this chapter are made use. In the second approach, multicast routing based on *optical burst/label switching* is proposed [27]. In this method, the multicast data is transmitted as optical burst at the rate of capacity of a wavelength. Here, the overheads are control packets and guard bands. To minimize these overheads, efficient multicast schemes are proposed in [29]. This method may make use of *fiber delay lines* (FDL). These FDLs are used for creating a delay in the optical burst so that the header can be checked for routing information. If FDLs do not exist, then the multicast data may be dropped due to heavy load at the receiver. In [30], burst recovery schemes for reliable delivery of packets are proposed.

References

[1] B. Mukherjee, *Optical Communication Networks*, (McGraw-Hill Companies, Inc., 1997).

[2] T. E. Stern and K. Bala, *Multiwavelength Optical Networks- A Layered Approach*, (Addison Wesley Longman, Inc., 1999).

[3] R. Ramaswami and K. N. Sivarajan, *Optical Networks- A Practical Perspective*, (Morgan Kaufmann Publishers, Inc., 1998).

[4] M. Borella and B. Mukherjee, Multicasting in a WDM local lightwave networks, *In Proceedings of ITC Sponsored Seminar on Teletraffic Analysis and Methods for Current and Future Telecom Networks*, pp. 71-77, November 1993.

[5] M. Borella and B. Mukherjee, A reservation-based multicasting protocol for WDM local lightwave networks, *In Proceedings of ICC'95*, pp. 1277-1281, 1995.

[6] G. Rouskas and M. Ammar, Multidestination communication over tunable-receiver single-hop WDM networks, *IEEE Journal on Selected Areas in Communications*, vol. 15, no. 3, pp. 501-511, April 1997.

[7] J. Jue and B. Mukherjee, The advantages of partitioning of multicast transmissions in a single-hop optical WDM networks, *In Proceedings of ICC'97*, 1997.

[8] Z. Ortiz, G. N. Rouskas, and H. G. Perros, Maximizing multicast throughput in WDM networks with tuning latencies using the virtual receiver concept, *European Transactions on Telecommunications*, vol. 11, no. 1, pp. 63-72, January-February 2000.

[9] D. Gusfield, Connectivity and edge disjoint spanning trees, *Information Processing Letters*, vol. 16, pp. 87-89, February 1983.

[10] K. Bala, K. Petropoulos, and T. E. Stern, Multicasting in a linear lightwave network, *In Proceedings of IEEE INFOCOM 1993*, pp. 1350-1358, 1993.

[11] H. Harai, M. Maruta, and H. Miyahara, Multicast routing method in optical switching networks, *Journal of Electronics and Communications*

in Japan (Translated from Denshi Joho Taushin Gakkai Ronbunshi), part. 1, vol. 79, no. 8, pp. 12-23, 1996.

[12] R. Ramaswami, Multiwavelength lightwave networks for computer communication, *IEEE Communications Magazine*, vol. 31, pp. 78-88, February 1993.

[13] X. Zhang, J. Wei, and C. Qiao, Constrained multicast routing in WDM networks with sparse light splitting, *In Proceedings of INFOCOM 2000*, March 2000.

[14] I. Chalmtac, A. Ganz, and G. Karmi, Lightpath communications: An approach to high bandwidth optical WANs, *IEEE Transactions on Communications*, vol. 40, no. 7, pp. 1171-1182, July 1992.

[15] S. Subramaniam, M. Azizoglu, and A. Somani, All-optical networks with sparse wavelength conversion, *IEEE/ACM Transactions on Networking*, vol. 4, pp. 544-557, August 1996.

[16] L. H. Sahasrabuddhe and B. Mukherjee, Light-trees: Optical multicasting for improved performance in wavelength routed networks, *IEEE Communications Magazine*, pp. 67-73, February 1999.

[17] G. Sahin and M. Azizoglu, Routing and wavelength assignment in all-optical networks with multicast traffic, *European Transactions on Telecommunications*, vol. 11, no. 1, pp. 55-62, January-February 2000.

[18] M. Ali and J. S. Deogun, Power-efficient design of multicast in wavelength routed networks, *IEEE Journal on Selected Areas in Communications*, vol. 18, no. 10, pp. 1852-1862, October 2000.

[19] R. K. Pankaj, Wavelength requirements for multicasting in all-optical networks, *IEEE/ACM Transactions on Networking*, vol. 7, no. 3, pp. 414-424, June 1999.

[20] R. Malli, X. Zhang, and C. Qiao, Benefit of multicasting in all-optical networks, *In Proceedings of SPIE , All Optical Networking*, pp. 209-202, November 1998.

[21] N. Sreenath, G. Mohan, and C. Siva Ram Murthy, Virtual source based multicast routing in WDM optical networks, *Photonic Network Communications*, vol 3, no. 3, pp. 217-230, July 2001.

[22] R. Sriram, G. Manimaran, and C. Siva Ram Murthy, Multimedia multicasting: A survey of issues and solutions, *Technical Report, Department of Computer Science and Engineering, Indian Institute of Technology, Madras, India*, 1998.

[23] H. Takahashi and A. Matsuyama, An approximation solution for the steiner problem in graphs, *Math. Japonica*, vol. 24, pp. 573-577, 1980.

[24] K. B. Kumar and J. M. Jaffe, Routing to multiple destinations in computer networks, *IEEE Transactions on Communications*, vol. 31, no. 3, pp 343-351, March 1983.

[25] V. P. Kompella and C. Joseph, Multicast routing in multimedia communication, *IEEE/ACM Transactions on Networking*, vol. 1, no. 3, pp. 286-292, June 1993.

[26] N. Sreenath, N. Krishna Mohan Reddy, G. Mohan, and C. Siva Ram Murthy, Virtual source based multicast routing in WDM networks with sparse light splitting, *In Proceedings of IEEE Workshop on High Performance Switching and Routing*, May 2001.

[27] C. Qiao, M. Jeong, A. Guha, X. Zhang, and J. Wei, WDM multicasting in IP-over-WDM networks, *In Proceedings of International Conference on Network Protocols*, pp. 89-96, November 1999.

[28] X. Zhang, J. Wei, and C. Qiao, On fundamental issues in IP-over-WDM multicast, *In Proceedings of International Conference on Computer Communications and Networks*, October 1999.

[29] M. Jeong, Y. Xiong, H. C. Cankaya, M. Vandenhoute, and C. Qiao, Efficient multicast schemes for optical burst-switched WDM networks, *In Proceedings of ICC 2000*, pp. 1289-1294, June 2000.

[30] M. Jeong, C. Qiao, and Y. Xiong, Reliable WDM Multicast in optical burst-switched networks, *In Proceedings of OPTICOMM 2000*, pp. 153-166, October 2000.

[31] M. Salvador, S. Heemstra de Groot, and D. Dey, Supporting IP dense mode multicast routing protocols in WDM all-optical networks, pp. 179-190, *In Proceedings of OPTICOMM 2000*, October 2000.

[32] T. Ballardie, P. Francis, and J. Crowcroft, Core based trees (CBT)- An architecture for scalable inter-domain multicast routing, *In Proceedings of SIGCOMM 93*, pp. 85-95, 1993.

[33] Z. Oritz, G. N. Rouskas, and H. G. Perros, Scheduling combined unicast and multicast traffic in broadcast WDM networks, *Photonic Network Communications*, vol. 2, no. 2, pp. 135-153, May 2000.

[34] M. Listanti, A. Cervelli, and R. Sabella, Strategies and algorithms for routing both unicast and multicast paths in WDM networks, *European Transactions on Telecommunications*, vol. 11, no. 1, pp. 43-54, January/February 2000.

OPTICAL NETWORKS - RECENT ADVANCES
L. Ruan and D.-Z. Du (Eds.) pp. 271 - 297
©2001 Kluwer Academic Publishers

Architecture and Analysis of Terabit Packet Switches Using Optoelectronic Technologies

Ti-Shiang Wang
Nokia Research Center
5 Wayside Road, Burlington, MA 01803, U.S.A
E-mail: ti-shiang.wang@nokia.com

Sudhir Dixit
Nokia Research Center
5 Wayside Road, Burlington, MA 01803, U.S.A
E-mail: sudhir.dixit@nokia.com

Contents

References

Abstract

To support all traffic types, especially data-centric traffic, carrier-based terabit packet switches are required to satisfy the ever-increasing bandwidth explosion. In this paper, we list briefly the recent activities on optical packet switches. Then, we present several architectures of a terabit-level packet switch using optoelectronic technologies. As a near term solution for all-optical switching, two optical bufferless interconnection networks, implemented by small modular structures to provide capacity in the range of terabit per second, are proposed and discussed. Alternatively, as a long-term approach, a novel optical packet switch with small quantity of optical buffers is presented and its performance with respect to buffer size is discussed.

1   Introduction

The explosive growth of traffic in the Internet has put tremendous pressure on the telecommunication networks for more capacity. As the data traffic increases dramatically, packet switching is poised to take over circuit switching since it can deliver services far more cost-effectively and flexibly. On one hand, electronic packet switches and routers, including the accompanying technologies of Internet Protocol (IP) and asynchronous transfer mode (ATM), will get faster and process hundreds of thousands of flows with service-level-guarantees. On the other hand, the huge bandwidth of optical fiber ($\sim$Tb/s) and advanced WDM technologies, such as erbium-doped fiber amplifier (EDFA), planar lightwave circuit (PLC) and optoelectronic integrated circuit (OEIC), have been considered as the emerging solution for the next generation Internet [1]. That is, the increasing growth of Internet traffic has led to a dramatic increase in the demand for data transmission capacity. As shown in Figure 1, EDFA played a pivotal role in accelerating the deployment of point-to-point wavelength division multiplexing (WDM) in long haul applications. With the advent of optical add/drop multiplexers (OADMs), which enable adding and dropping of the local traffic, WDM deployment in the metropolitan area network (MAN) is currently receiving a lot of attention. With migration toward an IP optical networking backbone, optical packet switched networks will support the very high bandwidth demand of future networks. For packet-based applications, packets will be individually routed and switch reconfiguration speed will become quite important.

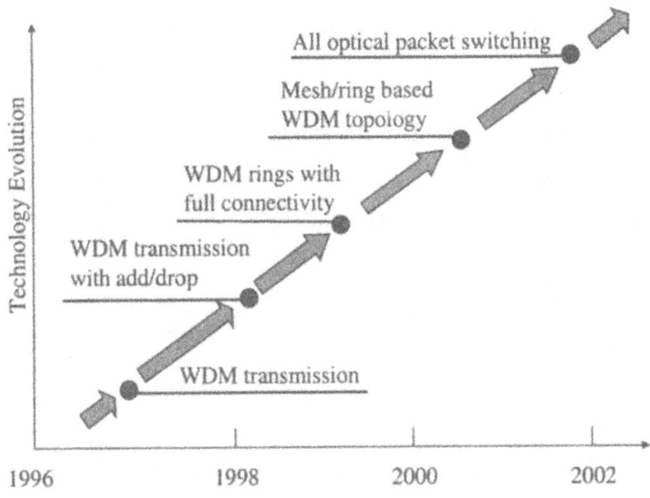

Figure 1: Evolution scenario of WDM networks.

In the present Internet, IP services are provided using a wide range of electronic multiplexing and switching equipment. A typical network may include many different electronic multiplexing and switching layers. For example, as shown in Figure 2, Internet packets may be carried using ATM, which, in turn, are carried over Synchronous Optical Networks (SONET) or Synchronous Digital Hierarchy (SDH) transport frames. The multitude of layers produce bandwidth inefficiencies, add to the latency of connections, and inhibit quality of service assurances. Worse, the layers are largely unaware of each other, causing duplication of network services and generating unnecessary overhead at each layer. Alternatively, one could use the simplified protocol stack where IP traffic is carried by SONET/SDH over the WDM or fiber layer, or IP traffic is directly mapped to the WDM layer. Thus, all optical switches open the possibility for new network architectures that are transparent to packet data rate and format, and extend the success of point-to-point links to an all-optical switched network. Thus, optical packet switches open the possibility for new network architectures to manage the abundance of bandwidth and make it possible that the IP or ATM traffic is directly mapped on the WDM layer. The question of how to build high capacity packet switching systems in the range of terabits using the huge bandwidth of optical fiber and optoelectronic technologies remains an open challenging issue for the industry.

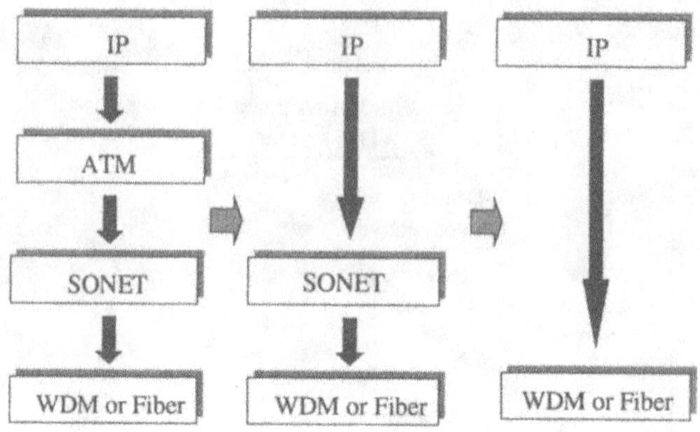

Figure 2: Evolution of IP over WDM or fiber.

2 Transparant vs. Opaque

Basically, optical packet switches can be categorized as transparent or opaque depending on whether the payload is processed optically or electronically, as shown in Figure 3. Transparent optical switches are a major advance on bit rate/signal format independence, network integration capacity, and operational simplicity. That is, in the transparent optical switches, packets are allowed to bypass extensive electronic signal processing at intermediate switching nodes. On the other hand, in opaque optical switches, the payload of each incoming packet is regenerated electrically at each switching node. Transponders, which perform optical-to- electrical (O/E) and electrical-to-optical (E/O) function, clean up the signal and provide wavelength interchange. However, in the switching core, either optical or electronic switching fabric can be used to perform the switching function in the opaque switches. As output contention occurs when multiple packets are destined to the same output port or wavelength, the switching nodes without buffering either rely on some deflection approach or on wavelength conversion scheme to solve the contention problem [2]. Deflection makes the contending packet(s) to be rerouted to other available output port(s). In addition, deflection routing, because of its inherent limited-time buffering, can eliminate the need for optical amplifiers in the optical memory. However, the sequencing and/or the service requirements of packets are controlled by the higher layer protocol (e.g., TCP layer). Alternatively, wavelength conversion technique is used to switch all the contending packets to different wavelengths so that these packets do not experience wavelength conflict at the same output port. In the opaque optical switches, the transponders can, at the same time, perform the wavelength conversion and reshape the incoming packet electronically. On the other hand, in the transparent optical switches, all optical wavelength converters are required to perform the wavelength conversion in the switching node. All optical wavelength conversion techniques based on semiconductor optical amplifiers (SOA) are proving to be most promising for wavelength conversion. In particular, nonlinearities in SOA, when applied to wavelength conversion, are cross-gain modulation (XGM), cross-phase modulation (XPM) and four-wave mixing (FWM) [3].

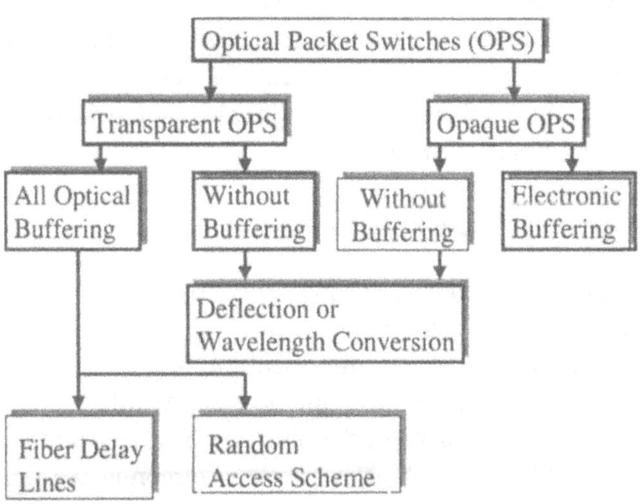

Figure 3: Categories of optical packet switches.

3 Electronic vs. Optical Buffers

Another approach to deal with the problem of contention is to use buffers to store the contending packet(s). Due to the opacity of the opaque optical switch, electronic buffering can be integrated with the transponder and placed at either the input or the output of the switching node. On the other hand, unlike electronic buffers or memory with mature logical function and signal processing, optical buffers can only be implemented by delaying the transmission time of optical packets by fiber lines. For transparent optical switches, two kinds of optical buffering schemes have been proposed: fiber delay line and random access. A comprehensive survey of optical buffering schemes and their comparison is presented in [4]. Optical buffering schemes based on fiber delay lines are used to stagger the incoming contended packets into different fiber delay lines, which represent different packet times to send out the packets to the same output port. In optical buffering based on random access scheme, all the packets on different wavelengths share the same fiber delay line and packets (or wavelengths) can be stored-in or read-out by controlling optical switches in the optical buffer.

4   An Overview of Related Work

In this section, we review several approaches to build a large-capacity packet switch and discuss their advantages and disadvantages. Basically, these switch architectures are classified, depending upon the type of the storage used to resolve the contention, into optical-buffered packet switches and electronic-buffered packet switches.

4.1   Optical-Buffered Packet Switches

In all optical packet switches, both logical control and contention resolution is assumed to be handled by an electronic controller. However, the payload of a packet is carried and stored in optical buffers.

Haas proposed an "almost-all" optical switch, called Staggering Switch which uses fiber delay lines for packet storage [5]. This architecture consists of two rearrangeable nonblocking networks interconnected by optical delay lines with different amounts of delays and an electronic controller. The scheduling stage distributes the packets to the optical delay lines in such a way that no two packets arrive at the switching stage destined for the same output in any packet time slot. Output collisions are solved by delaying the colliding packets with different number of slots. The switching stage permutes the outputs of the delay lines so the packets emerge at appropriate output ports. However, the implementation of the scheduling algorithm to preserve the packet sequence is complicated, and it degrades the performance of packet loss probability. Furthermore, the buffering scheme needs space division approach to route packets to the delay lines. Thus, large number of control signals to these two rearrangeable nonblocking networks complicate the electronic controller implementation. In addition, significant power loss will be experienced in these two rearrangeable nonblocking networks.

D. Chiaroni et al. proposed a 16×16 optical packet switching architecture with a data rate of 10Gb/s at each input/output port [6]. This switching architecture uses wavelength converters and wavelength selectors to achieve packet routing and the exploitation of fiber delay lines accessed by fast optical gates to perform packet buffering. In this switch, in a packet time slot, the incoming packets are carried with different wavelengths and sent to the fiber delay lines simultaneously. The fast switching gates, followed by the fiber delay lines, are controlled to select the WDM signals stored in the fiber delay lines destined to their destinations. Only one of the selected WDM signals will be selected and sent to the output port by a wavelength selector. This switch can perform the multicast function due to the broadcast and-select property of the switch. However, the capacity of this switch is limited by the size of the multiplexer/demultiplexer. Furthermore, the number of fast switching gates increases proportional to the buffer size and the wavelength number: it increases the optical component and interconnection complexity. The related and comprehensive work based on this architecture is described in [7].

Duan et al. introduced a packet switching architecture with tunable wavelength converters at input ports and fixed wavelength filters at output ports [8]. Once an incoming packet enters the switch, a small part of power will be used by the control unit to detect its output address. The control unit will then tune the wavelength converter to an appropriate wavelength, depending on the output address. Multiple packets destined to the same output port are assigned to different fiber delay lines and wavelengths are selected from the fiber delay lines to the destined output port. The optical delay lines are used to store packets temporarily if the star coupler is dealing with packets for the same destination. The control unit checks the number of packets going to the same output and then indicates the specified optical delay line to which those packet(s) must be forwarded. This switch cannot perform multicast functions because a fixed wavelength filter is used at each output port. In addition, the capacity is limited by the size of star coupler and passive couplers.

Sasayama et al. proposed a frequency router-based optical packet switch [9]. In this switch, at each input port, a tunable frequency converter assigns a frequency channel corresponding to the desired output port. That is, the header information is analyzed to obtain the destination address as the frequency-assignment signal. Based on the assignment signal, the optical packet is carried by the specified wavelength related to its destined output port. Two or more packets coming from different input ports destined to the same output port are always carried by different wavelengths and the WDM signal is stored in the output buffer at the output port. This frequency routing function is achieved by using an arrayed-waveguide grating router (AWG). However, the capacity of the switch is limited by the size of the frequency router and this switch cannot perform multicast functions.

Choa et al. introduced a star coupler-based optical packet switch [10]. This switch uses integrated optical devices and highly parallel processing and pipeline control electronic circuits. The switch handles signals at the packet level and can achieve very high-speed and high-throughput operation. At each input port, a waveguide taper is used to bring down a small part of the optical signal. After O/E conversion, the header information is extracted from the packet for processing while the packet payload is delayed by a fiber to compensate for the packet processing time. The synchronizer at each input port is used to synchronize the timing between the system clock and the input signal. After the synchronization, a new virtual channel identifier (VCI) is written into the packet to replace the old VCI. After the VCI overwriting, the packet coming from each input port will be wavelength-converted to the corresponding wavelength port by a wavelength converter. Then, it is selected by the optical concentrator to the corresponding output. Multiple packets destined to the same output port are stored in the optical buffer with different wavelengths. However, the capacity of the switch is limited by the size of the star coupler.

4.2   Electronic-Buffered Packet Switches

For the electronic-buffered packet switches, all the logical control, contention resolution, and packet storage are handled electronically. That is, the contended packets are stored in the electronic buffers and optical switches are used for interconnection and transmission between electronic input and output ports.

Arthurs et al. proposed a 64Gb/s optoelectronic hybrid packet switching system (HYPASS) [11]. It is based on an input buffered, output controlled arbitration protocol. The HYPASS is formed from two networks: a data transport network that connects the input ports to the output ports, and a control network in which the output ports transmit their status to the input ports. Packets arriving at an input port are held at the port controller until a "go" signal from the destined output port, is received, indicating that the packet is to be sent to the output port. In this switch, electronic components are used for buffering, logic function, and the control of optical components for routing and transport. A unique optical wavelength is associated with each of the output ports and, therefore, addressing with the output ports is achieved by transmission on that unique wavelength. However, the capacity of this switch is limited by the size of the star couplers. Furthermore, throughput is limited to 31.8% because of the use of a multi-access arbitration protocol with collisions. This switch design cannot provide multicast functions due to fixed wavelength receivers used at the output ports.

Lee et al. introduced an optical multicast switch system, called Star-Track [12]. It is based on a two-phase contention resolution algorithm. This switch consists of two internal networks, an optical "star" transport network, and one or more electronic "tracks" surrounding the star and sequentially connecting the input ports and output ports. Packets arriving from input ports are stored in electronic buffers. There are two control phases, which take place during a transmission cycle. The input ports write information into tokens indicating to which output ports their packets are to be sent (reserving these output ports) in the first control phase. The output ports read the tokens, and tune their receivers to the appropriate input port wavelengths during the second phase. As the token is emitted from the token generator, transmission of the packet is commenced at the input ports. A novel "call splitting" algorithm that is proposed in the paper enhances the multicast performance of the switch. The major disadvantage of this switch is that it is difficult to scale up the switch size, which is limited by the size of star coupler and the electronic processing time. Though the overall throughput can be improved by using a multiple major track scheme, this approach increases its control complexity.

Cisneros et al. proposed a packet switch based on memory switches and star couplers [13]. This architecture consists of input modules, output modules, optical star couplers, and a contention resolution device (CRD). Packets stored in the input memory modules are given new routing header information. Operation begins with the input modules transmitting request to the CRD for the head-of-queue packets. The information provided to the CRD by input modules is the output module number. The CRD resolves the output contention and responds to the input modules. Packets that won output contention are routed through the same star coupler with different wavelengths to the output modules. The output modules then route packets to their destined output ports. However, this switch cannot perform multicast functions due to the use of fixed wavelength filters at the output modules. Additional optical amplifiers are necessary to compensate for the power loss caused by star couplers so the complexity of the switch is increased. Furthermore, this architecture is input-buffered, which limits the maximum throughput to 58.6%. The arbitration scheme used to solve the contention problem is very complex and may limit the scalability of the switch.

Munter et al. introduced a high-capacity packet switch based on advanced electronic and optical technology [14]. The main components of the switch are input buffer modules, output buffer modules, a high-speed switching core, and a central control unit. The core switch contains a cross-connect network using optical links running at . The central control unit receives requests from input buffer modules and returns grant messages. Each request message indicates the number of queued packets in the input buffer module, which is later used to determine the size of burst allowed for transmitting to the switch fabric. A connection can only be made when both input and output ports are free. A control bus is used by the free input ports to broadcast their requests, and by the free output ports to return grant messages. An arbitration frame consists of 16 packet time slots for a core switch. In each slot, the corresponding output port polls all 16 inputs. For example, in time slot 1, output port 1 (if it is idle) will choose the input that has the longest queue destined for output port 1. If the input is busy, another input port that has the second longest queue will be examined; this operation repeats until a free input port is found. If a match is found (free input, free output, and outstanding request), a connection is made for the duration corresponding to the number of packets queued for this connection. So, the switch is a burst switch, not a packet switch. In time slot 2, output port 2 repeats the above operation. The switch capacity is limited by the speed of the central control unit. Packet streams can experience a long waiting time in the input buffer modules under a high traffic load.

Nakahira et al. introduced an optical packet switch architecture in which an electronic buffer is provided in the input side and an optical buffer is provided in the output side [15]. The optical buffer scheme consists of fiber delay lines so that packets destined to the same output port are stored in the different delay line positions. This switch consists of input group modules (IGM's), one switch module, and several output group modules (OGM's). A complex arbitration is necessary to solve the output contention not only for those packets in the same IGM but also for the packets in the different IGM's. A scheduling algorithm is necessary to preserve the packet sequence and to assign the proper wavelengths to the incoming packets so that there are no two packets with the same wavelengths entering the star coupler in the IGM. This will increase the control complexity. Furthermore, there are many O/E and E/O converters necessary in the switching path: it also limits the user data rate and reduces the capacity that optical components can supply.

5   Generic Architecture of Optical Packet Switch

In Figure 4, we show a generic optical packet switch architecture. Basically, each optical switching node consists of four subsystems: input port interface (IPI), optical switching matrix, output port interface (OPI), and electronic controller. Depending on the architecture, both IPI and OPI perform one or more of the following functions: buffering, synchronization, packet delineation, header processing and updating, and optical or optoelectronic conversion. Optical switching matrix is also controlled by the electronic controller to perform the switching function and to route optical packets to their destinations.

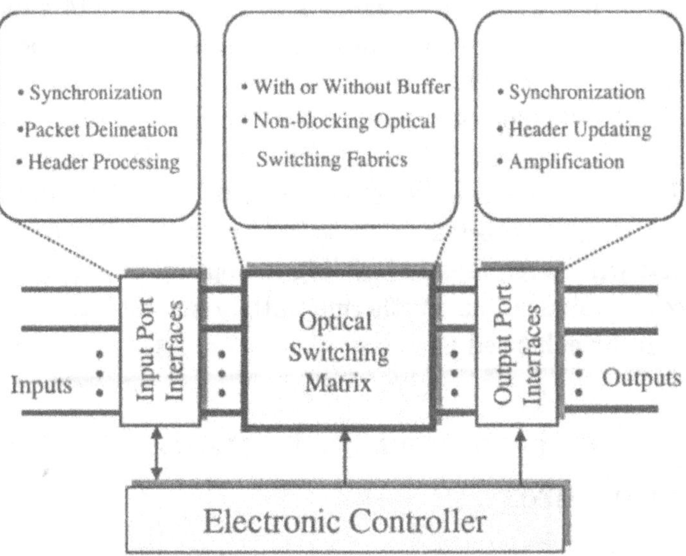

Figure 4: Generic optical packet switch architecture.

In the IPI, packet synchronization circuits are used to perform time alignment of packets at multiple input ports in order to correlate the packet positions with actual switching events. Packet delineation circuits are used to find the beginning of each packet. Both of these two functions can be done in both transparent and opaque optical switches. In the opaque optical switch, electronic buffers or memory inherently provide these functions. On the other hand, extra circuits with some fiber delay lines and optical switching components are required to perform these functions in the transparent optical switches. In addition, the header information of each packet is tapped at the IPI for processing and updated by a new header information at the OPI. In the OPI, additionally, amplifiers are used to ensure that the packets are routed through multiple switches in terms of routing, contention, timing and proper signal level. For transparent optical switches, both SOA and EDFA can be integrated with other subsystems in the switch node to perform amplification optically without regenerating the packets in their electrical format. In the opaque optical switches, the amplification can be done in the electronic domain if necessary.

Optical switching matrix plays an important role in routing the packets and resolving the contention in real time. The design of optical switching matrix depends on the switch requirements, the maturity in technology of optical devices, and the contention approach used in the switch. Some optical switching devices with fast switching speeds, such as SOA gates, lithium niobates and optical external modulators, have been considered for use in the future optical packet switching applications. Thanks to current advanced OEIC and PLC technologies, optical active components (e.g., laser diodes, SOA gates, optical modulators) and passive devices (e.g., optical couplers, optical multiplexers/demultiplexers) that can be monolithically integrated in the same package or substrate. These technologies enable wide variety of functions that will be required in building the optical switching matrix and providing cost minimization. Furthermore, the reliability in optical switching matrix can be enhanced as well.

6 Optical Interconnection Networks (OIN) Based Architectures

Due to the lack of mature technology for optical buffering, in packet switching with electronic buffering, the optical switching matrix performs the interconnection function between input and output ports. To perform switching function at high data rates of up to 10 Gb/s and to avoid the electromagnetic interference, the concept of optical interconnection networks (OIN) has been introduced. Basically the OIN is optical bufferless and performs the interconnection between electronic IPI and OPI. The electronic buffers can be placed at either IPI or OPI, or at both places. Depending on the placement of buffers, extra arbitration scheme is required for solving contention if buffers are placed at IPI. Fast packet switches or routers (e.g., ATM switches, IP routers) perform the standard switching function and aggregate the traffic in the IPI and/or OPI. They are interconnected to each other by an OIN. In the OIN, the specification of optical devices, optical performance and complexity will have to be considered so as to meet the requirements for packet switching. For example, to achieve over 90% transmission efficiency for a 64-byte packet with 10 Gb/s data rate, the reconfiguration speed of a switching node or OIN is less than about 5 nanosecond. For OIN, SOA gates can perform this fast switching speed and provide some optical amplification simultaneously [16].

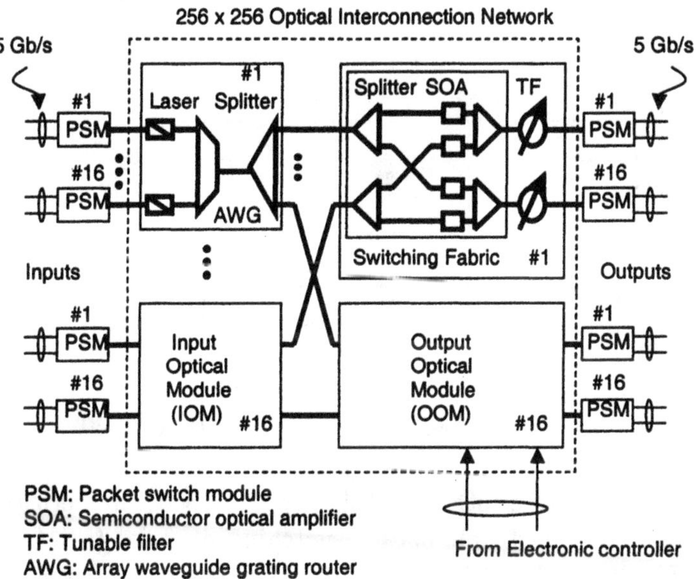

Figure 5: 3-Stage Clos type optical interconnection network.

Figure 5 shows a 128 × 128 bufferless optical interconnection network (OIN) for interconnecting multiple electronic packet switch modules (PSM's) with fibers running at 10 Gb/s [4]. Basically, OIN consists of three kinds of optical modules: transmit module (TM), core routing module (CRM) and receive module (RM). Transmit modules carry the packets from input packet switch modules (IPSM's) in term of wavelengths and send packets to the core routing modules. CRM's connect these transmitter modules with receive modules. Each receive module performs the packet(s) selection optically and then sends them to the output packet switch modules (OPSM's). As far as the electronic controller is concerned, it performs the control of OIN and the resolution of output contention if there are multiple packets from IPSM's destined to the same OPSM in a packet time slot. Since the proposed architecture function is similar to a 3-stage Clos network [17], to meet the nonblocking requirement, the number of wavelengths needed to connect each TM (or RM) with CRM's is at least equal to the sum of the group size of each TM and RM minus one.

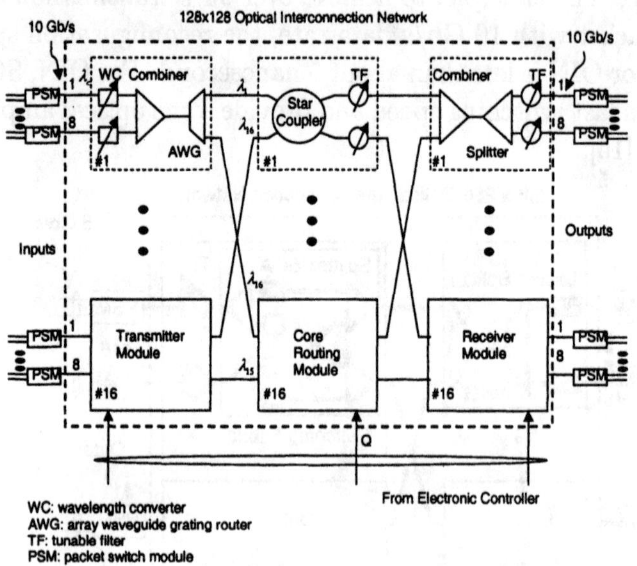

Figure 6: Broadcast-and-select type optical interconnection network.

Here, we also present another WDM-based 256×256 bufferless OIN to interconnect multiple electronic packet switching modules (PSM's) in Figure 6 [18]. This 256×256 OIN consists of two kinds of switching modules: input optical modules (IOM's) and output optical modules (OOM's). Basically, IOM's perform the function to modulate incoming electronic packets into optical formats and OOM's perform the switching function to deliver these packets (or wavelengths) to their requested output ports. Each IOM uses the same set of 16 different wavelengths and each of the 16 input links at an IOM is assigned a distinct wavelength from the set. This OIN can easily provide the multicast function due to its broadcast-and-select property. After solving the contention in the electronic controller at each packet time slot, at each IOM, up to 16 packets, which are from dedicated input packet switching modules (PSM's), are optically modulated and carried by different wavelengths. Then, they are optically multiplexed through an arrayed-waveguide grating (AWG) router or a combiner. This multiplexed WDM signal is then broadcast to all 16 OOM's by a passive splitter. This broadcast capability inherently provides multicast function in the OIN. The fast arbitration scheme to solve the output contention has been described in detail in [19].

At each OOM, a 16×16 switching fabric with a mesh connection performs the multicast switching function by properly controlling the SOA gates. In a 16 × 16 switching fabric, each SOA gate is controlled by the electronic controller and performs switching function at multiple wavelength level so that, at most, 16 out of 256 SOA gates can be turned ON simultaneously. Each tunable filter at each output PSM is used to select one and the only one wavelength, which is dedicated to its specific input PSM.

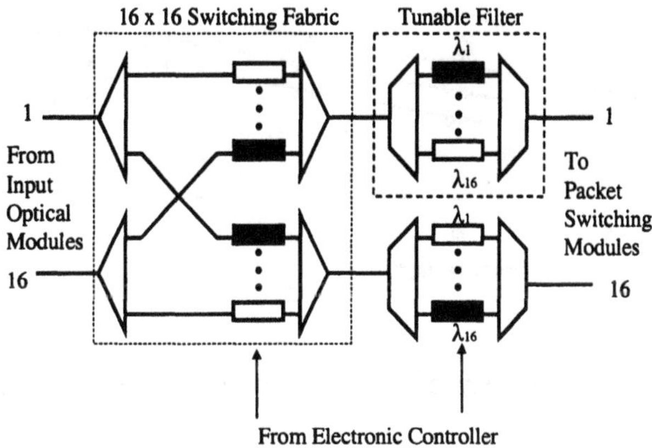

Figure 7: Operation of output optical module and tunable filters for N=256 and W=16.

To understand the operation of this OIN and tunable filter, let us consider an example. As shown in Figure 7, without loosing generality, we consider the operation of an OOM. Assuming that there is no output contention or the contention has been solved in the electronic controller, the $16th$ input PSM (i.e., with λ_{16}) in the first IOM and the first input PSM (i.e., with λ_1) in the $16th$ IOM are sent to the first and $16th$ output PSM in the OOM, respectively. In this case, in the switching fabric, only the bottom SOA gate is ON for the first output PSM and the top SOA gate is ON for the $16th$ output PSM. Due to dedicated wavelengths being assigned at each input PSM at each IOM, the selection of the tunable filter at each output PSM is controlled by both input and output address information. Here, for the first output PSM, only the top SOA gate in the tunable filter is ON and only the bottom SOA gate is ON for the 16*th* output PSM.

In Figures 8 and 9, we show the quantity of all optical devices needed in the proposed 3-stage Clos-based and broadcast-and-select based OINs with available wavelengths, respectively, although several of them can be monolithically integrated together in the same package or substrate using currently advanced PLC and OEIC technologies.

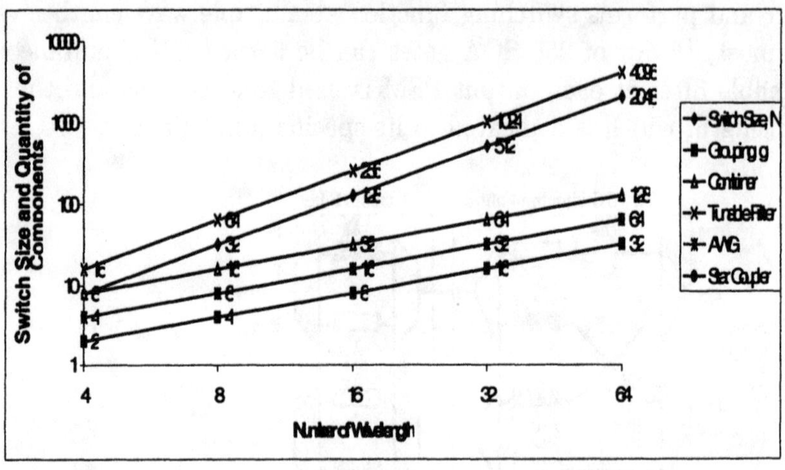

Figure 8: Component complexity of 3-stage Clos type optical interconnection network.

7   All-Optical Packet Switching (AOPS) Based Architecture

In the long term, while providing less complexity and high network performance (such as best throughput/delay performance), optical buffering schemes offer advantages in the design of all-optical packet switches by offering the benefits of transparency of data rate or format and data-aware services. To achieve the goal of all-optical packet switching, the concept of OIN can also be applied to all-optical and transparent packet switches [20]. As far as the OSN is concerned, Figure 10 depicts a generic architecture of the proposed $N \times N$ AOPS with W wavelengths at each incoming/outgoing fiber. Generally speaking, the optical switching fabric (OSF), which performs the switching function optically, consists of N wavelength converter

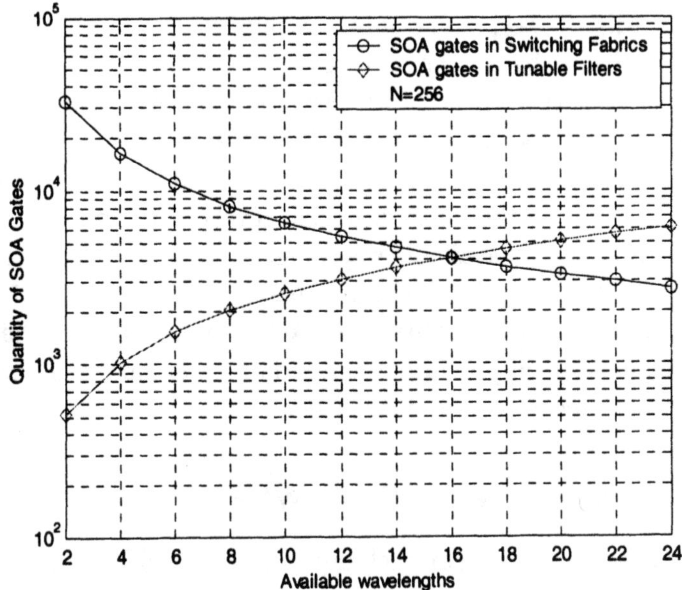

Figure 9: Quantity of SOA gates of broadcast-and-select type optical interconnection network with respect to available wavelengths.

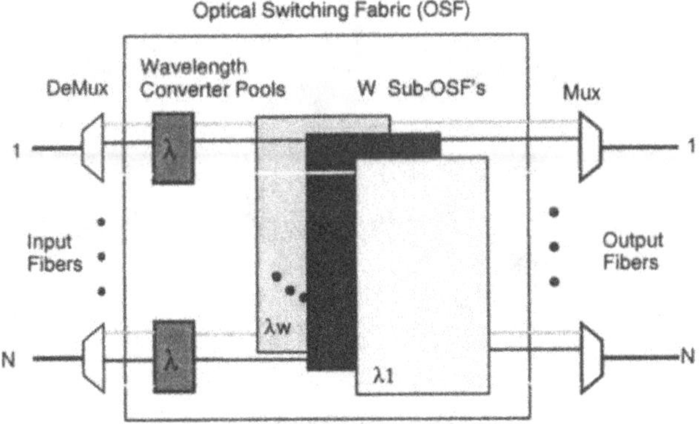

Figure 10: A generic proposed all-optical packet switch with W sub-optical switching fabrics.

pools (WCPs) and W sub-OSF's. Each WCP is controlled to balance the traffic load (or distribute the incoming wavelengths evenly) in the whole network and is shared by all the sub-OSF's. Each sub-OSF performs the switching function with respect to its corresponding wavelength, and consists of a non-blocking switching fabric (NSF) and optical delay lines as buffers (to be discussed later). This modular architecture has the advantage of scaling the capacity in the range of terabit levels. For example, assuming the switch size of $N = 16$, available wavelengths, $W = 32$, and data rate of each wavelength, $S = 2.5Gb/s$, the total capacity is $N \times W \times S = 1.28Tb/s$. Furthermore, multi-terabit capacity in packet switches can be achieved by advances in technology on providing higher data rate, more available wavelengths, or both to accommodate more users.

7.1   Wavelength Converter Pool (WCP)

All optical wavelength conversion techniques based on SOA gates are proving to be most promising for wavelength conversion [3]. Basically, each WCP performs not only the wavelength conversion function but also distributes the incoming traffic evenly to all sub-OSF's. That is, any incoming wavelength or traffic from any input fiber link can be wavelength- converted to another wavelength and switched in the sub-OSF.

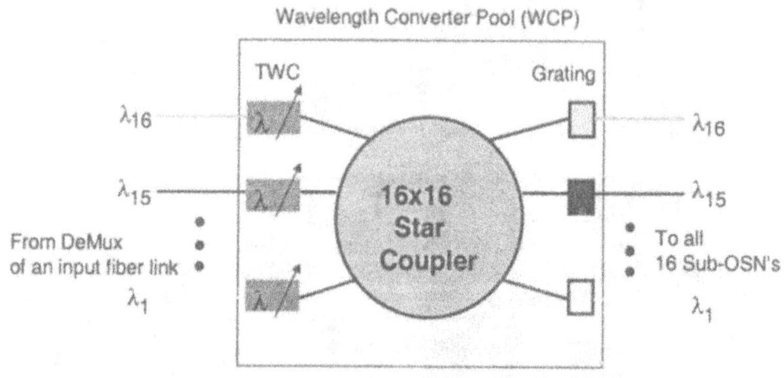

Figure 11: Star coupler based wavelength conversion pool (SC-WCP).

There are two types of WCPs presented in this paper. One is star-coupler based WCP (i.e., SC-WCP) and the other is arrayed-waveguide grating based WCP (i.e., AWG-WCP). Without losing generality, we consider $W = 16$ here. In the SC-WCP, as shown in the Figure 11, the incoming wavelength is wavelength-converted to its specific sub-OSF through its tunable wavelength converter, and then broadcast to all of the output ports of star coupler. Then, a fixed grating device used at the output port filters out the converted wavelength, followed by switching in the sub-OSF. Compared to SC-WCP, AWG-WCP incurs less power loss, as shown in Figure 12. For AWG-WCP, the incoming wavelength is wavelength-converted through a tunable wavelength converter at the input port to specific sub-OSF based on the routing connection of the AWG. At the output of the AWG, a wavelength is then converted by a fixed wavelength converter and sent to a sub-OSF.

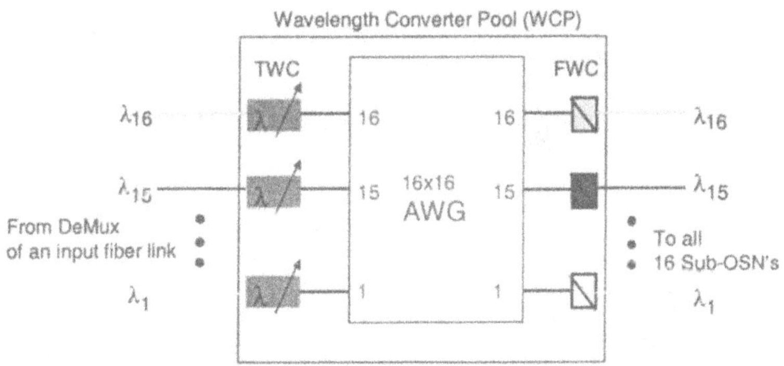

Figure 12: Arrayed-waveguide grating based wavelength conversion pool (AWG-WCP).

7.2 Sub-OSF Architecture and Size of Optical Buffers

Figure 13 shows a generic architecture of a sub-OSF. Basically, each sub-OSF provides broadcast-and-select function to all output switching modules (OSMs). Thus, the proposed switch is capable of providing multicast function as well. At each OSM, there is a non-blocking switching fabric (NSF) and several fiber delay lines used as optical buffers as shown in Figure 14. Because of output buffering, the proposed switch could achieve very good throughput/delay performance. Control signals from electronic controller are required to control the SOA gates in the NSF to stagger the incoming wavelengths (or packets) to optical buffers if multiple inputs are destined for the same output.

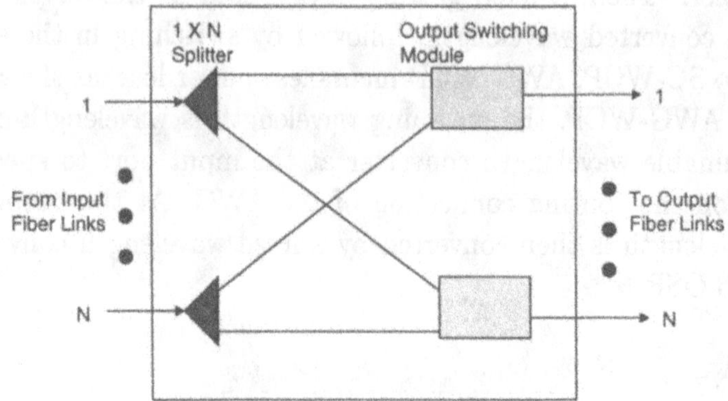

Figure 13: A generic architecture of a sub-OSF.

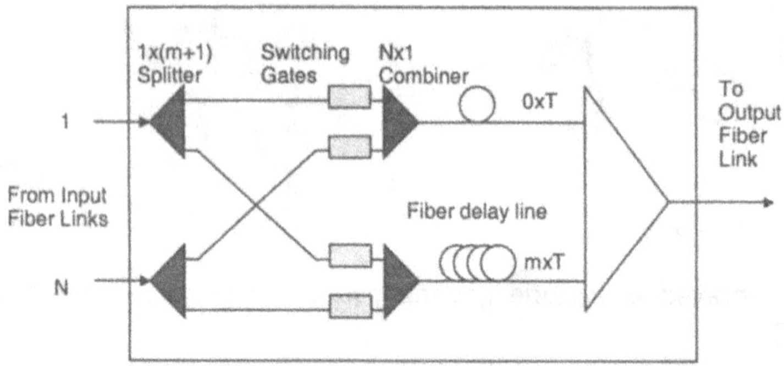

Figure 14: An architecture of an OSM with optical delay lines as buffers at each sub-OSF.

Figure 15 shows simulation results on packet loss probability with respect to buffer size (or the depth of fiber delay line) at each OSM. For simplicity, it is assumed that the packet arrivals to each input are independent and with identical Bernoulli process. Based on the simulation results, it is concluded that the probability of packet loss reduces as the number of available wavelengths increases while also decreasing the size of optical buffers at each OSM. For example, when switch size is 16 ports ($N = 16$), traffic load is 0.8 ($\rho = 0.8$) and available wavelengths are 4 ($W = 4$), an optical buffer of depth 9 (i.e., $m = 9$) is sufficient to achieve a packet loss rate of 10^{-10}. Furthermore, the size of SOA gates at each OSM is also reduced when the number of available wavelengths is increased. Here, we recommend gain-clamped SOA gates because they are capable of providing high speed switching function ($\sim$ns) to achieve high transmission efficiency and supporting optical amplification simultaneously.

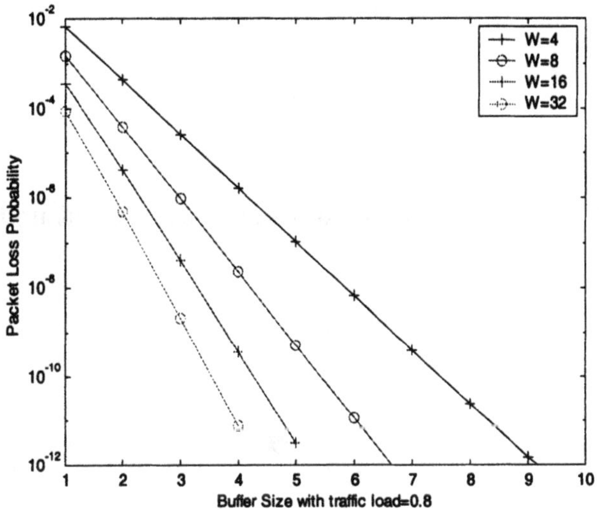

Figure 15: Plot of packet loss probability versus buffer size, assuming (N, ρ)=(16, 0.8), with different available wavelengths.

7.3 Header Replacement Unit (HRU)

To support future optical packet switching networks, a packet header replacement capability will be required. To perform this function optically, several proposals have been made [21-23]. However, in our opinion, very limited work has been done to integrate the header replacement function with packet switching function. In this section, a novel header replacement unit (HRU) is proposed and its operation integrated with the proposed switching function. In Figure 16, the proposed HRU is placed at the output of each OSM.

At each input of OSM, a small part of optical power is tapped and packet's header information is detected for processing in the electronic con-

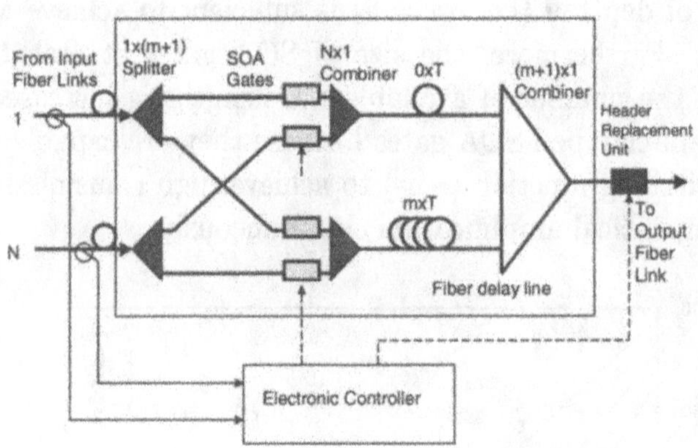

Figure 16: Placement of a proposed hearder replacement unit (HRU).

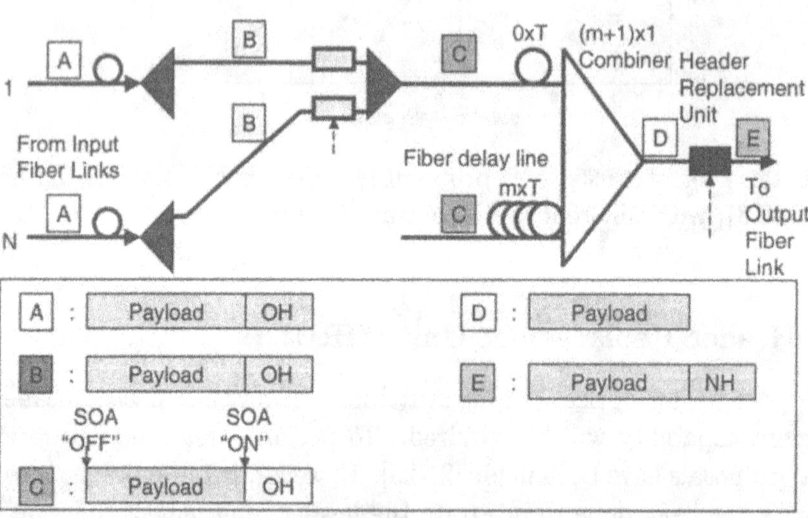

Figure 17: Operation of a proposed hearder replacement unit (HRU) at each output switching module (OSM).

troller. Then, based on the header information, electronic controller sends control signals to control SOA gates in OSM to deliver the packets with their old headers to the specific fiber delay lines. When packets are sent to output ports in a sequence of packet time slots, T, their new header information is updated through HRU. For example, in Figure 17, before performing the new header replacement function (as shown in position A and B), packet buffering/switching and erasure of old header can be performed simultaneously by controlling the SOA gates, as shown in position C of Figure 17. This logic function and clock control function can be easily implemented in the electronic controller. At each packet time slot, only payload of a packet is presented (as shown in position D) and its new header is updated in the HRU (as shown in position E).

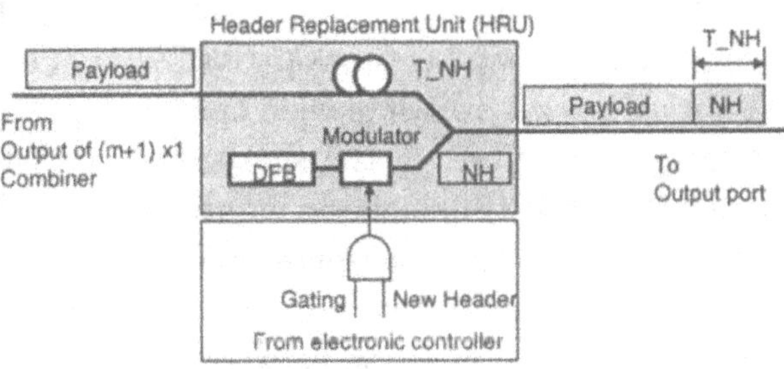

Figure 18: An architecture of a proposed header replacement unit (HRU).

Figure 18 shows an architecture of a proposed HRU where a small amount of fiber delay line (i.e., with the length of T_NH) is reserved for delaying the payload for updating the new header. A fixed laser associated with high-speed optical modulator (i.e., external modulator or SOA gate) is controlled to modulate the new header using the electronic controller.

Our proposed HRU at each OSM has the following advantages. First, each HRU is shared by all inputs destined to the same output port, thereby providing cost advantages. Second, with the robustness of logic function supported by the electronic controller, the erase function of the old header can be performed along with the packet switching function. It means that the complexity of the HRU is reduced and the overhead processing is avoided [22]. In addition, the simplicity of the proposed HRU architecture allows it to be implemented with current technology without the complexities of the wavelength converter [21][23]. The reader is referred to [24] for more details, which includes a comparison of the cascadability of the proposed

AOPS based on SC-WCP with AOPS based on AWG-WCP.

8   Conclusion

To meet the future needs of the traffic volume in the next generation broad-band Internet, optical technologies will enable development of all- optical packet switching systems. That is, the modular optical subsystems based on PLC and OEIC will keep progressing and will perform some key switching features optically, such as carrying high data rates of up to or beyond 10 Gb/s, and providing fast switching speeds in the order of nanosecond. These advanced technologies for WDM subsystems or devices will make all-optical switching systems possible in the near future. In spite of the major challenges, eventually they will provide less expensive and simpler solutions while also improving the end-to-end network performance. The race towards developing an all-optical packet switch is wide open and potential business implications enormous. Therefore, in the short-term, with less dense interconnectivity and without eletro-magnetic interference, WDM will play a key role in optically interconnecting high-performance electronic switches of the future. Due to limited resources of available wavelengths and optical device technology, optical interconnection networks should be designed and implemented in several small modular WDM subsystems to provide cost-effectiveness, reliability and flexibility in optical performance. In addition, OIN should be able to scale up in space, time and wavelength dimensions so as to be future proof. An optimum combination of these three dimensions is crucial to the success of OIN. Alternatively, in the long-term approach, advanced electronic technologies, matured optical switching matrix or OIN technologies and all-optical buffering will be critical for the success in building all-optical packet switching systems of terabit per second capacity.

References

[1] N. Ghani, S. Dixit and T. S. Wang, On IP-over-WDM Integration, *IEEE Communications Magazine*, pp. 72-84, March 2000.

[2] S. Yao, B. Mukherjee and S. Dixit, Advances photonic packet switching: an overview, *IEEE Communications Magazine*, Vol. 38, No. 2, pp. 84-94, February 2000.

[3] S. J. B. Yoo, Wavelength conversion technologies for WDM network applications, *J. Lightwave Technol.*, vol. 14, no. 6, pp. 955-966, June. 1996.

[4] T. S. Wang and S. Dixit, Optical interconnection networks for terabit packet switches, *Photonic Network Communications*, Kluwer Academic Publisher, vol. 2, no. 1, pp. 61-71, January-March 2000.

[5] Z. Hass, The "staggering switch": an electrically controlled optical packet switch, *J. Lightwave Technol.*, vol. 11, no. 5/6, pp. 925-936, May/June. 1993.

[6] D. Chiaroni et al, Sizeability analysis of a high-speed photonic packet switching architecture, *Proc. 21st Eur. Conf. on Opt. Comm. (ECOC'95)*, pp. 793-796, 1995.

[7] C. Guillemot et al, KEOPS optical packet switch demonstrator: architecture and testbed performance, *OFC'00 Technical Digest*, paper ThO1, pp. 204-206, 2000.

[8] G. H. Duan et al, Analysis of ATM wavelength routing systems by exploring their similitude with space division switching, *ICC '96*, vol. 3, pp. 1783-1787, 1996.

[9] Koji Sasayama et al, FRONTIERNET: frequency-routing-type time- division interconnection network, *J. Lightwave Technol.*, vol. 15, no. 3, pp. 417-429, March. 1997.

[10] F. S. Choa and H. J. Chao, On the optically transparent WDM ATM multicast (3M) switches, *Fiber and Integrated Optics*, vol. 15, pp. 109-123, 1996.

[11] E. Arthurs et al, HYPASS: an optoelectronic hybrid packet switching system, *IEEE Journal on Selected Areas in Communications*, vol. 6, no. 9, pp. 1500-1510, December. 1988.

[12] T. T. Lee et al, STAR-TRACK: a broadband optical multicast switch, *Bellcore Technical Memorandum Abstract.*, TM Number. TM-ARH-013510, 1989.

[13] A. Cisneros and C. A. Brackett, A large ATM switch based on memory switches and optical star couplers, *IEEE Journal on Selected Areas in Communications*, vol. 9, no. 8, pp. 1348-1360, October. 1991.

[14] E. Munter et al, A high-capacity ATM switch based on advanced electronic and optical technologies, *IEEE Communications Magazine*, pp. 64-71, November. 1995.

[15] Y. Nakahira et al, Evaluation of photonic ATM switch architecture-proposal of a new switch architecture, *International Switching Symposium*, pp. 128-132, 1995.

[16] S. L. Danielsen et al, Detailed experiment and theoretical investigation and comparison of the cascadability of semiconductor optical amplifier gates and gain-clamped semiconductor optical amplifier gates, *Technical Digest of Optical Fiber Communication Conference (OFC'98)*, vol. 2, pp. 41-42, 1998.

[17] C. Clos, A study of non-blocking switching networks, *Bell Syst. Tech. J.*, vol. 32, pp. 406-424, March 1953.

[18] T. S. Wang and S. Dixit, A scalable and modular optical interconnection network for a terabit packet switch, *International Conference on Applications of Photonic Technology*, pp. 66-75, June 2000.

[19] H. J. Chao, X. Guo, C. Lam and T. S. Wang, A terabit IP switch router using optoelectronic technology, *Journal of High Speed Networks*, pp. 35-38, January 1999.

[20] T. S. Wang and S. Dixit, All-optical terabit packet switches: issues and solutions, *European Conference on Networks and Optical Communications*, vol. 1, pp. 78-84, June 2000.

[21] E. Park, D. Norte and A. E. Willner, Simultaneous all-optical packet header replacement and wavelength shifting for a dynamically- reconfigurable WDM network, *IEEE Photonic Technology Letters*, vol. 7, no. 7, July 1995.

[22] J. Spring, R. M Fortenberry and R. S. Tucker, Photonic header replacement for packet switching, *Electronics Letters*, vol. 29, no. 17, August 1993.

[23] X. Jiang, X. P. Chen and A. E. Willner, All optical wavelength independent packet header replacement using a long CW region generated directly from the packet flag, *IEEE photonic technology Letters*, vol. 10, no. 11, November 1998.

[24] T. S. Wang and S. Dixit, A scalable and high capacity all-optical packet switch: design, analysis, and control, accepted for publication in *Photonic Network Communications*, Kluwer Academic Publisher, vol. 3, No. 1/2, pp. 103-112, January-June 2001.

OPTICAL NETWORKS - RECENT ADVANCES
L. Ruan and D.-Z. Du (Eds.) pp. 299 - 345
©2001 Kluwer Academic Publishers

Allocation of Wavelength Converters in All-Optical Networks

Gaoxi Xiao
Center for Advanced Telecommunications Systems and Services
University of Texas at Dallas, Richardson, TX75083
E-mail: **gxiao@utdallas.edu**

Yiu-Wing Leung
Department of Computer Science
Hong Kong Baptist University
Kowloon Tong, Hong Kong
E-mail: **ywleung@comp.hkbu.edu.hk**

Contents

Abstract

All-optical networks deliver information in the optical domain so that the electronic bottleneck can be avoided. They can support high data rate and provide large network capacity. In this chapter, we survey the state-of-the-art technologies for all-optical networks. In particular, we focus on the problem of allocating wavelength converters in all-optical networks. We explain why an all-optical network can use wavelength converters to improve its performance. To maximize the performance, it is necessary to allocate wavelength converters to the network nodes optimally. We describe three approaches to tackle this allocation problem:

1. *Intuitive approach*: This approach applies intuitive ideas to allocate wavelength converters.

2. *Analytical approach*: In this approach, the network performance is derived analytically, and then optimization algorithms are designed to allocate wavelength converters based on the analytical results. This approach is very popular in the literature and it is adopted by many existing allocation methods. However, various models and assumptions have to be adopted in deriving the network performance, and the resulting allocation methods are only applicable to these specific models and assumptions.

3. *Simulation-based optimization approach*: In this approach, utilization statistics of wavelength converters are collected from computer simulation, and then optimization algorithms are designed to allocate wavelength converters based on the utilization statistics. This approach is widely applicable and it is not restricted to any particular model or assumption.

Finally, we identify some problems for further investigation.

1 Introduction to All-Optical Networks

1.1 Optical Networks

Traditional information networks communicate through air, copper wires or coaxial cables (e.g., TV/radio networks, telephone networks and computer networks). With the increasing demand for high-speed communications, the traditional communication media cannot provide enough bandwidth. Optical fibers can be used as the communication media and they can provide very large bandwidth. Theoretically, a single-mode optical fiber can support a data rate of 50 Tbps [1]. If an information network uses optical fibers as its communication media, it is called an *optical network*.

1.2 Wavelength Division Multiplexing (WDM)

It is difficult to directly utilize the available capacity of an optical fiber for two reasons:

1. The theoretical capacity of an optical fiber is much larger than the speed of electronic hardware.

2. It is very difficult or costly to realize optical transmitters/receivers operated at a very high speed (say, several Tbps).

To overcome the above difficulties, **wavelength division multiplexing (WDM)** was proposed. WDM divides the bandwidth of an optical fiber into multiple *wavelength channels* at distinct wavelengths with a separating frequency space between any two adjacent channels. Each wavelength channel is operated at a reasonably high-speed that matches with the speed of electronic hardware and optical transmitters/receivers. In this manner, an optical fiber can provide multiple wavelength channels for multiple concurrent transmissions, so that it can provide a large capacity.

Using the current WDM technologies, an optical fiber can provide many high-speed wavelength channels. In laboratory, an optical fiber can provide more than one thousand wavelength channels at 37 Mbps/channel [2] or a speed of higher than 3 Tbps per channel [3]. In real-world deployment, an optical fiber provides 40 channels at 10 Gbps/channel in the Lucent's system [4] and 160 channels at 10 Gbps/channel in the Nortel's system [5].

The WDM technologies are evolving into several subclasses:

- **Coarse wavelength division multiplexing (CWDM)** [6, 7, 8, 9]: Using CWDM, an optical fiber can provide a small or moderate number of wavelength channels (e.g., 4-16) with a relatively large separating frequency space between two adjacent channels. CWDM provides a moderate capacity at a relatively low cost.

- **Dense wavelength division multiplexing (DWDM)** [1], [10]-[12]: Using DWDM, an optical fiber can provide a large number of wavelength channels (e.g., 8-160) with a relatively small separating frequency space (typically 50 GHz).

- **Ultra DWDM** [2]: Ultra DWDM is the latest WDM technology. Its main goal is to reduce the separating frequency space (e.g., to 10 GHz [2]) so that an optical fiber can provide more wavelength channels.

1.3   Types of Optical Networks

There are two major types of optical networks using WDM. The first type is called **single-hop networks** [10]. In single-hop networks, there is only one hop between any two nodes. As a result, network control and management are usually simpler, but each network can only accommodate a limited number of nodes. The single-hop networks can be further divided into two classes:

- **Point-to-Point Networks** [11, 12]: These networks are usually used for long-distance communication between two remote endpoints [12].

- **Broadcast-and-Select Networks** [10]-[15]: Each network has a star topology in which all the nodes communicate through a broadcast medium such as a *passive star coupler* (see Figure 1). Broadcast-and-select networks are suitable for local area networks (LANs) and metropolitan area networks (MANs). They may also be suitable for constructing high performance packet switches [16, 17].

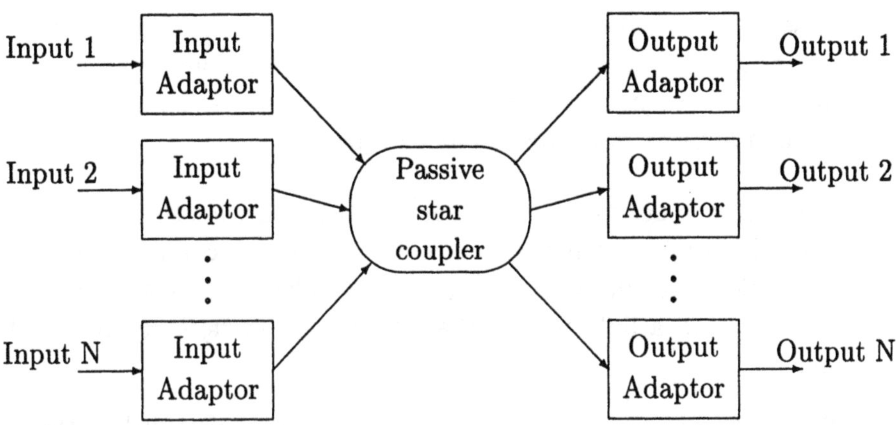

Figure 1. WDM broadcast-and-select networks [14].

The second type of optical networks using WDM is called **multihop networks**. In these networks, there can be one or more intermediate nodes between a source node and a destination node. Multihop networks are suitable for LANs, MANs, and wide area networks (WANs). They can be further divided into two classes:

- **Optical networks with O/E/O conversions:** In each intermediate node, information is converted from the optical domain into the electronic domain for processing and buffering, and then it is converted back to the optical domain for transmission. In other words, O/E and E/O conversions are performed in each intermediate node. These conversions, however, limit the speed of information flow.

- **All-optical networks:** Information is delivered in the optical domain within the networks. It is not necessary to perform O/E or E/O conversion in any intermediate node. Consequently, the speed of information flow can be higher, and the processing overhead in the intermediate nodes is significantly smaller.

1.4   All-Optical Networks

All-optical networks can support high-speed communications and provide large network capacity. They have been a hot research topic in recent years.

It is generally believed that all-optical networks will be widely deployed as backbone networks in the near future [1].

When two nodes want to communicate through an all-optical network, it is necessary to determine a source-to-destination path such that information can be transmitted along this path in the optical domain without any electronic processing in the intermediate nodes. There are two approaches to maintain all-optical transmission along a path:

1. The path can occupy a channel at the same wavelength on each of its links. Figure 2(a) shows an example. The above condition is known as *wavelength continuous constraint* [18]. When this constraint cannot be fulfilled in a link, we say that there is *wavelength conflict* in this link.

2. The path can occupy a channel on each of its links. When there is wavelength conflict in a link, the corresponding node uses a **wavelength converter** to convert optical signals from one wavelength to another wavelength in order to maintain all-optical transmission along the path. Figure 2(b) shows an example. This approach is more flexible and has a better performance because it is not necessary to fulfill the wavelength continuous constraint, but it has to use additional optical devices (i.e., wavelength converters).

Using either approach, the source-to-destination path for all-optical transmission is known as a **lightpath** [18]. The problem of determining a lightpath is composed of two co-dependent subproblems: (1) determine a route between the source and destination nodes, and (2) assign a wavelength channel to each hop of the route. For this reason, the problem is known as *routing and wavelength assignment (RWA) problem* [18, 19]. Two main types of RWA problems are tackled in the literature:

1. **Static RWA** [18, 19]: The static RWA problem is to determine all the lightpaths required when the traffic load and traffic pattern are given. Common objectives include minimizing the number of wavelength channels needed to support all the required lightpaths or maximizing the number of lightpaths that can be established. Several heuristic algorithms have been proposed for static RWA [18, 19, 20].

2. **Dynamic RWA** [18], [20]-[28]: The dynamic RWA problem is to determine a lightpath upon a request. Dynamic RWA is more flexible

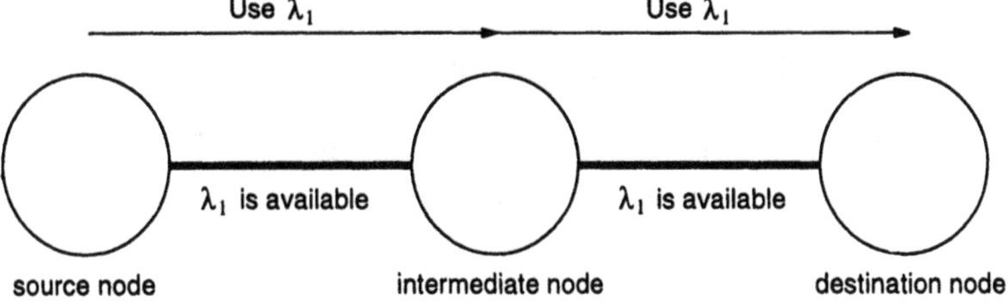

(a) A lightpath occupies the same wavelength on each of its links.

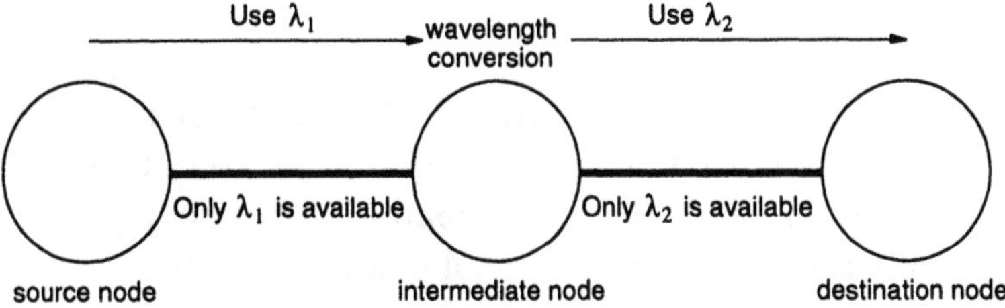

(b) A lightpath occupies different wavelengths on different links.
Wavelength conversion is performed in the intermediate
node to ensure all-optical transmission along the path.

Figure 2.   Examples of two approaches to maintaining all-optical
transmission along a path.

than static RWA because it can adapt to the changes in traffic load
or traffic pattern. When routing is fixed, several dynamic wavelength
assignment methods have been proposed:

- **Random** wavelength assignment: Identify the available wavelengths along the route, and randomly select one.

- **First-fit** wavelength assignment [18]: Identify the available wavelengths along the route, and select the one with the smallest index.

- **Least-used** wavelength assignment [29]: Identify the available wavelengths along given route, and select the one that is least used in the network.

- **Most-used** wavelength assignment [29]: It is the opposite of least-used wavelength assignment.

- **Min-product** wavelength assignment [30]: It is designed for the all-optical networks with multiple fibers per link. It minimizes the number of fibers being used in the network.

- **Least-loaded** wavelength assignment [31]: It is also designed for the multiple-fiber case. It selects the wavelength that has the largest residual capacity on the most-loaded link along the route.

When both routing and wavelength assignment are dynamic, the existing methods can be divided into two classes:

- **Fixed alternate routing** [32]-[36]: In the planning phase, several alternative routes are determined for every node pair. In the operation phase, one of the routes for this node pair is dynamically selected upon the arrival of a new request.

- **Adaptive routing:** The route between a node pair is determined dynamically based on the network state. The popular methods include

  (a) *Adaptive least-cost-path routing:* The route is determined by using the shortest path algorithm and a certain definition of link weight. Most of the existing adaptive routing methods belong to this class.

  (b) *Least congestion routing:* The route is determined by minimizing the maximum traffic load on every link along the route [37]. This can balance the load among the links.

For large all-optical networks, it is desirable that a RWA method uses distributed computation based on local information (i.e., it does not need any central controller and precise global information). Several distributed routing and wavelength assignment methods have been proposed, such as the deflection routing method [38, 39], the fixed-path least-congestion (FPLC) method [40], the hot-potato routing method [41, 42] and the alternate link routing method [43], etc. Several distributed signal mechanisms have also been proposed to define how the RWA is actually done in a distributed way (e.g., [44], [45]).

2　Wavelength Conversion in All-Optical Networks

2.1　Benefits of Wavelength Conversion

A wavelength converter is an optical device that converts optical signals from one wavelength to another wavelength. When a lightpath is to be established but wavelength conflict is encountered in a link, the corresponding node can use a wavelength converter to resolve this conflict in order to establish the lightpath (e.g., see Figure 2(b)). In general, wavelength converters can increase the probability of establishing lightpaths successfully. Therefore, they can improve the performance of all-optical networks.

The performance improvement due to wavelength converters depends on the following factors:

1. *Network size:* When the network size is larger, the performance improvement is larger. In a larger all-optical network, some routes may consist of more hops. The probability that there is wavelength conflict in at least one of these hops is larger. Then it is more likely that wavelength converters are used to resolve these conflicts. Therefore, the performance improvement is larger.

2. *Network topology:* After extensive empirical studies, it is commonly believed that the performance improvement is more significant for networks with regular or irregular mesh topology [46, 47].

3. *Number of channels per fiber:* The performance improvement is more significant when there are more wavelength channels per fiber [46, 47].

4. *Number of fibers per link:* The performance improvement is less significant when there are more fibers per link [30].

5. *Traffic load:* When the traffic load is neither very light nor very heavy, the performance improvement is significant. When the traffic load is very light, the performance improvement is not significant. It is because the probability of wavelength conflict is small under light traffic, and therefore it is unlikely that wavelength converters are needed to resolve conflict. When the traffic load is very heavy, the performance improvement is not significant. Under heavy traffic, the main issue is that the network does not have enough channels, and wavelength conflict is a relatively minor issue. Even when there are wavelength converters, the performance cannot be improved significantly.

6. *Traffic pattern:* When the traffic pattern is less bursty, wavelength converters can improve the performance more significantly [37, 48]. The benefit of wavelength conversion under various bursty traffic patterns is a topic of current research.

7. *Routing and wavelength assignment methods:* The performance improvement depends on the RWA method adopted, because different RWA methods have different possibility of encountering wavelength conflicts.

8. *Heterogeneous links:* If different links of a network have different number of wavelength channels, wavelength converters can improve the performance more significantly [49].

2.2   Wavelength Conversion Technologies

Wavelength converters can be classified from different viewpoints. First, they can be classified based on their conversion ranges as follows:

1. *Full-range wavelength converters* (FWCs) [19, 46, 47, 50, 51, 52, 53]: A FWC can convert an input wavelength to any output wavelength.

2. *Limited-range wavelength converters (LWCs)* [54, 55, 56]: A LWC can convert an input wavelength to an output wavelength belonging to a given subset of possible wavelengths. This subset defines a *conversion range.* In general, a LWC with a smaller conversion range is less costly.

Wavelength converters can also be classified based on their underlying physical mechanisms as follows:

1. *O/E/O converters* [1]: In an O/E/O converter, incoming optical signal is converted into electronic signal, which is then used to modulate an output laser to generate optical signal at the desired wavelength. The conversion range depends on the tunable range of: (1) the optical filter for the incoming optical signal and (2) the laser for generating the outgoing optical signal. Optical filters and lasers usually have large tunable ranges, and therefore O/E/O converters can have large conversion ranges. However, O/E/O converters have the bottleneck due to O/E and E/O conversions (e.g., limited speed, high energy consumption, unqualified transparency, etc.).

2. *All-optical converters*: An all-optical converter performs wavelength conversion in the optical domain. Therefore, it can avoid the bottleneck due to O/E and E/O conversions. They can be further classified into two major types:

   - *Coherent effect wavelength converters:* These wavelength converters make use of the *wave-mixing effect* (i.e., the nonlinear response of a medium subject to more than one interacting input at different wavelengths) to generate optical signal at another wavelength (as a linear combination of the phases and frequencies of the interacting waves). They have the advantages of strict format transparency and bit-rate-independence. However, they have several shortcomings: they require accurate polarization control and phase matching, they have limited conversion ranges, and high energy consumption. This type of wavelength converters includes those using *four wave mixing* (FWM) [57]-[61] and *difference frequency generation* (DFG) [62, 63].

   - *Cross modulation wavelength converters:* These wavelength converters utilize active semiconductor optical devices such as *semiconductor optical amplifier* (SOA) and *semiconductor laser*.

     - When SOA is used, it can be operated in either one of two modes: cross-gain modulation [57, 64] and cross-phase modulation [57, 64, 65, 66]. Each mode has its own strength and weakness, and a good discussion can be found in [57].

     - When semiconductor laser is used, input optical signal is used to modulate the lasing mode intensity through lasing mode gain saturation. The converted signal is inverted compared to the input signal. This mechanism has been employed in a multielectrode distributed feedback (DFB) laser [67] and a distributed Bragg reflector (DBR) laser [68] to convert signals at a speed of up to 10 Gbps.

3 Allocating Wavelength Converters in All-Optical Networks

3.1 Problem Definition

When a node of an all-optical network has n incoming wavelength channels, at most n lightpaths can be established across this node. If this node has n wavelength converters, it can always provide a free wavelength converter to any lightpath across it. This can resolve all possible wavelength conflicts across this node to achieve the ideal performance. However, each node requires many wavelength converters. Wavelength converters are expensive, and it is widely believed that they will still be expensive at least in the next several years [50, 57]. Therefore, it is desirable if each node can use a smaller number of wavelength converters to give near-ideal performance.

In an all-optical wide area network, some of the nodes are required to handle heavier volumes of traffic than the others. It is because the topology of a wide area network is usually irregular, and the traffic is often non-uniform. It is desirable to allocate more wavelength converters to the nodes handling heavier volumes of traffic, so that the wavelength converters can be efficiently utilized to resolve wavelength conflicts.

The allocation problem is to allocate a given number of wavelength converters to the nodes of an all-optical network to maximize the performance. A common performance measure for all-optical networks is the blocking probability [69]-[71]. In an equivalent allocation problem, the objective is to minimize the number of wavelength converters required to achieve a given performance target.

Extensive researches have been done to solve the allocation problem and its variants. We classify the existing solutions into three approaches: *intuitive approach*, *analytical approach*, and *simulation-based optimization approach*.

3.2 Intuitive Approach to Allocating Wavelength Converters

The intuitive approach applies intuitive ideas to allocate wavelength converters. The following two studies adopt this approach:

- Subramaniam, Azizoglu and Somani [47] proposed to select some of the nodes randomly and equip them with sufficient number of FWCs for *complete wavelength conversion* (i.e., FWCs are always available when they are needed). The remaining nodes do not have any FWC. When the number of nodes with complete wavelength conversion is large enough, this allocation method can give nearly the same blocking probability as the ideal case (i.e., all the nodes can support complete wavelength conversion).

- Lee and Li [51] proposed to equip every node with the same and limited number of FWCs for *partial wavelength conversion* (i.e., FWCs may not be available when they are needed). When the number of FWCs per node is large enough, this allocation method can give nearly the same blocking probability as the ideal case.

The above two studies [47, 51] were not aimed to optimize the allocation of FWCs, but they reached the important conclusion that an all-optical network can use a smaller number of FWCs to give near-ideal performance. This conclusion triggered many subsequent researches on allocating wavelength converters.

3.3 Analytical Approach to Allocating Wavelength Converters

The analytical approach is adopted by many allocation methods [69]-[74]. In general, this approach is composed of two main steps:

1. Derive the blocking probability analytically.

2. Based on the analytical results, optimize the allocation of wavelength converters to reduce the blocking probability as far as possible.

The main difficulty in the analytical approach is to derive the blocking probability. Therefore, various models and assumptions have to be adopted so that the blocking probability can be derived for subsequent minimization. For example, the following are some common assumptions [69]-[74]:

- Traffic arrival process is Poisson.

- Holding time of a lightpath is exponentially distributed.

- Wavelength assignment is random.

- Routing is static.

- Occupancy of a wavelength channel is statistically independent of that of the other wavelength channels on the same link and the other links.

These assumptions are restrictive and undesirable. For example, random wavelength assignment is not efficient and it is not preferred in practice; dynamic routing is preferred for better performance; etc. Using the analytical approach, the resulting allocation methods are only applicable to the specific models and assumptions adopted.

4   Simulation-Based Optimization Approach to Allocating Wavelength Converters

In a recent study, we propose a new approach to allocating wavelength converters in all-optical networks. This approach is called *simulation-based optimization approach*, and it is composed of two main steps:

1. Collect utilization statistics of wavelength converters from computer simulation.

2. Based on the utilization statistics, optimize the allocation of wavelength converters.

Since computer simulation is adopted in the first step, it is not necessary to use any restrictive model or assumption. As a result, the simulation-based optimization approach is widely applicable and it is not restricted to any particular model or assumption. In this section, we describe an allocation method using this approach.

4.1   Network Model

We consider an all-optical wide area network. The network consists of N nodes and the network topology can be irregular. Every network link consists of a set of distinct wavelength channels. We consider the node configuration proposed in [51] because it requires the least number of FWCs to obtain the same blocking performance. This node configuration is called *share-per-node* [51] and is shown in Figure 3. After demultiplexing, the wavelength channels are fed to the first switch. If a wavelength channel does not need wavelength conversion, it is switched to an appropriate multiplexer for outgoing transmission. Otherwise, it is switched to the FWC bank in which it is converted to another wavelength by a FWC, and then it is switched to an appropriate multiplexer for outgoing transmission. We use the shortest path routing.

4.2   Allocating Full-Range Wavelength Converters

Using the share-per-node structure, we need to determine the number of FWCs in the FWC bank of every node. Our objectives are to reduce (1) the overall blocking probability (i.e., better mean quality of service) and (2) the maximum of the blocking probabilities experienced at all the source node

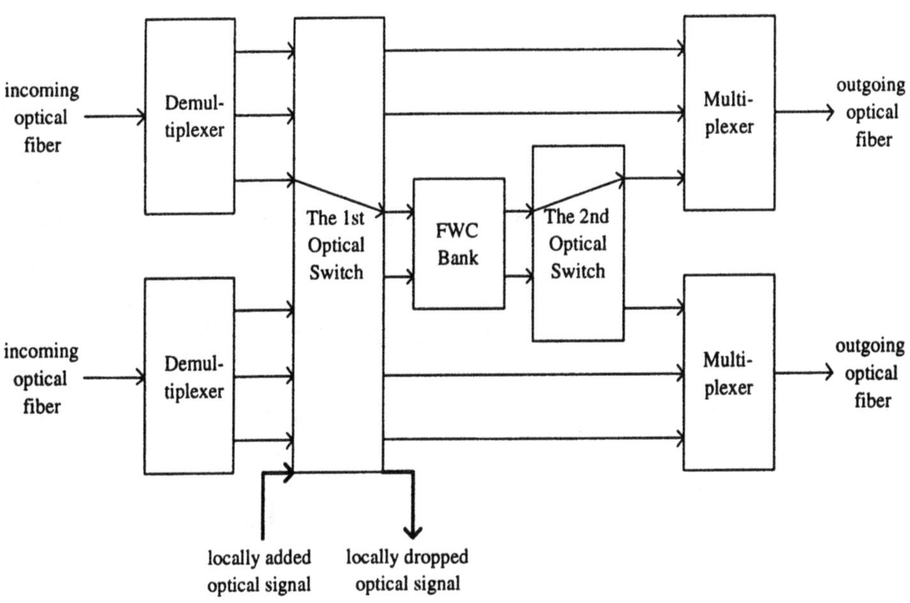

Figure 3.   Node configuration [51].

(i.e., better fairness). Equivalently, for a given blocking requirement, our objective is to reduce the required number of FWCs.

The blocking probability for all-optical networks is available in analytical form only under some simplifying assumptions, specific traffic models, or specific routing and wavelength assignment methods ([32, 72]). Therefore, we adopt simulation-based optimization approach, so that our method is widely applicable and is not restricted to any specific model or assumption.

Our main idea is as follows. First, when there is complete wavelength conversion, we record the utilization statistics of FWCs in every node by computer simulations. Specifically, we let node i require M_i FWCs for complete wavelength conversion, where M_i is equal to the number of outgoing fibers of node i times the number of channels per fiber, and $M = \max\{M_1, M_2, \ldots, M_N\}$. We measure the *utilization matrix*

$$\mathbf{U} = [U_{i,j}]_{(1 \leq i \leq N) \times (0 \leq j \leq M)}$$

in computer simulations, where $U_{i,j}$ is the percentage of time that j FWCs are being utilized simultaneously in node i. Second, when there are a limited number of full-range wavelength converters, we optimize the allocation of the given number of FWCs based on $\mathbf{U}$. This is an approximate approach because when there is partial wavelength conversion, the utilization matrix is changed. Nevertheless, this approximation is good for the following reason. In a well-engineered network, the traffic load handled by each node should not approach or exceed its capacity, so that the blocking probability can be kept at a reasonably low value (say, 0.01). Even if node i has M full-range wavelength converters for complete wavelength conversion, it is likely that only some of them are being used at a time while the others are left idle. Therefore, $U_{i,M}, U_{i,M-1}, \cdots$ are relatively small. For this reason, when node i is equipped with a fewer number of FWCs, the utilization matrix is only changed slightly. In section 4.4, we will present simulation results to demonstrate that when there are a small number of FWCs, our approximate approach can already result in blocking probabilities close to that with complete wavelength conversion.

With the above idea, we divide the problem into the following two subproblems:

1. Record the utilization matrix via computer simulations.

2. Based on the utilization matrix, optimize the allocations of the full-range wavelength converters.

In the following subsections, we design algorithms to solve these subproblems.

4.2.1 Recording Utilization Matrix

We record the utilization matrix via simulation experiments. One important issue is that, when there is wavelength conflict, we need to determine where we should perform wavelength conversion. Different methods can lead to different utilization matrices. In our study, we design and adopt one possible method to resolve wavelength conflict that gives good results. However, our simulation-based optimization methodology is also applicable to any other conflict resolution method.

For any given call duration statistics, we can generate the duration for each transmission. Therefore, when a new request arrives, we can determine its finish time. We use this feature when we record the utilization matrix as follows:

Recording Algorithm

1. When a transmission request arrives, identify those transmissions that have been finished since the last transmission request arrives, record their duration and the number of FWCs that have been used on each node of their source-to-destination paths, then release all the FWCs used by them.

2. If at least one wavelength (the same one) is available on every hop of the source-to-destination path (this wavelength channel is called a *clear channel*), admit this transmission request and assign a clear channel to it on a first-fit basis. Otherwise, go to step 3.

3. If there is at least one free wavelength channel (at any wavelength) on every hop of the source-to-destination path, execute the following steps:

   (a) Execute *Conflict Resolution Algorithm* (to be explained) to (i) select a wavelength requiring the smallest number of FWCs, and (ii) when there is more than one choice, select the one that minimizes the maximum number of FWCs being used on every node of the path.

   (b) After selecting the nodes that perform wavelength conversion (called the *tuning nodes*), select a wavelength between any two consecutive tuning nodes on a first-fit basis.

   Otherwise, the transmission request is blocked.

4. Repeat the above steps for the next and new transmission request. $\square$

In step 1, we perform recording only when a transmission request arrives. In this manner, it is not necessary to monitor the finish time of all the ongoing transmissions, so that the algorithm can be simpler. In step 3(a), we resolve wavelength conflict to fulfill two objectives: (i) the resulting source-to-destination path requires the smallest number of FWCs, and (ii) when there is more than one choice, we select the one that minimizes the maximum number of FWCs being used on every node of the path. In [51], Lee and Li proposed a *graph transformation* method to minimize the required number of FWCs (i.e., to fulfill the first objective). To tackle the second objective, we modify and enhance the graph transformation method.

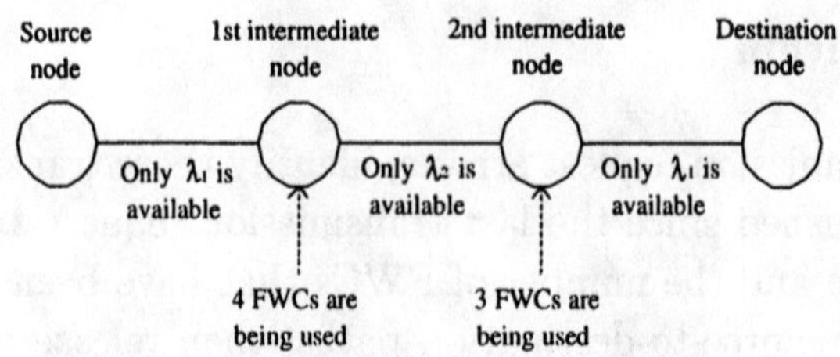

(a) A source-to-destination path with 2 intermediate nodes
and 10 wavelength channels per link.

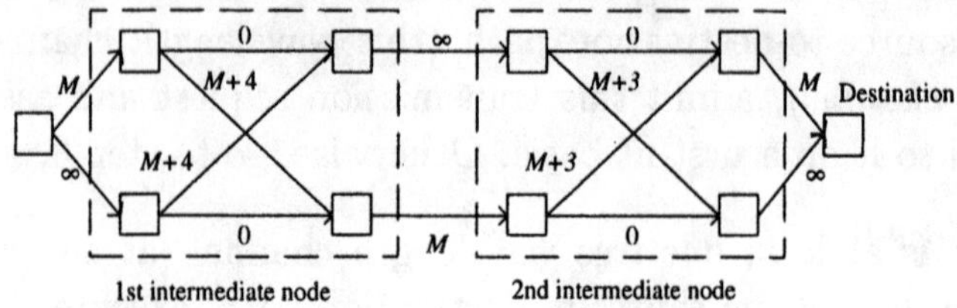

(b) A directed graph for the source-to-destination path shown in (a).

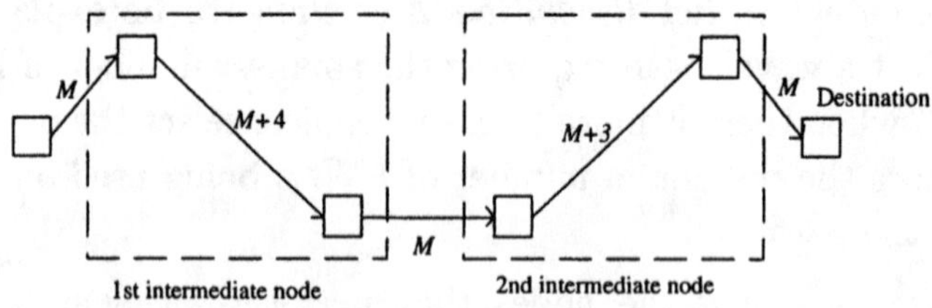

(c) The shortest path in the directed graph. The weight of the shortest path
is $5M+\max(3, 4)=5M+4$.

Figure 4.   An example to illustrate the graph transformation method
to resolve wavelength conflict.

The main idea is to transform the problem of resolving wavelength conflict into an equivalent shortest path problem in a directed graph, where the
length of a path in the directed graph is determined by (i) the total number
of FWCs used and (ii) the maximum number of FWCs being used on every
node of the source-to-destination path. By determining the shortest path in
this directed graph, we can fulfill both of our objectives. We construct the
directed graph as follows. Along the source-to-destination path in the network, the *intermediate nodes* (excluding the source and destination nodes)
are indexed from 1 to L. Let $W(l)$ denote the number of FWCs being used
on the l-th intermediate node. Now perform the following steps to construct

a directed graph.

1. For wavelength λ on the j-th hop of the path, let $c(\lambda, j)$ be the weight of this wavelength channel. If it is available, then $c(\lambda, j) = M$; otherwise, $c(\lambda, j) = \infty$.

2. For the l-th intermediate node ($1 \leq l \leq L$), we create a vertex $v_i(\lambda_i, l)$ for every incoming wavelength channel λ_i, and create a vertex $v_o(\lambda_o, l)$ for every outgoing wavelength channel λ_o. The weight of the edge connecting vertex $v_i(\lambda, l)$ to vertex $v_o(\lambda, l)$ for any λ is zero. The edge connecting $v_i(\lambda_i, l)$ and $v_o(\lambda_o, l)$ for any $\lambda_i \neq \lambda_o$ is called a *converter edge*, and its weight is $M + W(l)$.

Figure 4(a) and 4(b) illustrate the construction of the directed graph.

We recall that the length of a path in the directed graph must reflect two quantities: (i) the total number of FWCs used and (ii) the maximum number of FWCs used on the network nodes (e.g., see Figure 4(c)). The traditional shortest path algorithms can only tackle the first quantity. To tackle both quantities, we modify the Dijkstra's algorithm to solve our problem as follows:

Conflict Resolution Algorithm

1. Initially, label the source vertex s, and leave all the other vertices unlabeled. Let $d(s) = 0$ and $d(y) = \infty$ for all $y \neq s$. Let $x = s$.

2. Let $a(x, y)$ be the weight for the edge from vertex x to vertex y. For each unlabeled vertex y, if $a(x, y) \leq M$, compute $d(y)$ as follows:

$$d(y) = \min\{d(y), d(x) + a(x, y)\} \tag{1}$$

If $M < a(x, y) < \infty$ (i.e., the edge from vertex x to vertex y is a converter edge), compute $d(y)$ as follows:

$$d(y) = \min\{d(y), [d(x) - d'(x)] + [M + \max(d'(x), a'(x, y))]\} \tag{2}$$

where $d'(x)$ is the maximum of the number of FWCs being used in all the pervious nodes up to x:

$$\begin{cases} d'(x) = \mathrm{mod}(d(x), M) \\ a'(x, y) = \mathrm{mod}(a(x, y), M) \end{cases} \tag{3}$$

In other words, $a'(x, y)$ is the number of FWCs being used on the intermediate node; and $d(x)$ is equal to M times the number of hops up to x, plus M, and plus the maximum number of FWCs being used on all the previous nodes up to x. Label the unlabeled vertex y with the smallest value of $d(y)$.

3. If the destination vertex has been labeled (i.e., a path has been determined), identify the tuning nodes, increment $W(l)$ of these nodes by 1, and then stop. Otherwise, repeat step 2. $\square$

The above algorithm is modified from the Dijkstra's algorithm, and equations (2) and (3) are tailored for our problem. Specifically, we use these equations to minimize the number of FWCs needed in the path, for once a converter edge is included, the cost of the path has to be increased by at least M units. When there is a tie, we use equations (2) and (3) to minimize the maximum number of FWCs being used on every intermediate node, for $d(x)$ is also determined by the maximum number of FWCs that have been used on all the previous nodes up to x. The time complexity of the above algorithm can be found to be $O(n^2)$, where n is the number of vertices in the directed graph.

4.2.2 Allocating Full-Range Wavelength Converters

In this subsection, we optimize the allocations of a given number of full-range wavelength converters based on the utilization matrix $\mathbf{U}$.

After allocating a certain number of FWCs to a node, we can get from $\mathbf{U}$ the percentage of time that this node has sufficient FWCs to serve the transmission. For convenience, we call this quantity the *total utilization*. For example, if node i has J wavelength converters, the total utilization is $\sum_{j=0}^{J} U_{i,j}$. To optimize the allocation of a given number of FWCs without any assumption about the traffic pattern, we consider three different objective functions:

1. Maximize the sum of total utilizations of all the nodes, so that the overall utilization of FWCs can be improved. As a result, the overall blocking probability can be smaller and hence the mean quality of service is better.

2. Maximize the product of the total utilizations of all the nodes. In this manner, the overall utilization of FWCs can be improved (i.e., better

mean quality of service) and the allocation of FWCs to the nodes can be more fair.

3. Maximize the minimum value of total utilization of the N nodes, so that the allocation of FWCs to the nodes can be more fair.

A. Maximize the Sum of Total Utilizations

Let T denote the total number of available FWCs. For the trivial case that T is large enough to provide complete wavelength conversion (i.e., $T \geq MN$), the optimal allocation of FWCs is simple: allocate M FWCs to every node so that every node can perform complete wavelength conversion. In the following, we consider the case $T < MN$.

We define $x_{i,j}$ as follows:

$$x_{i,j} = \begin{cases} 1 & \text{if } j \text{ or more WCs are allocated to node } i \\ 0 & \text{otherwise} \end{cases} \tag{4}$$

The total utilization of node i is $\sum_{j=1}^{M} U_{i,j} x_{i,j}$. The problem of maximizing the sum of the total utilization of all the nodes can be formulated as follows:

$$\begin{cases} \text{maximize}_{x_{i,j}} & \sum_{i=1}^{N} \sum_{j=1}^{M} U_{i,j} \cdot x_{i,j} \\ \text{subject to} & 1.\ \sum_{i=1}^{N} \sum_{j=1}^{M} x_{i,j} = T \\ & 2.\ x_{i,j} \geq x_{i,j+1},\ i = 1, 2, \ldots, N,\ j = 1, 2, \ldots, M-1 \\ & 3.\ x_{i,j} \in \{0, 1\} \end{cases} \tag{5}$$

The objective function is the sum of the total utilization of all the nodes. The first constraint ensures that the total number of FWCs allocated to the nodes is T. The second and the third constraints ensure that $x_{i,j} \geq x_{i,j+1}$ and $x_{i,j}$ is binary. The variables to be optimized are $x_{i,j}$ for all $1 \leq i \leq N$ and $1 \leq j \leq M$. After optimization, if $x_{i,1} = x_{i,2} = \cdots = x_{i,n} = 1$ and $x_{i,n+1} = x_{i,n+2} = \cdots = x_{i,M} = 0$, then the optimal number of FWCs allocated to node i is n.

We solve the above optimization problem as follows. Observing that $0 \leq U_{i,j} \leq 1$ and $T < N \cdot M$, we can transform the above optimization problem into the following problem which has the same optimal solutions as those of (5).

$$\begin{cases} \text{minimize}_{x_{i,j}} & \sum_{i=1}^{N} \sum_{j=1}^{M} (1 - U_{i,j}) \cdot x_{i,j} \\ \text{subject to} & 1.\ \sum_{i=1}^{N} \sum_{j=1}^{M} x_{i,j} = T \\ & 2.\ x_{i,j} \geq x_{i,j+1},\ i = 1, 2, \ldots, N,\ j = 1, 2, \ldots, M-1 \\ & 3.\ x_{i,j} \in \{0, 1\} \end{cases} \tag{6}$$

If we remove the second constraint $x_{i,j} \geq x_{i,j+1}$, $i = 1, 2, \ldots, N$, $j = 1, 2, \ldots, M - 1$, the above problem reduces to the knapsack problem [75]. Therefore, the above problem can be regarded as a constrained knapsack problem. It is well known that the knapsack problem can be solved by the dynamic programming method [75] which transforms the knapsack problem to an equivalent shortest path problem in directed graphs. To tackle the constraint $x_{i,j} \geq x_{i,j+1}$, we modify the dynamic programming method and construct a directed graph for our problem as follows. Every vertex is indexed as (i_1, j_1, k_1) where for each $1 \leq i \leq N$, $1 \leq j \leq M$, $i_1 = (i-1) \cdot M + j$, $j_1 = \sum_{p=1}^{i} \sum_{q=1}^{j} x_{p,q}$, and $k_1 = x_{i,j}$. Therefore, $0 \leq i_1 \leq MN$, $0 \leq j_1 \leq T$ and $0 \leq k_1 \leq 1$. Since $0 \leq k_1 = x_{i,j} \leq 1$, the third constraint of (6) has been fulfilled. To fulfill the second constraint, there is an edge connecting vertex (i_1, j_1, k_1) to vertex (i_2, j_2, k_2) if and only if (i) $i_2 = i_1 + 1$ and $j_2 = j_1 + k_2$ for $k_1 \geq k_2$ or (ii) $\mathrm{mod}(i_1, M) = 0$ and $j_2 = j_1 + k_2$. The weight of this edge is set to $1 - U_{i,j}$ (where $i_2 = (i - 1)M + j$) if $j_2 = j_1 + 1$ and set to 0 if $j_2 = j_1$. In this manner, we have incorporated the objective function of (6) into the shortest path problem. To fulfill the first constraint of (6), we find the shortest paths (i) from vertex (0,0,0) to vertex $(MN, T, 0)$ and (ii) from vertex (0,0,0) to vertex $(MN, T, 1)$, and then select the shorter one between them. Figure 5 shows an illustrative example. The optimization details are given below.

Optimization Algorithm 1

1. Construct a directed graph $G = (V, A)$ where the set of vertices V is:

$$
\begin{aligned}
V \;=\; & \{0, 1, \ldots, M \cdot N\} \times \{0, 1, \ldots, T\} \times \{0, 1\} \\
& - \{0, 1, \ldots, M \cdot N\} \times \{0\} \times \{1\}
\end{aligned}
\tag{7}
$$

   There is an edge in A connecting vertex (i_1, j_1, k_1) to vertex (i_2, j_2, k_2) if and only if (i) $i_2 = i_1 + 1$ and $j_2 = j_1 + k_2$ for $k_1 \geq k_2$ or (ii) $\mathrm{mod}(i_1, M) = 0$ and $j_2 = j_1 + k_2$. The weight of this edge is set to $1 - U_{i,j}$ (where $i_2 = (i-1)M + j$) if $j_2 = j_1 + 1$ and set to 0 if $j_2 = j_1$.

2. Find the shortest path (i) from vertex $(0, 0, 0)$ to vertex $(M \cdot N, T, 0)$ and (ii) from vertex (0,0,0) to vertex $(M \cdot N, T, 1)$, and select the shorter one between them. If the shortest path passes through the edge with weight $1 - U_{i,j}$, the optimal value of $x_{i,j}$ is 1; otherwise, the optimal value is 0. $\square$

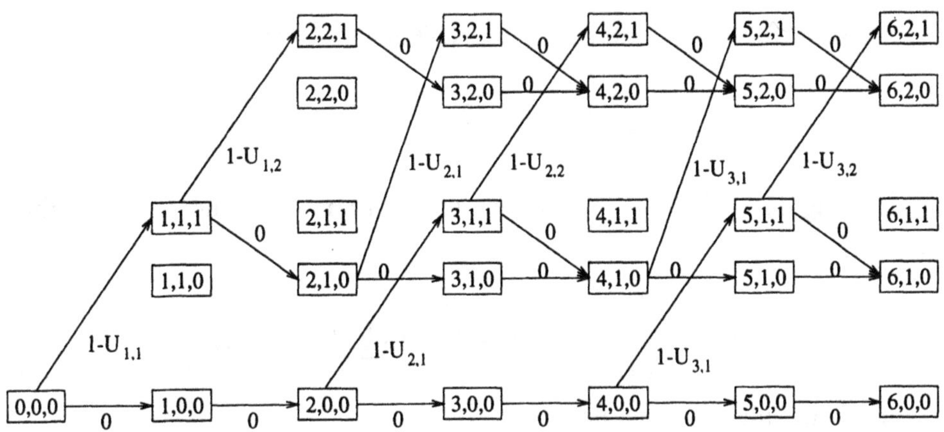

Figure 5.   An example to illustrate the directed graph used in Optimization Algorithm 1; $T = M = 2$, $N = 3$.

The time complexity of above algorithm can be analyzed as follows. From equation (7), we see that the directed graph has $O(MNT)$ vertices and hence we can use the Dijkstra's algorithm to find the shortest path in this directed graph in $O(M^2N^2T^2)$ time.

B. Maximize the Product of the Total Utilizations

In this subsection, we formulate and solve the problem of maximizing the product of the total utilizations of the nodes. Let $x_{i,j}$ be as defined in equation (4). The problem can be formulated as follows:

$$\begin{cases} \text{maximize}_{x_{i,j}} & \prod_{i=1}^{N}\{\sum_{j=0}^{M}(U_{i,j} \cdot x_{i,j})\} \\ \text{subject to} & 1.\ \sum_{i=1}^{N}\sum_{j=1}^{M} x_{i,j} = T \\ & 2.\ x_{i,0} = 1,\ i = 1,2,\ldots,N \\ & 3.\ x_{i,j} \geq x_{i,j+1},\ i = 1,2,\ldots,N,\ j = 1,2,\ldots,M-1 \\ & 4.\ x_{i,j} \in \{0,1\} \end{cases} \qquad (8)$$

The variables to be optimized are $x_{i,j}$ for all $1 \leq i \leq N$ and $1 \leq j \leq M$. The objective function is the product of the total utilizations of all the nodes. The first constraint ensures that the total number of FWCs allocated to the nodes is T. The second constraint ensures that the $U_{i,0}$ is always included in the objective function so that the objective function will not be zero. The third and fourth constraints ensure that $x_{i,j} \geq x_{i,j+1}$ and $x_{i,j}$ is binary.

We solve the above optimization problem as follows. Since

$$0 < \sum_{j=0}^{M} U_{i,j} x_{i,j} \le 1, \qquad 1 \le i \le N$$

maximizing $\prod_{i=1}^{N} \sum_{j=0}^{M} U_{i,j} x_{i,j}$ is equivalent to maximizing the following function:

$$\log\{\prod_{i=1}^{N} \sum_{j=0}^{M} (U_{i,j} x_{i,j})\} = \sum_{i=1}^{N} \log\{\sum_{j=0}^{M} (U_{i,j} \cdot x_{i,j})\} \tag{9}$$

We need the following lemma.

Lemma 4.1 *For any $1 \le i \le N$, if*

$$\begin{cases} x_{i,0} = 1 \\ x_{i,j} \ge x_{i,j+1} & j = 1, 2, \dots, M-1 \\ x_{i,j} \in \{0, 1\} & j = 1, 2, \dots, M \end{cases} \tag{10}$$

then

$$\log[\sum_{j=0}^{M} (U_{i,j} \cdot x_{i,j})] = \sum_{j=1}^{M} \left\{ [\log(\sum_{s=0}^{j} U_{i,s}) - \log(\sum_{s=0}^{j-1} U_{i,s})] \cdot x_{i,j} \right\} + \log(U_{i,0}) \tag{11}$$

Proof. (i) If $x_{i,1} = 0$, then for any $1 \le j \le M$, $x_{i,j} = 0$. The result is obviously correct.

(ii) If there exists $1 \le j_0 \le M$ such that $x_{i,j_0} = 1$ and $x_{i,j_0+1} = 0$, then from (10), we have $x_{i,1} = x_{i,2} = \dots = x_{i,j_0} = 1$ whereas $x_{i,j_0+1} = x_{i,j_0+2} = \dots = x_{i,M} = 0$. Thus,

$$\begin{aligned}
&\sum_{j=1}^{M} \left\{ [\log(\sum_{s=0}^{j} U_{i,s}) - \log(\sum_{s=0}^{j-1} U_{i,s})] \cdot x_{i,j} \right\} + \log(U_{i,0}) \\
&= \sum_{j=1}^{j_0} [\log(\sum_{s=0}^{j} U_{i,s}) - \log(\sum_{s=0}^{j-1} U_{i,s})] + \log(U_{i,0}) \\
&= [\log(\sum_{s=0}^{j_0} U_{i,s}) - \log(U_{i,0})] + \log(U_{i,0}) \\
&= \log \sum_{s=0}^{j_0} U_{i,s} \\
&= \log \sum_{s=0}^{j_0} (U_{i,s} \cdot x_{i,s}) \\
&= \log \sum_{j=0}^{M} (U_{i,j} \cdot x_{i,j})
\end{aligned} \tag{12}$$

(iii) If $x_{i,M} = 1$, then $x_{i,j} = 1$, $j = 1, 2, \dots, M$. Similar to case (ii), formula (11) can be proved to be correct. This completes the proof. $\square$

Using the above lemma, we can transform optimization problem (8) into an equivalent form of (5), so that we can apply Optimization Algorithm 1 to find the optimal solutions. The details are given in the following algorithm.

Optimization Algorithm 2

1. Use the following transformation

$$U'_{i,j} = \log(\sum_{s=0}^{j} U_{i,s}) - \log(\sum_{s=0}^{j-1} U_{i,s}), \qquad 1 \le i \le N, \ 1 \le j \le M \quad (13)$$

we transform problem (8) to the following form:

$$\begin{cases} \text{maximize}_{x_{i,j}} & \sum_{i=1}^{N}\sum_{j=1}^{M}(U'_{i,j} \cdot x_{i,j}) \\ \text{subject to} & 1. \ \sum_{i=1}^{N}\sum_{j=1}^{M} x_{i,j} = T \\ & 2. \ x_{i,j} \ge x_{i,j+1}, \ i = 1, 2, \ldots, N, \ j = 1, 2, \ldots, M-1 \\ & 3. \ x_{i,j} \in \{0, 1\} \end{cases}$$
$$(14)$$

2. Apply Optimization Algorithm 1 to solve (14). $\square$

C. Maximize the Minimum Total Utilization

In this subsection, we formulate and solve the problem of maximizing the minimum total utilization of all the N nodes. Let $x_{i,j}$ be as defined in equation (4). This problem can be formulated as follows:

$$\begin{cases} \text{maximize}_{x_{i,j}} & \left(\min_{1 \le i \le N}\{\sum_{j=0}^{M} U_{i,j} x_{i,j}\} \right) \\ \text{subject to} & 1. \ \sum_{i=1}^{N}\sum_{j=1}^{M} x_{i,j} = T \\ & 2. \ x_{i,j} \ge x_{i,j+1}, \ i = 1, 2, \ldots, N, \ j = 1, 2, \ldots, M-1 \\ & 3. \ x_{i,j} \in \{0, 1\} \end{cases}$$
$$(15)$$

The variables to be optimized are $x_{i,j}$ for all $1 \le i \le N$ and $1 \le j \le M$. The objective function is the minimum total utilization of all the N nodes, and the first constraint ensures that the total number of FWCs allocated to the N nodes is T.

The above optimization problem can be solved by the following greedy algorithm.

Optimization Algorithm 3

1. Initially, all the N nodes have no FWC.

2. Among the N nodes, find the node having the smallest total utilization, then allocate one more FWC to this node, and then update the total utilization for this node based on $\mathbf{U}$.

3. Repeat step 2 until all the T FWCs have been allocated to the nodes. $\square$

Theorem 4.2 *Optimization Algorithm 3 can find optimal solutions to optimization problem (15).*

Proof. By contradiction. Assume that Optimization Algorithm 3 cannot find the optimal solution to optimization problem (15). We let $W^*(i)$ be the optimal number of FWCs allocated to node i, and $W_a(i)$ be the number of FWCs allocated to node i by Optimization Algorithm 3. In addition, let $S^*(i)$ and $S_a(i)$ be the total utilization of node i corresponding to $W^*(i)$ and $W_a(i)$ respectively. In the optimal allocation of FWCs, suppose node i_0^* has the smallest total utilization; in the allocation by Optimization Algorithm 3, node i_0 has the smallest total utilization. In other words, we have

$$\begin{cases} S^*(i_0^*) & = & \min\{S^*(1), S^*(2), \ldots, S^*(N)\} \\ S_a(i_0) & = & \min\{S_a(1), S_a(2), \ldots, S_a(N)\} \end{cases} \tag{16}$$

Since

$$S^*(i_0) \geq S^*(i_0^*) \tag{17}$$

We have three possible cases:

$$\begin{cases} \text{case (a)} & W^*(i_0) > W_a(i_0) \\ \text{case (b)} & W^*(i_0) = W_a(i_0) \\ \text{case (c)} & W^*(i_0) < W_a(i_0) \end{cases} \tag{18}$$

Since the total utilization of a node is monotonically increasing with the number of FWCs allocated to it, both case (b) and case (c) will lead to a contradiction to the assumption that the solution got by Optimization Algorithm 3 is not optimal. For case (a), since the total number of available FWCs is fixed, there must exist at least one node (called node k_0) such that

$$W^*(k_0) < W_a(k_0) \tag{19}$$

If $S^*(k_0) \leq S_a(i_0)$, then

$$S^*(i_0^*) \leq S^*(k_0) \leq S_a(i_0) \tag{20}$$

which leads to a contradiction. If $S^*(k_0) > S_a(i_0)$, then we have

$$S_a(k_0) > S^*(k_0) > S_a(i_0) \tag{21}$$

Combined with (19), we see that Optimization Algorithm 3 has allocated more FWCs to the node with larger total utilization (node k_0), but not the node with the smallest total utilization (node i_0), which leads to contradiction again. This completes the proof. $\square$

4.3 Routing and Wavelength Assignment Algorithm

Routing and wavelength assignment (RWA) for all-optical network is a hot research topic and several RWA algorithms have been proposed (see section 1.4). These algorithms were designed for the networks having either complete wavelength conversion or no wavelength conversion. In this section, we design a new RWA algorithm for the networks in which different nodes may be allocated different number of FWCs.

The critical problem is that when a certain number of FWCs have been allocated to each node, how should we select the tuning nodes for every transmission request in order to get good blocking performance? Our main ideas for this problem are as follows:

1. Once a transmission request arrives, select the set of tuning nodes such that the required number of FWCs is minimized.

2. When there is more than one choice, select the one that maximizes the minimum number of free FWCs in each tuning node of the source-to-destination path. For simplicity, we call the tuning node with minimum number of free FWCs as the *critical node*.

3. When there is more than one choice, select the one that has the maximum number of FWCs installed on the critical node. When there is still more than one choice (though this rarely happens), randomly select one choice.

Based on the above ideas, we design the following algorithm for routing and wavelength assignment. For the i-th intermediate node ($1 \leq i \leq L$) let $N_t(i)$ and $N_a(i)$ be the total number of FWCs and the number of FWCs being used in this node respectively. Therefore, the number of free FWCs in this node is $N_t(i) - N_a(i)$. The details of our RWA algorithm are as follows:

RWA Algorithm

1. Check if there is at least one clear channel on the source-to-destination path. If there is one, assign this clear channel to the transmission request; if there is more than one channel, select one of them on a first-flt basis; if there is none, go to step 2.

2. If there is at least one free wavelength channel (at any wavelength) on every hop of the source-to-destination path, execute the following steps:

(a) Construct a directed graph in a manner similar to that in Conflict Resolution Algorithm. For each free wavelength channel on every hop, the weight of the corresponding edge is M. On every intermediate node l, the weight of the edge between the node $v_i(\lambda_i, l)$ and node $v_o(\lambda_o, l)$ is

$$c(\lambda_i, \lambda_o, l) = \begin{cases} M + S & \text{if } \lambda_i \neq \lambda_o \\ 0 & \text{if } \lambda_i = \lambda_o \end{cases} \qquad (22a)$$

where

$$S = \begin{cases} \frac{M}{N_t(l) - N_a(l)} + (1 - \frac{N_a(l)}{N_t(l)}) & \text{if } N_t(l) > N_a(l) \\ \infty & \text{if } N_t(l) = N_a(l) \end{cases} \qquad (22b)$$

Apply equations (1)-(3) in Conflict Resolution Algorithm to find the shortest path from the source to the destination.

(b) Determine the set of tuning nodes and increment $N_a(l)$ of each tuning node by 1.

Otherwise, the transmission request is blocked. $\square$

The RWA algorithm can minimize the number of FWCs required by each transmission because the weight of each converter edge is at least M. When there is more than one choice requiring the same number of tuning nodes, equation (22) ensures that we can select the choice that maximizes the minimum number of free FWCs in each tuning node. It is because a smaller value of $\frac{M}{N_t(l) - N_a(l)}$ implies a larger value of $N_t(l) - N_a(l)$ (i.e., a larger number of free FWCs in node l). When there is still more than one choice, equation (22) can also ensure that we can select the choice that has the maximum number of FWCs installed on the critical node. It is because, for the same value of $N_t(l) - N_a(l)$, a larger $N_t(l)$ can lead to a smaller value of $(1 - \frac{N_a(l)}{N_t(l)})$.

4.4   Numerical Results and Discussions

We use computer simulations to evaluate the performance of the proposed allocation method. The main steps are as follows:

1. Conduct a computer simulation for any given network with complete wavelength conversion and any given traffic load and pattern. During simulation, execute the Recording Algorithm to record the utilization matrix.

2. Based on the recorded utilization matrix, execute Optimization Algorithm 1 (or 2 or 3) to optimize the allocation of FWCs.

3. Conduct another computer simulation for the same network with the allocation of FWCs determined in step 2. During simulation, execute the RWA Algorithm to perform routing and wavelength assignment for each new request and record the blocking probability.

We have conducted extensive computer simulations to study the effectiveness of our algorithms. We consider a regular network (an 11×11 torus-mesh network with 121 nodes [46], see Figure 6) and an irregular network with 100 nodes. The irregular network is randomly generated. To ensure that the resulting network is not far from the reality, we adopt the following generation method:

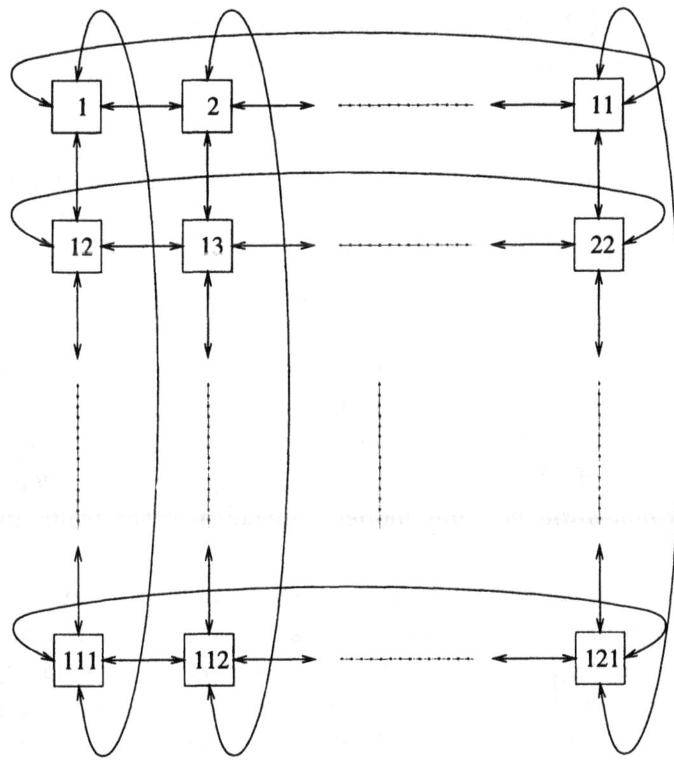

Figure 6. The 11×11 torus-mesh network, where every edge represents a link which is composed of oppo- two separate fibers going in site directions and each fiber has 10 channels.

1. Start from the 10×10 mesh network with 100 nodes and 180 bi-directional links.

2. Randomly delete 20 links from the network while ensuring that the resulting network is not disconnected.

3. Randomly add 30 links to the network as follows. For the j_1-th node on the i_1-th row and the j_2-th node on the i_2-th row, we define the *distance* between them as follows:

$$d_{(i_1,j_1),(i_2,j_2)} = \sqrt{(i_1 - i_2)^2 + (j_1 - j_2)^2} \qquad (23)$$

To ensure that a node is not directly connected to a very far-away node, we randomly select two nodes, and add a link between them if and only if (1) there is no existing link between them, and (2) their distance is not larger than $3\sqrt{2}$. This step is repeated until 30 links have been added.

We execute the above steps to get a sample network for our simulation experiments. This network is irregular with 100 nodes and 190 bi-directional links. The path length between any two nodes varies from 1 to 11 and the average is 5.1628 hops. The number of links connected to a node varies from 2 to 6 and the average

In both the regular and irregular networks, each link is composed of two separate fibers going in opposite directions and each fiber has 10 channels. The torus-mesh network has been adopted by many researchers for performance evaluation of all-optical network (e.g., see [46, 47]). We consider both *uniform* and *non-uniform* traffic models. Specifically, the traffic models are defined as follows:

The arrivals of transmission requests follow a Poisson process and the total arrival rate is λ_T. The duration of each transmission is exponentially distributed. The *traffic matrix* is $\mathbf{I} = [I_{i,j}]_{N \times N}$ where $I_{i,i} = 0$ and $I_{i,j}$ $(i \neq j)$ denotes the probability that there is a transmission request from node i to node j [76]. Therefore, the arrival rate from node i to node j is $\lambda_T I_{i,j}$. When all the non-diagonal entries of $\mathbf{I}$ are equal to each other, the traffic is uniform; otherwise, the traffic is non-uniform.

The non-uniform traffic on the regular network is as follows. Based on the network topology shown in Figure 6, we divide the network into 9 parts as follows:

1 (4×4)	2 (4×3)	3 (4×4)
4 (3×4)	5 (3×3)	6 (3×4)
7 (4×4)	8 (4×3)	9 (4×4)

Let $I'_{i,j}$ (where i could be equal to j) denotes the probability that there is a transmission request between two different nodes in the i-th part and the j-th part respectively. The non-uniform traffic is defined as follows:

$$\begin{cases} I'_{1,3} = \frac{5}{25784} \\ I'_{1,7} = I'_{3,9} = \frac{3}{25784} \\ I'_{1,9} = I'_{3,7} = \frac{8}{25784} \\ I'_{i,j} = \frac{1}{25784} \qquad \text{otherwise} \end{cases} \tag{24}$$

where 25784 is a normalization constant and it ensures that probabilities sum to one.

The non-uniform traffic on the irregular network is as follows. We divide the network into the two parts where the nodes in the upper 5 rows belong to part **1** and those in the lower 5 rows belong to part **2**. The non-uniform traffic is defined as follows:

$$\begin{cases} I'_{1,2} = \frac{x}{4900+5000 \cdot x} \\ I'_{1,1} = I'_{2,2} = \frac{1}{4900+5000 \cdot x} \end{cases} \tag{25}$$

where x is a parameter such that a larger x specifies a more non-uniform traffic, and $(4900 + 5000 \cdot x)$ is a normalization constant.

We consider two performance measures: (1) overall blocking probability (i.e., the average of the blocking probabilities experienced at all the source nodes) and (2) maximum blocking probability (i.e., the maximum of the blocking probabilities experienced at all the source nodes). The first performance measure can measure the mean quality of service, while the second one can measure the fairness. To make comparisons, we apply the blocking probability with complete wavelength conversion to evaluate the performance of the proposed algorithms under partial wavelength conversion.

In subsection 4.4.1, we compare the performance of our algorithms with the allocation method proposed in [51]. Though the method in [51] was not proposed for minimizing the number of FWCs required, among all the existing methods *not* based on theoretical analysis of network blocking probability, to the best of our knowledge, it is still the one requiring the least number of FWCs. For simplicity, we refer to it as the *"best existing allocation"* in the rest part of this chapter. In subsection 4.4.2, we demonstrate that our algorithms are robust under simulation and estimation uncertainty.

4.4.1 Performance

Figure 7 shows the performance of Optimization Algorithms 1, 2 and 3. When each node has one or more FWCs on average, we see that all the three algorithms can result in blocking probabilities close to those with complete wavelength conversion. In addition, we see that these algorithms have similar performance for our regular and irregular networks. Since Optimization Algorithm 3 has the smallest time complexity, it can be regarded as the most efficient one for these two networks. Therefore, unless otherwise specified, we consider Optimization Algorithm 3 in the remaining part of this section.

Figure 8 shows the performance of our allocation and the best existing allocation [51]. From this figure, we observe the following points:

- When each node has one or more FWCs on average, our method can already result in blocking probabilities close to those with complete wavelength conversion. This demonstrates that our approximate approach is very good.

- When the network topology is regular and the traffic is uniform, Figure 8(a) shows that our method and the best existing allocation have the same performance. It is because every node handles the same amount of traffic in this special case and hence the optimal allocation is to allocate the same number of FWCs to every node. Therefore, the best existing allocation can be regarded as a special case of our allocation.

- When the network topology is irregular and the traffic is uniform, Figure 8(b) shows that our method can give significantly better performance than the best existing allocation, especially when the number of available FWCs is small. For example, when the number of available FWCs is 100, the overall blocking probabilities of our method and the best existing allocation are 2.916% and 4.244% respectively (i.e., our method can reduce the overall blocking probability by 31.3%). In addition, the maximum blocking probabilities of our method and the best existing allocation are 7.158% and 10.460%, respectively (i.e.,

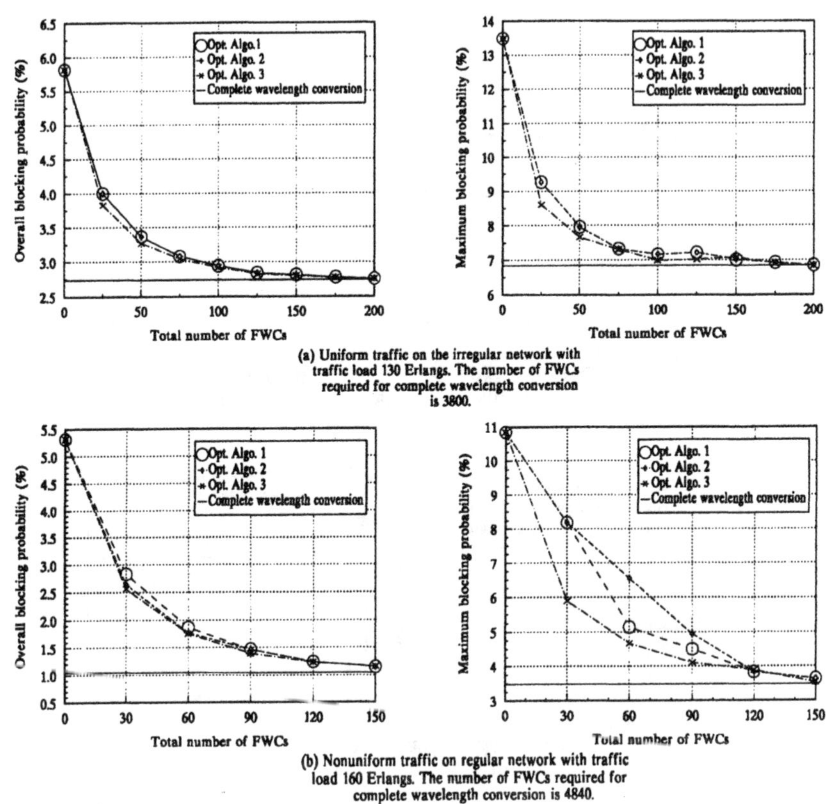

Figure 7.   Performance of Optimization Algorithms 1, 2 and 3.

our method can reduce the maximum blocking probability by 31.6%). From another point of view, when we want to achieve a given blocking performance, our method requires significantly fewer WCs than those required by the best existing allocation. For example, if we want to ensure that the overall blocking probability is about 3%, the number of FWCs required by our method and the best existing allocation are 100 and 300 respectively.

- When the network topology is regular and the traffic is non-uniform, Figure 7(c) shows that our method can also give significantly better performance than the best existing allocation. For example, when the number of available FWCs is 121, our method can reduce the overall and maximum blocking probability of the best existing allocation by

59.0% and 53.5%, respectively.

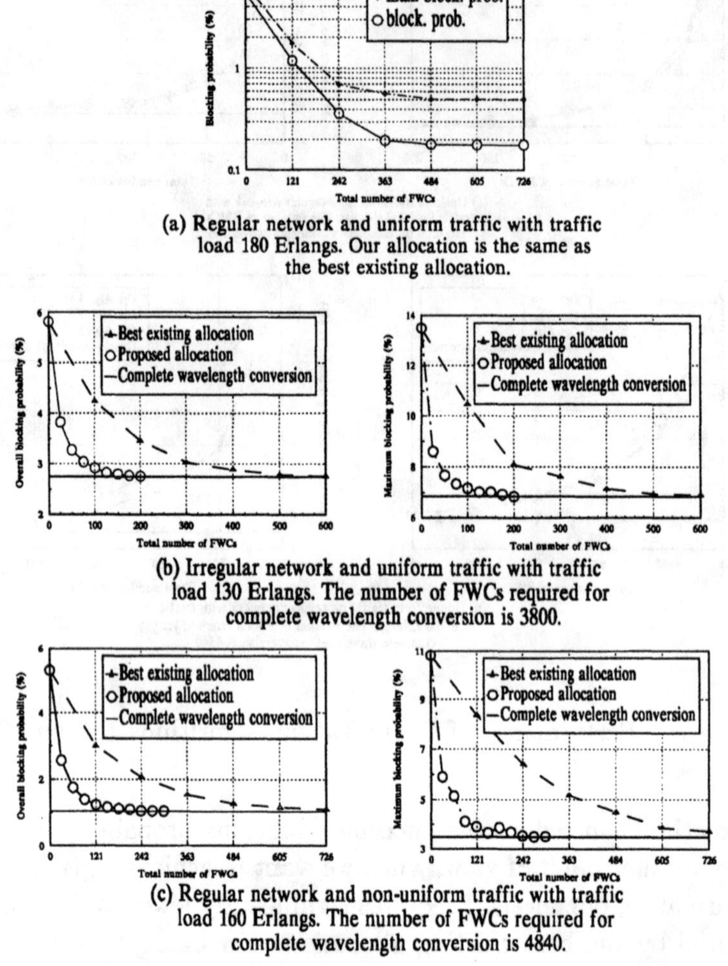

(a) Regular network and uniform traffic with traffic
load 180 Erlangs. Our allocation is the same as
the best existing allocation.

(b) Irregular network and uniform traffic with traffic
load 130 Erlangs. The number of FWCs required for
complete wavelength conversion is 3800.

(c) Regular network and non-uniform traffic with traffic
load 160 Erlangs. The number of FWCs required for
complete wavelength conversion is 4840.

Figure 8.   Performance of the proposed allocation and the best
existing allocation.

Figures 9 and 10 compare the performance of the proposed method with
the best existing allocation when the traffic becomes more non-uniform and
the traffic load becomes heavier. We see that our method is significantly
better than the best existing allocation, and its performance is quite close
to that with complete wavelength conversion.

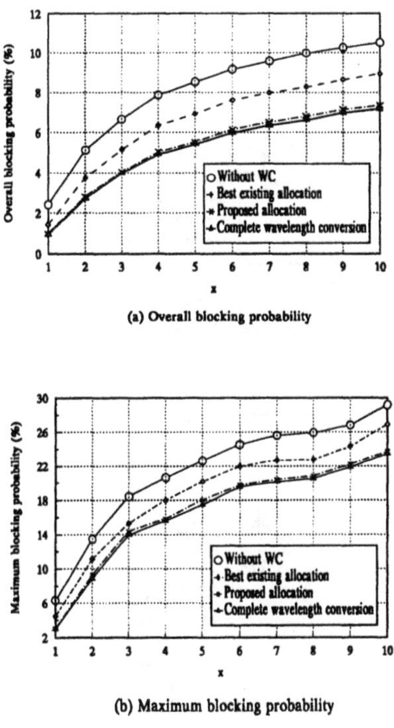

Figure 9. Performance of the proposed method and the best existing allocation versus the traffic parameter x, where a larger x specifies a more nonuniform traffic. Network topology is irregular, traffic load is 100 Erlangs, and there are 100 FWCs. The number of FWCs required for complete wavelength conversion is 3800.

Figure 11 shows the performance of the proposed allocation and the best existing allocation for low blocking probability. We observe similar results: (1) the proposed allocation can result in blocking probabilities close to that with complete wavelength conversion, and (2) the proposed allocation is significantly better than the best existing allocation.

4.4.2 Robustness

In computer simulations, uncertainty is unavoidable. In this subsection, we demonstrate that our simulation-based optimization method is robust under

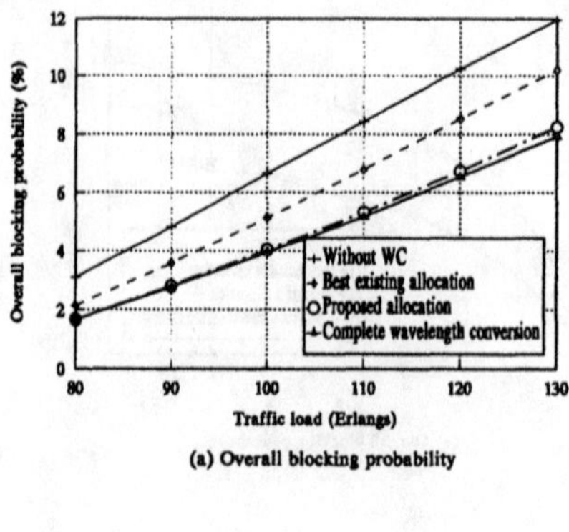

(a) Overall blocking probability

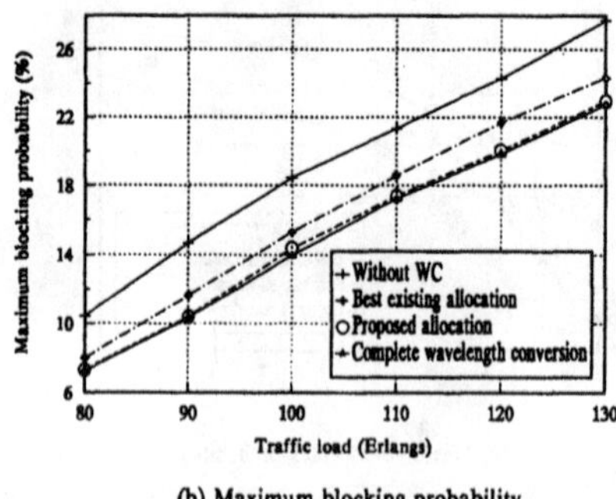

(b) Maximum blocking probability

Figure 10. Performance of the proposed method and the best existing allocation under different traffic load. Network is irregular, traffic is non-uniform with $x = 3$, and there are 100 FWCs. The number of FWCs required for complete wavelength conversion is 3800.

(1) simulation uncertainty and (2) estimation uncertainty of traffic pattern and traffic load.

To study the stability of Optimization Algorithms 1, 2 and 3 under simulation uncertainty, we conducted 10 independent simulation experiments on the irregular network using different kinds of random number generators and different seeds. Figure 12 shows the results. We see that our method is robust, and the blocking probability is relatively less sensitive to the uncertainty than the maximum blocking probability.

Figures 13 and 14 show that our method is robust under estimation uncertainty of traffic pattern and traffic load respectively. In particular, the

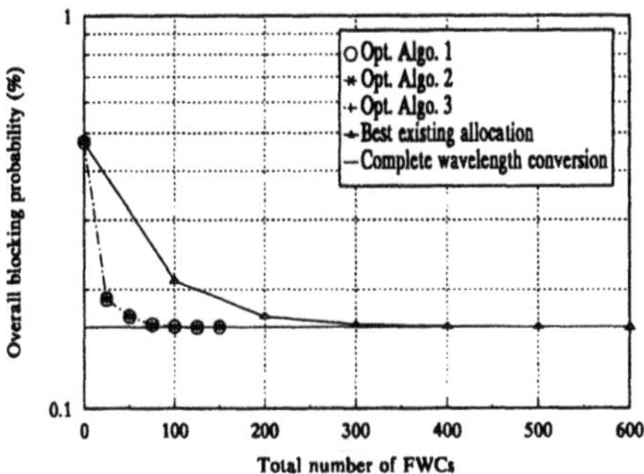

(a) Irregular network and uniform traffic with traffic
load 70 Erlangs. The number of FWCs required for
complete wavelength conversion is 3800.

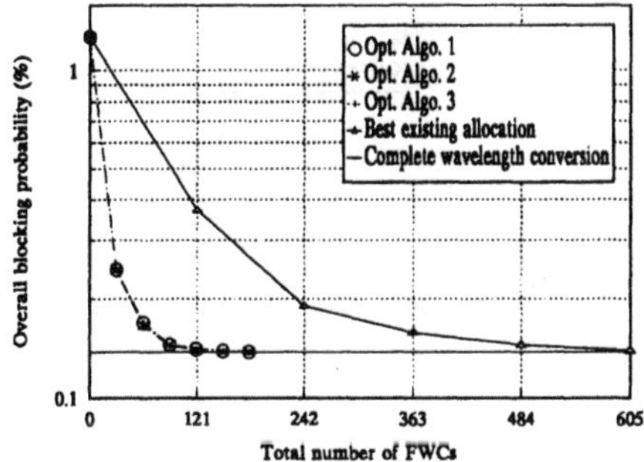

(b) Nonuniform traffic on regular network with traffic
load 110 Erlangs. The number of FWCs required for
complete wavelength conversion is 4840.

Figure 11.   Performance of the proposed method and the best existing
allocation for low blocking probability.

blocking probability is relatively less sensitive to the uncertainty than the
maximum blocking probability.

5   Summary and Future Research

In this chapter, we surveyed the state-of-the-art technologies for all-optical
networks. In particular, we focused on the problem of allocating wavelength converters in all-optical networks. We explained why an all-optical
network can use wavelength converters to improve its performance. Then
we stated the problem of allocating wavelength converters to the nodes of

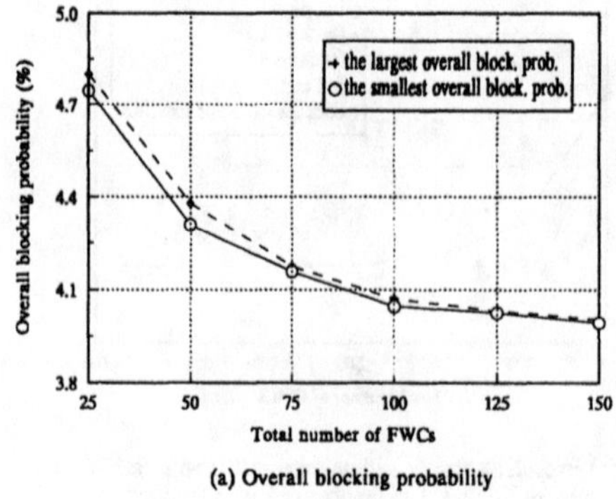

(a) Overall blocking probability

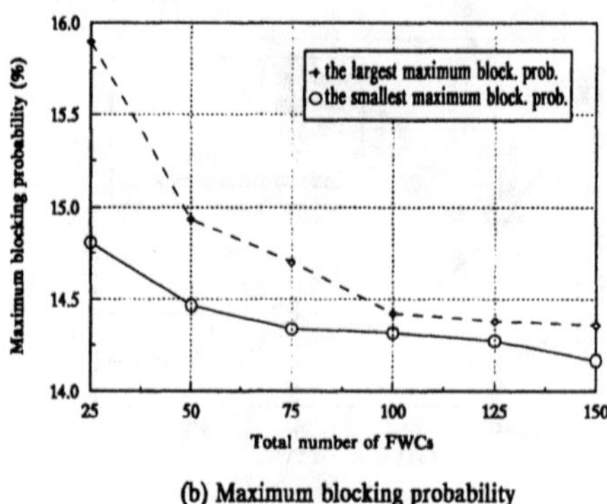

(b) Maximum blocking probability

Figure 12.   Performance of the proposed method and the best existing allocation for low blocking probability.

an all-optical network. We described three approaches to solve this allocation problem: *intuitive approach, analytical approach* and *simulation-based optimization approach*. In particular, the simulation-based optimization approach is widely applicable and it is not restricted to any particular model or assumption.

Many problems still need to be addressed. For example, if a wavelength converter has a smaller conversion range, it is usually cheaper. When limited-range wavelength converters are used, it is necessary to investigate how to modify the allocation method described in section 4. In particular, the main challenge is to modify the Recording Algorithm to handle the constraint on limited conversion range. In addition, when there are multiple

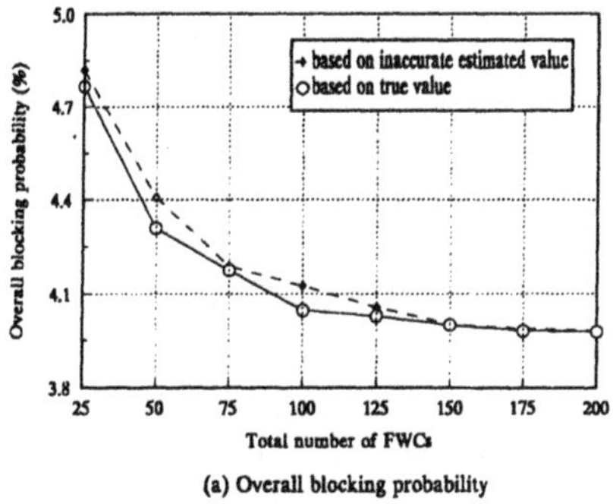

(a) Overall blocking probability

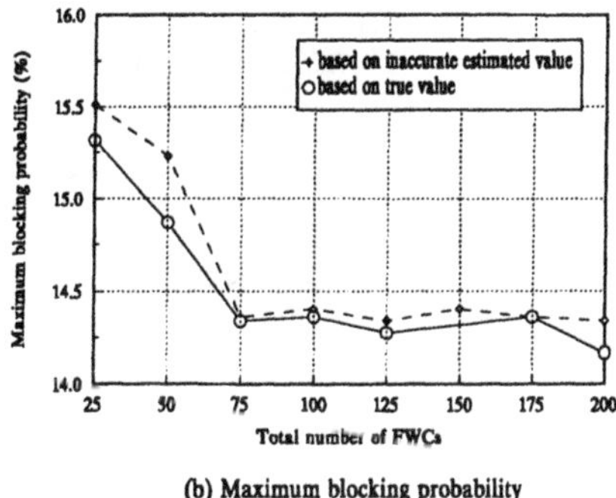

(b) Maximum blocking probability

Figure 13. Robustness of the proposed method under estimation uncertainty of traffic pattern. The network is irregular with traffic load 100 Erlangs. The traffic is non-uniform and the true value and the inaccurate estimated value of x are 3 and 2 respectively.

kinds of wavelength converters with different conversion ranges and different prices, it would be interesting to determine the most cost-effective choice. Finally, the simulation-based optimization approach is a general framework, and we believe that it is applicable to the design and optimization of many types of optical networks (e.g., allocation of optical buffer or bursty switches in IP over WDM networks). In this regard, there are many significant research opportunities.

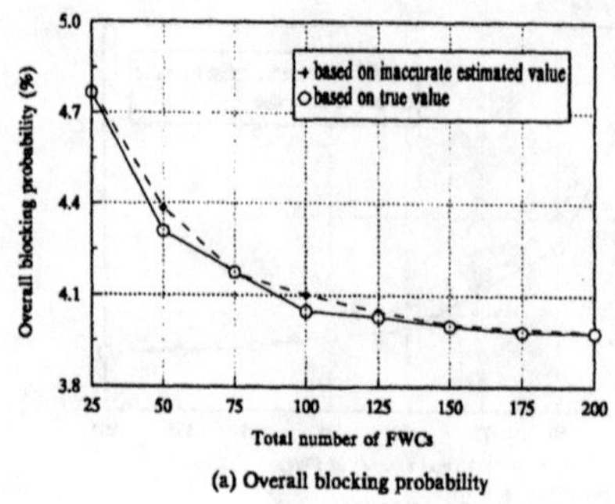

(a) Overall blocking probability

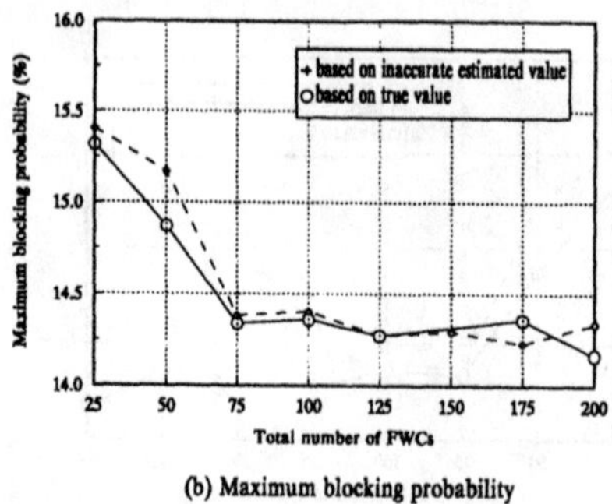

(b) Maximum blocking probability

Figure 14. Robustness of the proposed method under estimation uncertainty of traffic load. Network is irregular and traffic is non-uniform with $x = 3$. The true and the inaccurate estimated traffic load are 100 and 90 Erlangs respectively.

References

[1] R. Ramaswami and K. N. Savarajan, *Optical Networks: A Practical Perspective*, (Morgan Kaufmann Publishers, 1998).

[2] Bell Labs uses ultra-dense WDM to transmit 1,022 channels over fiber, *http://www.bell-labs.com/news/1999/november/10/2.html* (1999).

[3] Bell Labs scientists demo first long distance triple terabit transmission, *http://www.bell-labs.com/news/2000/march/16/2.html* (2000).

[4] Wavestar™ OLS 400G, *http://www.lucent-optical.com/solutions/products/ols400g/* (2000).

[5] MCI WorldCom to trial world's highest-capacity next-generation optical solution from Nortel Networks, *http://www.nortelnetworks. com/corporate/news/newsreleases/1999b/5_5_9999311_MCI_Worldcom. html* (1999).

[6] M. J. L. Cahill, G. J. Pendock, and D. D. Sampson, Demonstration of hybrid coherence multiplexing/WDM customer access network, *Proc. OFC'97* (1997) pp. 58-59.

[7] R. R. Patel, S. W. Bond, M. C. Larson, M. D. Pocha, H. E. Garrett, M. E. Lowry, and R. J. Deri, Multi-mode fiber coarse WDM grating router using broadband add/drop filters for wavelength re-use, *Proc. LEOS'99* Vol.2 (1999) pp. 826-827.

[8] B. St. Arnaud, Optical Networks for the rest of us: customer empowered networking, *Proc. Optical Network Workshop* (2000) http://catss.utdallas.edu/ONW2000/proceedings/.

[9] G. Finley, A CWDM-based Canadian regional advanced network, *Technical report, University of Calgary* (2000).

[10] B. Mukherjee, WDM-based local lightwave networks, Part I: single-hop systems, *IEEE Network* (May 1992) pp. 11-27.

[11] S. Melle, C. P. Pfistner, and F. Diner, Amplifier and multiplexing technologies expand network capacity, *Lightwave Magazine* (Dec. 1995) pp. 42-46.

[12] P. E. Green, Optical network update, *IEEE J. Select. Areas Commun.* Vol.14 (1996) pp. 764-779.

[13] G. R. Pieris and G. H. Sasaki, Scheduling transmission in WDM broadcast-and -select networks, *IEEE/ACM Trans. Network.* Vol.2 No.2 (1994) pp. 105-110.

[14] G. Xiao and Y.-W. Leung, Cost-effective WDM broadcast-and-select networks for all-to-all transmission schedules, *J. System. Archtec.* Vol.45 No. 2 (1998) pp. 115-129.

[15] C. A. Brackett, Foreword: Is there an emerging consensus on WDM networking? *J. Lightwave Tech.* Vol.14 No.6 (1996) pp. 936-941.

[16] A. Misawa and M. Tsukada, Broadcast-and-select photonic ATM switch with frequency division multiplexed output buffers, *J. Lightwave Tech*, Vol.15 No.10 (1997) pp. 1769-1777.

[17] M. Tsukada, D. Z. Wen, T. Matsunaga, M. Asobe, and T. Oohara, An ultrafast photonic ATM switch based on bit-interleave multiplexing, *J. Lightwave Tech.* Vol.14 No.9 (1996) pp.1979-1985.

[18] I. Chlamtac, A. Ganz, and G. Karmi, Lightpath communications: An approach to high bandwidth optical WAN's, *IEEE Trans. Commun.* Vol.40 No.7 (1992) pp. 1171-1182.

[19] R. Ramaswami and K. N. Sivarajan, Routing and wavelength assignment in all-optical networks, *IEEE/ACM Trans. Network.* Vol.3 No.5 (1995) pp.489-500.

[20] D. Banerjee and B. Mukherjee, A practical approach for routing and wavelength assignment in wavelength-routed optical networks, *IEEE J. Select. Areas Commun.* Vol.14 No.5 (1996) pp. 903-913.

[21] I. Chlamtac, A. Farago, and T. Zhang, Lightpath (wavelength) routing in large WDM networks, *IEEE J. Select. Areas Commun.* Vol.14 No.5 (1996) pp. 909-913.

[22] K.-C. Lee and V. O. K. Li, A wavelength rerouting algorithm in wide-area all-optical networks, *IEEE/OSA J. Light. Tech.* Vol.14 No.6 (1996) pp. 1218-1229.

[23] A. Mokhtar and M. Azizoglu, Adaptive wavelength routing in all-optical networks, *IEEE/ACM Trans. Network.* Vol.6 No.2 (1998) pp. 197-206.

[24] R. Ramaswami and A. Segall, Distributed network control for wavelength routed optical networks, *Proc. IEEE INFOCOM'96* (1996) pp. 138-147.

[25] H. Zang, L. Sahasrabuddhe, J. P. Jue, S. Ramamurthy, and B. Mukherjee, Connection management for wavelength-routed WDM networks, *Proc. IEEE GLOBECOM'99* (1999) pp.1428-1432.

[26] X. Yuan, R. Melhem, R. Gupta, Y. Mei, and C. Qiao, Distributed control protocols for wavelength reservation and their performance evaluation, *Photo. Network Commun.* Vol.1 No.3 (1999) pp. 207-218.

[27] K. Bala, T. E. Stern, D. Simchi-Levi, and K. Bala, Routing in a linear lightwave network, *IEEE/ACM Trans. Networking* Vol.3 No.4 (1995) pp. 459-469.

[28] Z. Zhang, D. Guo, and A. Acampora, Logarithmically scalable routing algorithms in large optical networks, *Proc. IEEE Infocom'95* Vol.3 (1995) pp. 1290-1299.

[29] S. Subramaniam and R. A. Barry, Wavelength assignment in fixed routing WDM networks, *Proc. IEEE ICC'97* (1997) pp. 406-410.

[30] G. Jeong and E. Ayanoglu, Comparison of wavelength-interchanging and wavelength-selective cross-connects in multiwavelength all-optical network, *Proc. IEEE Infocom'96* (1996) pp. 156-163.

[31] E. Karasan and E. Ayanoglu, Effects of wavelength routing and selection algorithms on wavelength conversion gain in WDM optical networks, *IEEE/ACM Trans. Network.* Vol.6 No.2 (1998) pp. 186-196.

[32] A. Birman, Computing approximate blocking probabilities for a class of all-optical networks, *IEEE J. Select. Areas Commun.* Vol.14 No.6 pp. 852-857, June 1996.

[33] H. Harai, M. Murata, and H. Miyahara, Performance of alternate routing methods in all-optical switching networks, *Proc. IEEE Infocom'97* Vol.2 (1997) pp. 517-525.

[34] S. Ramamurthy and B. Mukherjee, Fixed-alternate routing and wavelength conversion in wavelength-routed optical networks, *Proc. IEEE GLOBECOM'98* (1998) pp. 2295-2303.

[35] R. A. Barry and S. Subramaniam, The MAX-SUM wavelength assignment algorithm for WDM ring networks, *Proc. OFC'97* (1997).

[36] X. Zhang and C. Qiao, Wavelength assignment for dynamic traffic in multi-fiber WDM networks, *Proc. IC3N'98* (1998) pp. 479-485.

[37] K. Chan and T. P. Yum, Analysis of least congested path routing in WDM lightwave networks, *Proc. IEEE INFOCOM'94* Vol.2 (1994) pp. 962-969.

[38] F. Borgonovo, L. Fratta, and J. Bannister, On the design of optical deflection-routing networks, *Proc. IEEE INFOCOM'94* Vol.1 (1994) pp. 120-129.

[39] E. A. Varvarigos, The 'packing' and the 'scheduling' packet switch architectures for almost all-optical lossless networks, *IEEE/OSA J. Lightwave Tech.* Vol.16 No.10 (1998) pp. 1757-1767.

[40] L. Li and A. K. Somani, Dynamic wavelength routing using congestion and neighborhood information, *IEEE/ACM Trans. Network.* Vol.7 No.5 (1999) pp. 779-786.

[41] Z. Zhang and A. S. Acampora, Performance analysis of multihop lightwave networks with hot potato routing and distance-age-priorities, *IEEE Trans. Commun.* Vol.42 No.8 (1994) pp. 2571-2581.

[42] A. Bononi, G. A. Castanon, and O. K. Tonguz, Analysis of hot-potato optical networks with wavelength conversion, *IEEE/OSA J. Lightwave Tech.* Vol.17 No.4 (1999) pp. 525-534.

[43] J. P. Jue and G. Xiao, An adaptive routing algorithm for wavelength-routed optical networks with a distributed control scheme, *Proc. IEEE IC3N'00* (2000) pp. 192-197.

[44] Y. Mei and C. Qiao, Efficient distributed control protocols for WDM all-optical networks, *Proc. IEEE IC3N'97* (1997) pp. 150-153.

[45] Y. Mei and C. Qiao, Distributed control schemes for dynamic lightpath establishment in WDM optical networks, *Proc. Optical Network Workshop* (2000) http://catss.utdallas.edu/ONW2000/proceedings/.

[46] M. Kovacevic and A. Acampora, Benefits of wavelength translation in all-optical clear-channel networks, *IEEE J. Select. Areas Commun.* Vol.14 No.5 (1996) pp. 868-880.

[47] S. Subramaniam, M. Azizoglu, and A. K. Somani, All-Optical Networks with Sparse Wavelength Conversion, *IEEE/ACM Trans. Network.* Vol.4 No.4 (1996) pp. 544-557.

[48] S. Subramaniam, A. K. Somani, M. Azizoglu, and R. A. Barry, The benefits of wavelength conversion in WDM networks with non-Poisson traffic, *IEEE Commun. Lett.* Vol.3 No.3 (1999) pp. 81-83.

[49] J. M. Yates, M. P. Rumsewica, and J. P. Lacey, Wavelength conversion in networks with differing link capacities, *Proc. IEEE GLOBECOM'98* (1998) pp. 2315-2320.

[50] B. S. Glance, J. M. Wiesenfeld, U. Koren, and R. W. Wilson, New advances in optical components needed for FDM optical networks, *IEEE/OSA J. Lightwave Technol.* Vol.11 No.5/6 (1993) pp. 882-890.

[51] K.-C. Lee and V. O. K. Li, A wavelength-convertible optical network, *IEEE/OSA J. Lightwave Technol.* Vol.11 No.5/6 (1993) pp. 962-970.

[52] H. Harai, M. Murata, and H. Miyahara, Allocation of wavelength convertible nodes and routing in all-optical networks, *Proc. SPIE, Vol.30, All-Optical Communication Systems: Architecture, Control and Network Issues III* (1997) pp. 277-287.

[53] S. Subramaniam, M. Azizoglu, and A. K. Somani, On the optimal placement of wavelength converters in wavelength-routed networks, *Proc. INFOCOM'98* Vol.2 (1998) pp. 902-909.

[54] J. Yates, J. Lacey, D. Everitt, and M. Summerfield, Limited-range wavelength translation in all-optical networks, *Proc. IEEE INFOCOM'96* Vol.3 (1996) pp. 954-961.

[55] V. Sharma and E. A. Varvarigos, Limited wavelength translation in all-optical WDM mesh networks, *Proc. IEEE INFOCOM'98* Vol.2 (1998) pp. 893-901.

[56] T. Tripathi and K. Sivarajan, Computing approximate blocking probabilities in wavelength routed all-optical networks with limited-range wavelength conversion, *Proc. INFOCOM'99* Vol.1 (1999) pp. 329-336.

[57] J. M. H. Elmirghani and H. T. Mouftah, All-optical wavelength conversion technologies and applications in DWDM networks, *IEEE Commun. Mag.* Vol.38 No.3 (2000) pp. 86-92.

[58] D. Nesset, T. Kelly, and D. Marcenac, All-optical wavelength conversion using SOA nonlinearities, *IEEE Commun. Mag.* Vol.36 No.12 (1998) pp. 56-61.

[59] M. F. C. Stephens, D. Nesset, R. V. Penty, I. H. White, and M. J. Fice, Wavelength conversion at 40 Gbit/s via four wave mixing in semiconductor optical amplifier with integrated pump laser, *Elect. Letter.* Vol.35 No.5 (1999) pp. 420-421.

[60] R. W. Tkach, A. R. Chraplyvy, F. Forghieri, A. H. Gnauck, and R. M. Derosier, Four-photon mixing and high-speed WDM systems, *IEEE/OSA J. Lightwave Tech.* Vol.13 No.5 (1995) pp. 841-849.

[61] J. Zhou, N. Park, K. J. Vahala, M. A. Newkirk, and B. I. Miller, Four-wave mixing wavelength conversion efficiency in semiconductor traveling-wave amplifiers measured to 65 nm of wavelength shift, *IEEE. Photo. Tech. Letter.* Vol.6 No.8 (1994) pp. 984-987.

[62] S. J. B. Yoo, Wavelength conversion technologies for WDM network applications, *IEEE/OSA J. Lightwave Tech.* Vol.14 No.6 (1996) pp. 955-966.

[63] N. Antoniades, S. J. B. Yoo, K. Bala, G. Ellinas, and T. E. Stern, An architecture for a wavelength-interchanging cross-connect utilizing parametric wavelength converters, *IEEE/OSA J. Lightwave Tech.* Vol.17 No.7 (1999) pp. 1113-1125.

[64] T. Durhuus, B. Mikkelsen, C. Joergensen, S. L. Danielsen, and K. E. Stubkjaer, All-optical wavelength conversion by semiconductor optical amplifiers, *IEEE/OSA J. Lightwave Tech.* Vol.14 No.6 (1996) pp. 942-954.

[65] J. P. R. Lacey, G. J. Pendock, and R. S. Tucker, Gigabit-per-second all-optical 1300-nm to 1550-nm wavelength conversion using cross-phase modulation in a semiconductor optical amplifier, *Proc. OFC'96* Vol.2 (1996) pp. 125-126.

[66] M. Eisenberg, W. Peiper, and H. G. Weber, Decision gate for all-optical retiming using a semiconductor laser amplifier in a loop mirror configuration, *Elect. letter.* Vol.29 No.1 (1993) pp. 107-109.

[67] H. Kawaguchi, K. Magari, H. Yasaka, M. Fukuda, and K. Oe, Tunable optical wavelength conversion using an optically triggerable multielectrode distributed feedback laser diode, *IEEE J. Quantum Electron.* Vol.24 No.11 (1998) pp. 2153-2159.

[68] H. Yasaka, H. Sanjon, H. Ishii, Y. Yoshikuni, and G. Oe, Repeated wavelength conversion of 10 Gb/s signals and converted signal gating using wavelength-tunable semiconductor lasers, *IEEE/OSA J. Lightwave Tech.* Vol.14 No.6 (1996) pp. 1042-1047.

[69] S. Subramaniam, M. Azizoglu, and A. K. Somani, On optimal converter placement in wavelength-routed networks, *IEEE/ACM Trans. Network.* Vol.7 No.5 (1999) pp. 754-766.

[70] A. S. Arora and S. Subramaniam, Converter placement in wavelength routing mesh topologies, *Proc. IEEE ICC'00* Vol.3 (2000) pp. 1282-1288.

[71] S. Thiagarajan and A. K. Somani, An efficient algorithm for optimal wavelength converter placement on wavelength-routed networks with arbitrary topologies, *Proc. IEEE INFOCOM'99* Vol.2 (1999) pp. 916-923.

[72] R. Barry and P. Humblet, Models of blocking probability in all-optical networks with and without wavelength changers, *IEEE J. Select. Areas Commun.* Vol.14 No.5 (1996) pp. 858-867.

[73] P.-J. Wan, L. Liu, and O. Frieder, Optimal placement of wavelength converters in tree and tree of rings, *Proc. IEEE IC3N'99* (1999) pp. 392-397.

[74] K. R. Venugopal, M. Shivakumar, and P. S. Kumar, A heuristic for placement of limited range wavelength converters in all-optical network, *Proc. IEEE INFOCOM'99* Vol.2 (1999) pp. 908-915.

[75] A. Schrijver, *Theory of Linear and Integer Programming*, (John Wiley & Sons, 1986).

[76] A. Ganz and Y. Gao, Time-wavelength assignment algorithm for high performance WDM star based system, *IEEE Trans. Commun.* Vol.42 No.2/3/4 (1994) pp. 1827-1836.

[77] G. Xiao and Y.-W. Leung, Algorithms for allocating wavelength converters in all-optical networks, *IEEE/ACM Trans. Network.* Vol.7 No.4 (1999) pp. 545-557.

[78] G. Xiao, Y.-W. Leung, and K. W. Hung, Two-Stage Cut Saturation Algorithm for Designing All-Optical Networks, *IEEE Trans. Commun.* in press.

[79] Y. W. Leung, G. Xiao, and K. W. Hung, Design of Node Configuration for All-Optical Networks, *IEEE Trans. Commun.* to appear.

Network Theory and Applications

Kluwer Academic Publishers – Dordrecht / Boston / London

Printed in the USA
CPSIA information can be obtained
at www.ICGtesting.com
LVHW010556161023
761183LV00006B/558